Principles of Engineering Physics 2

This textbook is a follow-up to the volume *Engineering Physics 1* and aims for an introductory course in engineering physics. It provides a balance between theoretical concepts and their applications. Beginning with the fundamental concepts of crystal structure including lattice directions and planes, atomic packing factor, diffraction by crystal, reciprocal lattice and intensity of diffracted beam, the authors discuss various relevant topics including defects in crystals, X-rays, bonding in solids and magnetic properties of materials. The book also covers topics related to superconductivity, optoelectronic devices, dialectic materials, semiconductors, electron theory of solids and energy bands in solids.

All chapters are interspersed with rich pedagogical features such as solved problems, unsolved exercises and multiple choice questions with answers. It will help undergraduate students of engineering acquire skills for solving difficult problems in quantum mechanics, electromagnetism, nanoscience, energy systems and other engineering disciplines.

Md. N. Khan is Associate Professor at the Department of Physics, Indira Gandhi Institute of Technology (IGIT), Odisha. He has more than 22 years of teaching experience and has taught courses on engineering physics, physics of semiconductor devices and materials science. His areas of interest include X-ray scattering and materials science.

S. Panigrahi is Senior Professor at the Department of Physics and Astronomy, National Institute of Technology (NIT), Rourkela. He has more than two decades of teaching and research experience in the field of solid state physics, materials science and ferroelectrics.

Principles of
Engineering Physics 2

Md. N. Khan
S. Panigrahi

CAMBRIDGE
UNIVERSITY PRESS

Shaftesbury Road, Cambridge CB2 8EA, United Kingdom

One Liberty Plaza, 20th Floor, New York, NY 10006, USA

477 Williamstown Road, Port Melbourne, VIC 3207, Australia

314–321, 3rd Floor, Plot 3, Splendor Forum, Jasola District Centre, New Delhi – 110025, India

103 Penang Road, #05–06/07, Visioncrest Commercial, Singapore 238467

Cambridge University Press is part of Cambridge University Press & Assessment,
a department of the University of Cambridge.

We share the University's mission to contribute to society through the pursuit of
education, learning and research at the highest international levels of excellence.

www.cambridge.org
Information on this title: www.cambridge.org/9781316635650

© Cambridge University Press & Assessment 2016

First published 2016

A catalogue record for this publication is available from the British Library

ISBN   978-1-316-63565-0   Paperback

Additional resources for this publication at www.cambridge.org/9781316635650

To all our beloved people who have sacrificed their lives for the betterment of the world through science, technology and social service.

Contents

Preface

From time immemorial, mankind has manipulated specific properties of materials for specific self-benefits. A clear understanding of the basic principles of materials science is essential for technological development. The rapid development of materials science resulted in the invention of miniature electronic devices. All modern technologically advanced devices are directly related to an understanding of materials at the atomic and sub-atomic levels. Accordingly, the technical universities throughout the world include materials science as an essential ingredient in their course curricula.

Materials science is an interdisciplinary subject relying heavily on basic principles of physics and chemistry. Electrical and thermal conductivity, dielectric constant, magnetization, optical reflection and refraction, strength and toughness etc. are properties that originate from the internal structures of the materials. The present book, entitled *Principle of Engineering Physics 2,* contains chapters mostly related to materials science. It is designed as a textbook, keeping in view the engineering physics and materials science course curricula prescribed by most technical universities of India. It begins with 'Crystal Structure' and ends with 'Nano Structure & Thin Films', containing altogether thirteen chapters. The book is written in a logical and coherent manner for easy understanding by students. It presumes a working knowledge of quantum mechanics, optics, electricity and magnetism. Emphasis has been given to an understanding of the basic concepts and their applications to a number of engineering problems. Each topic is discussed in detail both conceptually and mathematically, so that students will not face comprehension difficulties. Derivations and solutions of numerical examples are also provided in detail. Each chapter contains a large number of solved numerical examples, unsolved numerical problems with answers, practical applications, theoretical questions, and multiple choice questions with answers. Certain topics and derivations which are not present in university syllabi have been included in the book for the sake of continuity and completeness. The scope of the book has thus been expanded beyond the basic needs of undergraduate engineering students. We hope, this book will be helpful not only to the students but also to the teachers.

In spite of utmost care, some typographical errors might have inadvertently crept into the book. Readers would be highly appreciated if they convey these errors to the authors. The authors sincerely request the readers for their constructive criticisms via emails *mdnkhan1964@yahoo.com* and *spanigrahi@nitrkl.ac.in* for future modification of the book.

Acknowledgment

It is a pleasure to express our deep appreciation to the engineering students (both continuing and passed out) of IGIT Sarang and NIT Rourkela who have borne with us in our class teachings. Many suggestions from our colleagues, students, and reviewers have gone a long way in the development of this book. Our sincere thanks are due to them. We gratefully acknowledge the ideas received from a number of standard books on solid state physics/ materials science as given in the bibliography. We sincerely thank the editorial team of Cambridge University Press, India, for the keen interest in publishing the book in a nice format. We particularly wish to thank Gauravjeet Singh Reen for many helpful suggestions and improvements.

1 Crystal Structure

1.1 Introduction

Solids consist of atoms, molecules or ions packed very closely together. The forces that hold them in place give rise to distinctive properties of the various kinds of solid. In a broader sense, solids are classified into two categories: Crystalline and non-crystalline or amorphous. A crystal may be defined as a solid in which atoms, molecules or ions are arranged in a periodic pattern in three dimensions. That means crystals have a regular internal structure. An amorphous solid may be defined as a solid in which atoms or molecules are arranged arbitrarily in three dimensions, i.e., amorphous solids have no regular internal structure. A few examples of amorphous substances are glass, plastic, and gel, whereas the list of crystalline solids is very large; most metals are crystalline. Crystals that are composed of two elements are called binary crystals. There are thousands of binary crystals; some examples are sodium chloride (NaCl), alumina (Al_2O_3) and ice (H_2O). A polycrystalline solid is made up of an aggregate of a large number of tiny single crystals called grains oriented in different directions and separated by well-defined boundaries called grain boundaries.

1.2 Geometry of Crystals

For the systematic study of crystals, we should first know the geometry of crystals in which actual atoms or molecules composing the crystal are ignored and their positions in space are taken into consideration. The positions of the atoms or molecules in the crystal define a set of points called point lattice. The point lattice may be regarded as the skeleton on which the actual crystal is built.

1.3 Fundamental Terms

i. *Point lattice*   The point lattice is defined as an array of points in space so arranged that every point has surroundings identical to that of every other point in the array. By identical surroundings we mean that, when we look in a particular direction putting ourselves at a lattice point, the same scenery is visible as that of any other point when we look in the same direction. A two-dimensional point lattice having infinite extension is shown in Fig. 1.1(a) and a three-dimensional point lattice assumed to have infinite extension is shown in Fig. 1.1(b). Point lattice, lattice or space lattice, are synonymously used.

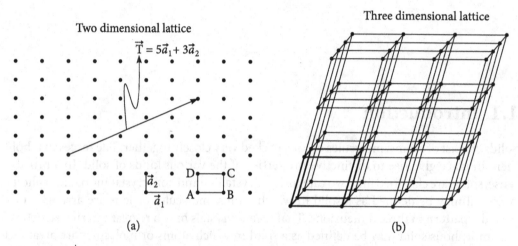

Figure 1.1 | (a) A two-dimensional lattice. (b) A three-dimensional lattice. Observe that each point has identical surroundings. ABCD represents a unit cell selected in a two-different ways in a two-dimensional lattice and the heavily outlined one is the unit cell in a three-dimensional lattice

ii. *Unit cell*   As can be seen in Fig. 1.1(a), the entire two-dimensional lattice can be produced by translating the cell ABCD along the horizontal as well as vertical directions. Hence, ABCD is a unit cell. As has been illustrated in Fig. 1.1(b), the entire three-dimensional lattice can be produced by translating the heavily outlined cell in space in all possible directions. Therefore, the heavily outlined cell of Fig. 1.1(b) is a unit cell. Thus, the unit cell is defined as the smallest cell, translation of which generates the entire lattice. A three-dimensional general unit cell is shown in Fig. 1.1(b) as heavily outlined.

iii. *Crystallographic axes*   The vectors $\vec{a}_1$, $\vec{a}_2$, and $\vec{a}_3$ that define the unit cell in Fig. 1.2 are called crystallographic axes of the unit cell. Thus, we can define the crystallographic axes of a unit cell as the three vectors defining the unit cell of a lattice. The crystallographic axes $\vec{a}_1$, $\vec{a}_2$, and $\vec{a}_3$ are also called primitive lattice vectors or basis vectors or fundamental translation vectors. Depending upon the magnitudes and directions of the basis vectors $\vec{a}_1$, $\vec{a}_2$, and $\vec{a}_3$, different types

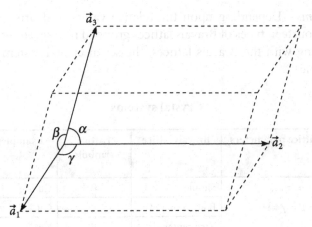

Figure 1.2 | A generalized unit cell. The vectors $\vec{a}_1$, $\vec{a}_2$, and $\vec{a}_3$ defining a unit cell are called crystallographic axes or basis vectors or primitive lattice vectors or fundamental translation vectors

(total seven in number) unit cells are formed. The volume of a unit cell V defined by basis vectors $\vec{a}_1$, $\vec{a}_2$, and $\vec{a}_3$ is given by

$$V = \left| \vec{a}_1 . \vec{a}_2 \times \vec{a}_3 \right| = \left| \vec{a}_2 . \vec{a}_3 \times \vec{a}_1 \right| = \left| \vec{a}_3 . \vec{a}_1 \times \vec{a}_2 \right| \qquad (1.1)$$

iv. *Lattice parameters* The magnitudes of the crystallographic axes a_1, a_2, and a_3 along with the interfacial angles α (angle between $\vec{a}_2$ and $\vec{a}_3$), β (angle between $\vec{a}_3$ and $\vec{a}_1$), and γ (angle between $\vec{a}_1$ and $\vec{a}_2$), define the unit cell of Fig. 1.2. The magnitudes of the crystallographic axes a_1, a_2, and a_3 along with the interfacial angles α, β, and γ are called lattice constants or lattice parameters of the unit cell.

v. *Lattice translation vector* Any two lattice points can be connected with each other by a vector of the form

$$\vec{T} = n_1 \vec{a}_1 + n_2 \vec{a}_2 + n_3 \vec{a}_3 \qquad (1.2)$$

The vector defined by Eq. (1.2) is called lattice translation vector $\vec{T}$ which specifies the position of a lattice point in a lattice. Here, n_1, n_2, and n_3 are integers, may be negative, zero or positive. To be particular, actually, n_1, n_2, and n_3 are the projections of the vector $\vec{T}$ along $\vec{a}_1$, $\vec{a}_2$, and $\vec{a}_3$ respectively.

vi. *Bravais lattice* A three-dimensional space lattice is generated by the repeated translation of basis vectors $\vec{a}_1$, $\vec{a}_2$, and $\vec{a}_3$. It turns out that there are only fourteen distinguishable ways of arranging points in three-dimensional space such that each arrangement confirms to the definition of a space lattice. These fourteen space lattices are known as Bravais lattices in honour of their originator, the French crystallographer Auguste Bravais.

vii. *Crystal systems*   Depending upon the relative values and orientation of the basis vectors, the fourteen types of Bravais lattices grouped into seven sets are called crystal systems. Along with the Bravais lattices, the seven crystal systems are listed in the following table.

Crystal systems

Systems	Lattice parameters	Bravais lattice	Lattice symbols	Examples	No. of lattice points per unit cell
Cubic	$a_1 = a_2 = a_3$,	Simple	P	Cu, Ag	1
	$\alpha = \beta = \gamma = 90°$	Body centered	I	CsCl	2
		Face centered	F	NaCl	4
Tetragonal	$a_1 = a_2 \neq a_3$,	Simple	P	β- Sn	1
	$\alpha = \beta = \gamma = 90°$	Body centered	I	TiO_2	2
Orthorhombic	$a_1 \neq a_2 \neq a_3$,	Simple	P	Ga	1
	$\alpha = \beta = \gamma = 90°$	Body centered	I	$Pbco_3$	2
		Base centered	C	α- S	2
		Face centered	F	K_2SO_4	4
Rhombohedral (Also called trigonal)	$a_1 = a_2 = a_3$, $\alpha = \beta = \gamma \neq 90°$	Simple	P	As, Bi, Sb, Calcite	4
Hexagonal	$a_1 = a_2 \neq a_3$, $\alpha = \beta = 90°, \gamma = 120°$	Simple	P	Ng, Zn	
Monoclinic	$a_1 \neq a_2 \neq a_3$,	Simple	P	Gypsum	1
	$\alpha = \gamma = 90° \neq \beta$	Base centered	C		2
Triclinic	$a_1 \neq a_2 \neq a_3$, $\alpha \neq \beta \neq \gamma \neq 90°$	Simple	P	$K_2Cr_2O_7$	1

viii. *Basis*   A group of atoms or molecules attached to a lattice point to form the crystal structure is called a basis.

ix. *Crystal structure*   A crystal structure is formed when a basis is attached identically to every lattice point. The space lattice is converted into a crystal structure when a basis is attached identically to every lattice point. The logical relation is

   Lattice points + basis = crystal structure

x. *Primitive unit cell*   The simplest unit cell is the primitive cell of the simple cubic unit cell of simple cubic crystals containing one atom which may be assumed to be at the origin.

Example 1.1

The fundamental lattice translation vectors of a hexagonal lattice may be defined as

$$\vec{a}_1 = \frac{\sqrt{3}}{2}a\hat{x} + \frac{1}{2}a\hat{y}, \vec{a}_2 = -\frac{\sqrt{3}}{2}a\hat{x} + \frac{1}{2}a\hat{y}, \vec{a}_3 = c\hat{z}.$$

Calculate the volume of the hexagonal unit cell.

Solution

$$\vec{a}_2 \times \vec{a}_3 = \left(-\frac{\sqrt{3}}{2}a\hat{x} + \frac{1}{2}a\hat{y}\right) \times c\hat{z} = \frac{1}{2}ac\hat{x} + \frac{\sqrt{3}}{2}ac\hat{y}$$

Thus, the volume of the unit cell V is calculated to be

$$V = \vec{a}_1.\vec{a}_2 \times \vec{a}_3 = \left(\frac{\sqrt{3}}{2}a\hat{x} + \frac{1}{2}a\hat{y}\right) \cdot \left(\frac{1}{2}ac\hat{x} + \frac{\sqrt{3}}{2}ac\hat{y}\right)$$

$$= \frac{\sqrt{3}}{4}a^2c + \frac{\sqrt{3}}{4}a^2c = \frac{\sqrt{3}}{2}a^2c$$

1.4 Lattice Directions and Planes

Certain physical properties of a crystal may depend on directions of measurement. The crystals whose certain properties depend on the direction of measurement are called anisotropic crystals and those properties are called anisotropic properties. For this reason, it is necessary to identify specific directions in the crystal. In a crystal, different lattice planes may pass through different lattice points in different orientations. For the study of crystal structure, it is very important to specify various lattice planes in the crystal. The lattice directions and lattice planes are also called crystal directions and crystal planes.

1.4.1 Lattice directions

The direction defined by the lattice translation vector $\vec{T}$ connecting two points in a lattice is given by

$$\vec{T} = n_1\vec{a}_1 + n_2\vec{a}_2 + n_3\vec{a}_3 \tag{1.3}$$

Here, n_1, n_2 and n_3 are the projections of the vector $\vec{T}$ along $\vec{a}_1$, $\vec{a}_2$ and $\vec{a}_3$ respectively. If one, two or all of n_1, n_2 and n_3 are fractions, they can be converted into smallest integral values by multiplying them by a suitable number (it may be the LCM of the denominators). These smallest integral values are called direction indices of the line represented by the vector $\vec{T}$ and are written within square brackets []. The direction indices of a line give the direction of the line in the crystal. It is important to remember that if the direction passes through the origin, to find the direction indices, the origin is first shifted to another lattice point and the direction indices are calculated with respect to this new origin. The following steps are followed in calculating the direction indices of a line.

i   If necessary shift the origin to any other lattice point.

ii.   The line whose direction indices are to be found out is represented by a vector of the form $\vec{T} = n_1\vec{a}_1 + n_2\vec{a}_2 + n_3\vec{a}_3$. If the coordinates of any two points on the line are known, then by using the principles of coordinate geometry we can represent the line by a vector of the form $\vec{T} = n_1\vec{a}_1 + n_2\vec{a}_2 + n_3\vec{a}_3$.

iii. The coefficients of $\vec{a}_1$, $\vec{a}_2$, and $\vec{a}_3$ i.e., n_1, n_2, and n_3 are written inside a square bracket like $[n_1\ n_2\ n_3]$. Commas, dots, are not to be put between the numbers.

iv.   If the bracketed terms are fractions, then multiply them with the LCM of their denominators to make them integers.

v.   If the bracketed terms are integers having a common multiple, then divide them by that common multiple to reduce them to the smallest integers.

vi.   The set of smallest integer written within a square bracket [] is called the direction indices of the line.

vii. If any one or all bracketed terms are negative, a bar is put over the integer(s).

The following example will elucidate the procedures and steps to find the direction indices of a line.

Case i: Direction indices of OA: The direction of OA is represented by the vector $\vec{T}_{OA} = 1\vec{a}_1 + 0 + 0$ as seen in Fig. 1.3. Hence, the direction indices of OA will be given by [1 0 0] (read as one zero zero).

Case ii: Direction indices of OB: The direction of OB is represented by the vector $\vec{T}_{OB} = 1\vec{a}_1 + 1\vec{a}_2 + 0$ as seen in Fig. 1.3. Hence, the direction indices of OB will be given by [1 1 0].

Case iii: Direction indices of OC: The direction of OC is represented by the vector $\vec{T}_{OC} = 1\vec{a}_1 + 1\vec{a}_2 + 1\vec{a}_3$ as seen in Fig. 1.3. Hence, the direction indices of OC will be given by [1 1 1].

Case iv: Direction indices of OE: The direction of OE is represented by the vector $\vec{T}_{OE} = 0 + 1\vec{a}_2 + 0$ as seen in Fig. 1.3. Hence, the direction indices of OE will be given by [0 1 0].

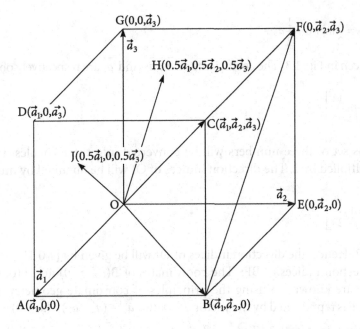

Figure 1.3 | Unit cell of a crystal system

Case v: Direction indices of OG: The direction of OG is represented by the vector $\vec{T}_{OE} = 0 + 0 + 1\vec{a}_3$ as seen in Fig. 1.3. Hence, the direction indices of OG will be given by [0 0 1].

Case vi: Direction indices of OH (H is the mid-point of the plane DCFG): The direction of OH is represented by the vector

$$\vec{T}_{OH} = \frac{1}{2}\vec{a}_1 + \frac{1}{2}\vec{a}_2 + 1\vec{a}_3$$

as seen in Fig. 1.3. The coefficients of $\vec{a}_1$, $\vec{a}_2$, and $\vec{a}_3$ are respectively obtained as

$$\left[\frac{1}{2} \quad \frac{1}{2} \quad 1\right]$$

This set of three numbers will be converted into three smallest integers when multiplied by 2. The direction indices of OH will be obtained by multiplying

$$\left[\frac{1}{2} \quad \frac{1}{2} \quad 1\right]$$

by 2. Hence, the direction indices of OH will be given by [1 1 2].

Case vii: Direction indices of OJ (J is the mid-point of the plane ADGO): The direction of OJ is represented by the vector

$$\vec{T}_{OJ} = \frac{1}{2}\vec{a}_1 + 0 + \frac{1}{2}\vec{a}_3$$

as seen in Fig. 1.3. The coefficients of $\vec{a}_1$, $\vec{a}_2$, and $\vec{a}_3$ are respectively obtained as

$$\left[\frac{1}{2} \; 0 \; \frac{1}{2}\right]$$

This set of three numbers will be converted into three smallest integers when multiplied by 2. The direction indices of OJ will be obtained by multiplying

$$\left[\frac{1}{2} \; 0 \; \frac{1}{2}\right]$$

by 2. Hence, the direction indices of OJ will be given by [1 0 1].

Case viii: Direction indices of BF: The coordinates of B(a_1, a_2, 0) and F(0, a_2, a_3) on the line are known, so using the principles of coordinate geometry, the direction of BF is represented by the vector $\vec{T}_{BF} = (0 - \vec{a}_1) + (\vec{a}_2 - \vec{a}_2) + (\vec{a}_3 - 0) = -1\vec{a}_1 + 0 + 1\vec{a}_3$

as seen in Fig. 1.3. The coefficients of and $\vec{a}_3$ are respectively obtained as –1, 0, and 1. Hence, the direction indices of BF will be given by [$\bar{1}$ 0 1] (read as bar one zero one).

From these discussions, it is clear that though

$$\left[\frac{1}{6} \; \frac{1}{2} \; \frac{2}{3}\right], \; [1 \; 3 \; 4], \; [2 \; 6 \; 8] \; \left[\frac{1}{2} \; \frac{3}{2} \; 2\right]$$

all represent the same direction, $\left[1 \; 3 \; 4\right]$ is the preferred form.

Example 1.2

If 0.2, 0.4, and 0.3 are the coordinates of a point on a line, determine the direction indices of the line.

Solution

The direction indices of the line containing the point (0.2, 0.4, 0.3) are [0.2, 0.4, 0.3]. The most preferred form is obtained by multiplying the indices by 10 as [2 4 3].

Family of lattice directions

In the previous discussion, we have represented the direction indices of the body diagonal OC by [111]. As can be checked, the body diagonal ED will be represented by direction indices [1$\bar{1}$1]. The following are the possible combinations including positive and negative values of the three numbers specifying the directions of all the body diagonals: [111], [$\bar{1}$11],

$[1\bar{1}1]$, $[11\bar{1}]$, $[\bar{1}\,\bar{1}1]$, $[1\bar{1}\,\bar{1}]$, $[\bar{1}1\bar{1}]$, $[\bar{1}\,\bar{1}\,\bar{1}]$. These combinations represent the direction of the body diagonal of the unit cell depicted in Fig. 1.3 and is called the family of lattice directions [111]. In symbols, the family of lattice directions [111] is written as

$$<111> = [111], [\bar{1}11], [1\bar{1}1], [11\bar{1}], [\bar{1}\bar{1}1], [1\bar{1}\bar{1}], [\bar{1}1\bar{1}], [\bar{1}\bar{1}\bar{1}].$$

These eight combinations give the direction indices of the body diagonals.

Similarly, the family of lattice directions of edge OA and face diagonal OB, are obtained respectively as $<100> = [100]\ [010]\ [001]\ [\bar{1}00]\ [0\bar{1}0]\ [00\bar{1}]$ and $<110> = [110]$, $[011]$, $[101]$, $[\bar{1}10]$, $[0\bar{1}1]$, $[\bar{1}01]$, $[1\bar{1}0]$, $[0\bar{1}1]$, $[10\bar{1}]$, $[\bar{1}\bar{1}0]$, $[0\bar{1}\bar{1}]$, $[\bar{1}0\bar{1}]$.

Linear density of atoms

The linear density of atoms in a lattice is the number of atoms per unit length in a particular direction in the crystal lattice. The number of atoms along the face diagonal of an FCC structure is 3 and the length of the face diagonal of the FCC is $\sqrt{2}$ a where a is the lattice parameter of FCC. Hence, the linear density of atoms along the face diagonal in FCC is $\dfrac{3}{\sqrt{2}a}$.

1.4.2 Crystal planes

In a crystal, the planes passing through the crystal in different orientations are called lattice planes or crystal planes. Crystal planes are known by their orientations with respect to crystallographic axes. The orientations of the crystal planes are specified by three parameters enclosed in lunar brackets $(hk\ell)$ called Miller indices. The method of finding Miller indices $(hk\ell)$ are explained here.

In general, the orientation of a given plane can be specified by knowing the three intercepts made by the plane with the crystallographic axes. These three intercepts will depend on the axial lengths a_1, a_2, and a_3. In order to make these three intercepts independent of the particular axial lengths involved in the given lattice, fractional intercepts are taken instead of intercepts. To avoid the introduction of infinity into the specification of orientation of crystal planes, the reciprocal of fractional intercepts are taken. Thus, we arrive at a workable symbolism for the orientation of a crystal plane called Miller indices. The working definition of Miller indices is given as the reciprocal of the fractional intercepts which the plane makes with the crystallographic axes. It is important to remember that if the crystal plane passes through the origin, the origin is shifted to another lattice point and the Miller indices is calculated with respect to this new origin.

The following steps are involved in the calculation of Miller indices of a crystal plane.

i.   If necessary, shift the origin to any other lattice point.

ii.   Write down the axial lengths a_1, a_2, and a_3 in order.

iii.   Write down the intercepts p, q, r in order.

iv.   Calculate the fractional intercepts, $\dfrac{p}{a_1}$, $\dfrac{q}{a_2}$, $\dfrac{r}{a_3}$.

[A fractional intercept means an intercept is a fraction of the corresponding axial length.]

v. Take the reciprocal of fractional intercepts, i.e., $\dfrac{a_1}{p}, \dfrac{a_2}{q}, \dfrac{a_3}{r}$.

vi. If $\dfrac{a_1}{p}, \dfrac{a_2}{q}, \dfrac{a_3}{r}$ are not the smallest integers, then by multiplying or by dividing by a single suitable number, they can be converted into a set of the three smallest integers $hk\ell$ to give the Miller indices.

vii. This set of three smallest integers $hk\ell$ written inside a lunar bracket $(hk\ell)$ gives the Miller indices of the given plane.

viii. If any one or all integers are negative, a bar is put over those integer(s).

The following example will elucidate the procedures and steps to find out the Miller indices of a plane. Let us consider the plane shown in Fig. 1.4.

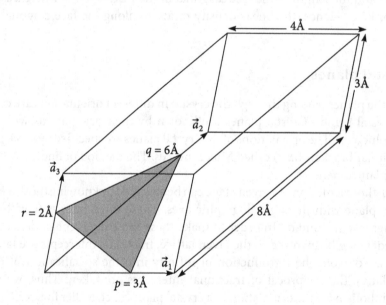

Figure 1.4 | The specification of the orientation of crystal planes by Miller indices. Re-draw the figure taking data from the example given here

As shown in Fig. 1.4, the axial length are given as 4Å, 8Å, and 3Å (a_1, a_2, a_3) respectively and axial intercepts are given as 3Å, 6Å, and 2Å (p, q, r) respectively. Our aim is to find the Miller indices of the plane shown in Fig. 1.4 by using the steps outlined earlier.

i.   Not necessary

ii. Axial lengths   4Å   8Å   3Å

iii. Intercept lengths   3Å   6Å   2Å

iv.	Fractional intercepts	$\dfrac{3}{4}$	$\dfrac{3}{4}$	$\dfrac{2}{3}$	
v.	Reciprocal of fractional intercepts	$\dfrac{4}{3}$	$\dfrac{4}{3}$	$\dfrac{3}{2}$	
vi.	Conversion to smallest integers	8	8	9	(multiplying by 6)

vii.   Miller indices of the plane (889). Step (viii) is not necessary in this case.

In some cases, intercepts are mentioned as pure numbers. In those cases, the intercepts are measured as multiples of the fundamental vectors $\vec{a}_1$, $\vec{a}_2$, and $\vec{a}_3$ or the intercepts are measured in the units of fundamental vectors $\vec{a}_1$, $\vec{a}_2$, and $\vec{a}_3$ which means that those numbers are fractional intercepts. Consider the case of the crystal plane shown in Fig. 1.5. According to the figure, the crystal plane makes intercepts of 4, 2, and 1 respectively.

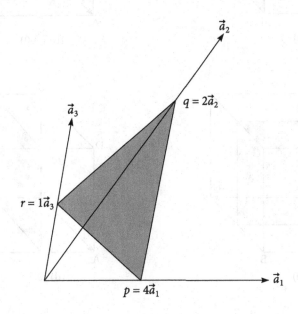

Figure.1.5   A crystal plane makes intercepts 4, 2, and 1 with crystallographic axes $\vec{a}_1$, $\vec{a}_2$, and $\vec{a}_3$ respectively

i.	Not necessary				
ii.	Axial lengths	a_1	a_2	a_3	
iii.	Intercept lengths	$4a_1$	$2a_2$	$1a_3$	
iv.	Fractional intercepts	4	2	1	
v.	Reciprocal of fractional intercepts	$\dfrac{1}{4}$	$\dfrac{1}{2}$	1	
vi.	Conversion to smallest integers	1	2	4	(multiplying by 4)

vii.   Miller indices of the plane (124). Step (viii) is not necessary in this case also.

Miller indices of different crystal planes of different orientations are illustrated in Fig. 1.6. In Fig. 1.6(b), (f), (h), and (j), it was necessary to shift the origin to another lattice point.

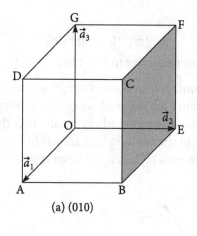

(a) (010)

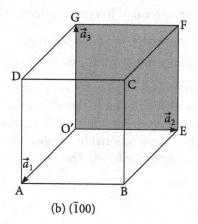

(b) ($\bar{1}$00)

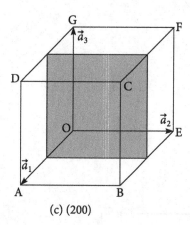

(c) (200)

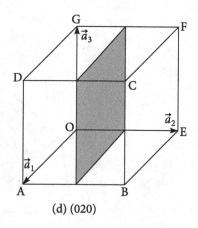

(d) (020)

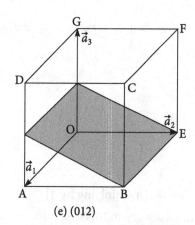

(e) (012)

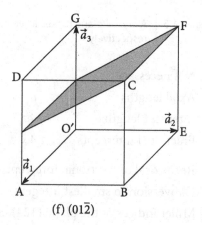

(f) (01$\bar{2}$)

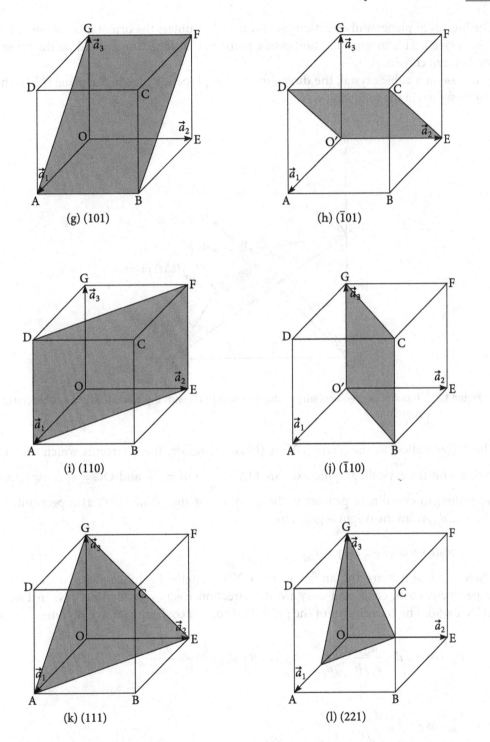

Figure 1.6 | Miller indices of different types of crystal planes. Origin is specified by *O* and where necessity arises to shift the origin, it is specified by *O'*

The indices of planes and directions are meaningless unless the orientation of the unit cell axes is given. This means that indices of a particular lattice plane depend on the origin of the unit cell chosen.

In case of a cubic crystal, the direction [$hk\ell$] is perpendicular to the plane ($hk\ell$) which can be shown in the following way.

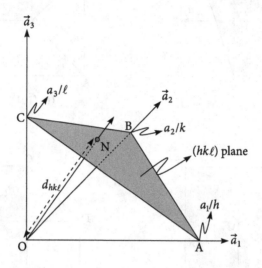

Figure 1.7 | Proof of perpendicularity of the direction [$hk\ell$] with the plane ($hk\ell$) in a cubic crystal system

The Miller indices of the plane ABC is ($hk\ell$). Therefore, the intercepts which the plane makes with the crystallographic axes are OA = $\dfrac{a_1}{h}$, OB = $\dfrac{a_2}{k}$ and OC = $\dfrac{a_3}{\ell}$, respectively. According to coordinate geometry, the equation of the plane ($hk\ell$) at a perpendicular distance $d_{hk\ell}$ from the origin is given by

$$x \cos\alpha + y \cos\beta + z \cos\gamma = d_{hk\ell}$$

where α, β, and γ are the angles which ON of length $d_{hk\ell}$ makes with $\vec{a}_1$, $\vec{a}_2$, and $\vec{a}_3$ respectively. cosα, cosβ, and cosγ are the direction cosines of the line ON. In Fig. 1.7, $\angle$ONA = 90. The coordinates of the point N, $d\cos\alpha$, $d\cos\beta$, and $d\cos\gamma$ becomes

$$d_{hk\ell} \cos\alpha = d_{hk\ell} \frac{d_{hk\ell}}{a_1/h} = \frac{d_{hk\ell}^2 h}{a_1}, \quad d_{hk\ell} \cos\beta = d_{hk\ell} \frac{d_{hk\ell}}{a_2/k} = \frac{d_{hk\ell}^2 k}{a_2}$$

and $\quad d_{hk\ell} \cos\gamma = d_{hk\ell} \dfrac{d_{hk\ell}}{a_3/\ell} = \dfrac{d_{hk\ell}^2 \ell}{a_3}$ 　　　　　　　　　　　　　(1.4)

For a cubic crystal, a_1, a_2 and a_3 are all equal. Let them be equal to a. Thus, in case of a cubic crystal, the coordinates of the point N on the line ON which is perpendicular to the plane ABC are

$$\frac{d_{hk\ell}^2 h}{a}, \frac{d_{hk\ell}^2 k}{a} \text{ and } \frac{d_{hk\ell}^2 \ell}{a}.$$

Therefore, the direction

$$\left[\frac{d_{hk\ell}^2 h}{a}, \frac{d_{hk\ell}^2 k}{a}, \frac{d_{hk\ell}^2 \ell}{a} \right]$$

is perpendicular to the plane ($hk\ell$). Or. In other words, the direction [$hk\ell$], obtained from

$$\left[\frac{d_{hk\ell}^2 h}{a}, \frac{d_{hk\ell}^2 k}{a}, \frac{d_{hk\ell}^2 \ell}{a} \right]$$

by dividing it by $\dfrac{d_{hk\ell}^2}{a}$, is perpendicular to the plane ($hk\ell$).

Example 1.3

In a triclinic or orthorhombic crystal, a plane makes intercepts 2.93 mm, 4.47 mm and 2.35 mm along three crystallographic axes having lengths 3.05 Å, 6.99 Å, and 4.90 Å respectively. Deduce the Miller indices of the plane.

Solution

Axial lengths:	3.05 Å	6.99 Å	4.90 Å
Intercepts:	29.3×10^6 Å	44.7×10^6 Å	23.5×10^6 Å
Fractional intercepts:	9.6×10^6	6.4×10^6	4.8×10^6
Reciprocal of fractional intercepts:	10×10^{-8}	15.6×10^{-8}	20.8×10^{-8}

Dividing these numbers by 5×10^{-8}, we get the Miller indices as (234)

Hexagonal lattice plane

In case of a hexagonal lattice, a slightly different system of plane indexing is used. The unit cell of a hexagonal lattice is defined by two coplanar vectors $\vec{a_1}$ and $\vec{a_2}$ of equal magnitude with an angle of 120° between them, and a third axis equal $\vec{a_4}$ at right angles to both $\vec{a_1}$ and $\vec{a_2}$ at their origin as shown in Fig. 1.8(a). The complete lattice is built up by the repeated translation of the points at the unit cell corners by the vectors $\vec{a_1}$, $\vec{a_2}$, and $\vec{a_4}$. The axis $\vec{a_3}$ lying on the basal plane formed by the axes $\vec{a_1}$ and $\vec{a_2}$ is inclined equally to both $\vec{a_1}$ and $\vec{a_2}$ so that it is used in conjunction with the other two. Thus, the indices of a crystal plane in a hexagonal crystal, called Miller–Bravais indices, refer to four axes and are written as $(hki\ell)$. The index i is the reciprocal of the fractional intercept on the $\bar{a}_3$ axis.

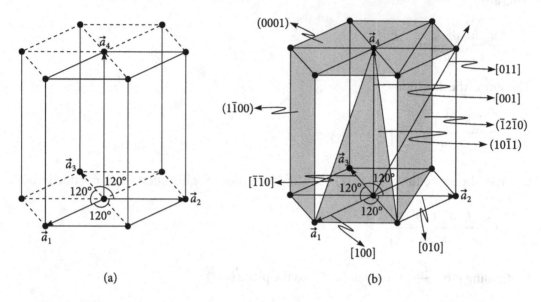

Figure 1.8 | Hexagonal systems. (a) The hexagonal unit cell (b) its indices of planes and directions

The intercepts of a plane on $\vec{a}_1$ and $\vec{a}_2$ determine the intercepts $\bar{a}_3$, the index i is determined by the indices h and k as

$$h + k = -i \quad \text{since } \vec{a}_1 + \vec{a}_2 = -\vec{a}_3 \qquad (1.5)$$

Since $h + k = -i$, the Miller–Bravais indices of a crystal plane is sometimes written as $(hk \cdot \ell)$ instead of $(hki\ell)$.

The directions in a hexagonal crystal are expressed with reference to three axes, vectors $\vec{a}_1$, $\vec{a}_2$, and $\vec{a}_4$. Figure 1.8(b) shows several examples of plane and direction indices of a hexagonal unit cell.

Family of planes

In Fig. 1.6, we have represented different crystal planes by Miller indices. It can be checked that the diagonal plane will be represented by Miller indices (110). The following are the possible combinations including positive and negative values of the three numbers specifying the diagonal planes. They are: (110), (101), (011), ($\bar{1}$10), ($\bar{1}$01), (0$\bar{1}\bar{1}$), (1$\bar{1}$0), (10$\bar{1}$), (01$\bar{1}$), ($\bar{1}\bar{1}$0), ($\bar{1}$0$\bar{1}$), (0$\bar{1}\bar{1}$). These combinations represent the diagonal planes of a simple unit cell and are called the family of planes (1 1 0). In symbols, the family of a lattice planes of a unit cell of a cubic crystal system (1 1 0) is written as

$$\{1\ 1\ 0\} = (1\ 10), (101), (011), (\bar{1}10), (\bar{1}01), (0\bar{1}\bar{1}), (1\bar{1}0), (10\bar{1}), (01\bar{1}), (\bar{1}\bar{1}0), (\bar{1}0\bar{1}), (0\bar{1}\bar{1}).$$

Similarly, the Miller indices of six faces of a unit cell of a cubic crystal are (100), (010), (001) ($\bar{1}$00), (0$\bar{1}$0), (00$\bar{1}$) and are called the family of lattice planes (100). In symbols, the family of facial planes of the unit cell of a cubic crystal system (1 0 0) is written as

$$\{100\} = (100), (010), (001) \, (\bar{1}00), (0\bar{1}0), (00\bar{1})$$

Salient features of Miller indices

We have now described the identification of crystal planes and directions by Miller indices. The conventions and implications of Miller indices are summarized here.

i. Unknown Miller indices are denoted by three letters h, k and ℓ.

ii. The family of directions is represented by $< hk\ell>$ and all the members of $< hk\ell>$ are not necessarily parallel to each other.

iii. By changing the signs of the indices of a crystal direction, we can obtain the anti-parallel or opposite direction. [1$\bar{2}$3] and [$\bar{1}2\bar{3}$] are anti-parallel or are in opposite directions.

iv. The family of planes is represented by $\{hk\ell\}$ and all the members of $\{hk\ell\}$ are not necessarily parallel to each other.

v. By changing the signs of the Miller indices of a crystal plane, we can obtain the mirror image of the plane about the origin. The plane [$\bar{1}4\bar{3}$] is located at the other side of the origin, at the same distance as the plane [1$\bar{4}$3] from the origin.

vi. Two-digit Miller indices are separated by commas for clarity like [5,12,16]

vii. The planes ($nh \, nk \, n\ell$) are parallel to the planes ($hk\ell$) and have $\dfrac{1}{n}$ th spacing. See Fig. 1.6(a) and (d).

viii. In a cubic crystal, the crystal plane ($hk\ell$) and direction [$hk\ell$] are perpendicular to each other.

Planar density of atoms

The planar density of atoms in a lattice is the number of atoms per unit area in a particular plane of the crystal lattice. The number of atoms on the facial plane of the FCC structure is 5 out of which 4 are corner atoms and one is a central atom. Each corner atom is shared by 4 faces and the share of one facial plane of an FCC is one atom. The central atom is not shared. Thus, in total, a facial plane contains 2 atoms. The area of the facial plane of FCC is a^2, where a is the lattice parameter of the FCC. Hence, the planar density of atoms on the crystal plane of an FCC will be $\dfrac{2}{a^2}$.

Interplanar spacing in terms of Miller indices

In a given crystal, the distance $d_{hk\ell}$ between any two consecutive crystal planes is called interplanar spacing. In general, the interplanar spacing depends on the Miller indices ($hk\ell$) of the crystal planes and the lattice parameters a_1, a_2, a_3, α, β, and γ of the crystal. The interplanar spacing $d_{hk\ell}$ of the $(hk\ell)$ plane in an orthogonal ($\alpha = \beta = \gamma = 90°$) crystal system, where the three crystal axes $\vec{a}_1$, $\vec{a}_2$, and $\vec{a}_3$ are mutually orthogonal is given as

$$\frac{1}{d_{hk\ell}^2} = \frac{h^2}{a_1^2} + \frac{k^2}{a_2^2} + \frac{\ell^2}{a_3^2} \tag{1.6}$$

This formula can be obtained in the following way. From Eq. (1.4), the direction cosines of the line ON (Fig. 1.7) of length $d_{hk\ell}$ are respectively

$$\cos\alpha = \frac{d_{hk\ell}h}{a_1}, \cos\beta = \frac{d_{hk\ell}k}{a_2}, \text{and } \cos\gamma = \frac{d_{hk\ell}\ell}{a_3}$$

Thus, we have

$$\frac{d_{hk\ell}^2 h^2}{a_1^2} + \frac{d_{hk\ell}^2 k^2}{a_2^2} = \frac{d_{hk\ell}^2 \ell^2}{a_3^2} = \cos^2\alpha + \cos^2\beta + \cos^2\gamma$$

Using the properties of direction cosines, from this equation, we have

$$\frac{d_{hk\ell}^2 h^2}{a_1^2} + \frac{d_{hk\ell}^2 k^2}{a_2^2} + \frac{d_{hk\ell}^2 \ell^2}{a_3^2} = 1$$

or $$\frac{1}{d_{hk\ell}^2} = \frac{h^2}{a_1^2} + \frac{k^2}{a_2^2} = \frac{\ell^2}{a_3^2}$$

The interplanar spacing of a few simpler crystal systems are given here

i.   Cubic system $\dfrac{1}{d_{hk\ell}^2} = \dfrac{h^2 + k^2 + \ell^2}{a^2}$ (1.7)

ii.   Tetragonal system $\dfrac{1}{d_{hk\ell}^2} = \dfrac{h^2 + k^2}{a_1^2} + \dfrac{\ell^2}{a_3^2}$ (1.8)

iii.   Orthorhombic system $\dfrac{1}{d_{hk\ell}^2} = \dfrac{h^2}{a_1^2} + \dfrac{k^2}{a_2^2} + \dfrac{\ell^2}{a_3^2}$ (1.9)

iv.   Hexagonal system $\dfrac{1}{d_{hk\ell}^2} = \dfrac{4}{3}\left(\dfrac{h^2 + hk + k^2}{a_1^2}\right) + \dfrac{\ell^2}{a_3^2}$ (1.10)

Example 1.4

The lattice constant of a cubic lattice is 4.50 Å. Calculate the spacing between {011}, {101}, and {112}.

Solution

For a cubic lattice, we have

$$\frac{1}{d_{hk\ell}^2} = \frac{h^2 + k^2 + \ell^2}{a^2}$$

or $d_{hk\ell} = \dfrac{a}{\sqrt{h^2 + k^2 + \ell^2}}$

Thus, the spacing between {011} is: $d_{011} = \dfrac{4.50\text{Å}}{\sqrt{0^2 + 1^2 + 1^2}} = 3.18\text{Å}$

Thus, the spacing between {101} is: $d_{101} = \dfrac{4.50\text{Å}}{\sqrt{1^2 + 0^2 + 1^2}} = 3.18\text{Å}$

Thus, the spacing between {112} is: $d_{112} = \dfrac{4.50\text{Å}}{\sqrt{1^2 + 1^2 + 2^2}} = 1.84\text{Å}$

Example 1.5

The density of KCl (sylvite) is 1.98 gm/cm³ and its molecular mass is 74.55. Find the distance between adjacent atoms in the crystal and between adjacent atoms of the same type.

Solution

The mass of 6.02×10^{23} KCl molecules is 74.55 gm.

Hence, the mass of one KCl molecule $= \dfrac{74.55}{6.02 \times 10^{23}}$ gm $= 12.4 \times 10^{-23}$ gm.

12.4×10^{-23} gm corresponds to one KCl molecule. 1.0 gm corresponds to $\dfrac{1}{12.4 \times 10^{-23}}$ molecules. Hence, 1.98 gm corresponds to $\dfrac{1.98}{12.4 \times 10^{-23}} = 1.60 \times 10^{22}$ molecules.

Thus, the unit volume of KCl contains 1.60×10^{22} molecules.

The unit volume, i.e., 1 cm³ of KCl contains $= 2 \times 1.60 \times 10^{22} = 3.20 \times 10^{22}$ atoms since KCl is diatomic. Therefore, a^3 volume contains $3.20 \times 10^{22} \times a^3$ atoms.

 KCl is a cubic crystal. Let d be the distance measured along the length of the cube, between the adjacent atoms in the crystal and let N be the number of atoms along the edge of a 1 cm cube. Then, length of an edge is Nd and the volume of this unit cube is $N^3 d^3$. 1 cm³ volume contains N^3 number of atoms, i.e., $N^3 = 3.20 \times 10^{22}$. Therefore, we have

$N^3 d^3 = 1$

or $d = \dfrac{1}{N} = \dfrac{1}{\left(3.20 \times 10^{22}\right)^{1/3}} cm = 3.14 \overset{\text{o}}{A}$

Thus, the interplanar spacing in KCl is 3.14 Å. The distance between two atoms of the same kind is twice of 3.14 Å = 6.28 Å. This is the length of a unit cell of KCl, i.e., lattice parameter.

Example 1.6

Show that in a simple cubic lattice, interplanar spacings of {111}, {110}, and {100} planes are in the ratio $\dfrac{1}{\sqrt{3}} : \dfrac{1}{\sqrt{2}} : 1$

Solution

For a simple cubic lattice, we have

$$d_{hk\ell} = \dfrac{a}{\sqrt{h^2 + k^2 + \ell^2}}$$

Thus, the spacing between {111} is: $d_{111} = \dfrac{a}{\sqrt{1^2 + 1^2 + 1^2}} = \dfrac{a}{\sqrt{3}}$

Thus, the spacing between {110} is: $d_{101} = \dfrac{a}{\sqrt{1^2 + 1^2 + 0^2}} = \dfrac{a}{\sqrt{2}}$

Thus the spacing between {100} is: $d_{100} = \dfrac{a}{\sqrt{1^2 + 0^2 + 0^2}} = a$

Therefore, we conclude that $d_{111} : d_{101} : d_{100} = \dfrac{1}{\sqrt{3}} : \dfrac{1}{\sqrt{2}} : 1$

1.5 Coordination Number

Every atom in a crystalline solid is surrounded by other atoms in a periodic manner. The coordination number of a particular atom in a crystalline solid is the number of its nearest atoms. More the coordination number, the more closely the atoms are packed and larger is the density of the solid. For example, the coordination number of carbon in methane is four, and it is five in protonated methane. In the following, the coordination number of atoms in SC, FCC, and BCC are calculated.

1.5.1 Simple cubic (SC) lattice

The unit cell of a simple cubic lattice contains atoms only at the corners. Let us calculate the coordination number of a corner atom. A corner atom at a distance of a on same plane has 04 nearest atoms and in addition to this, there is one atom each just above and below it at a distance of the same a. Hence, the coordination number of a corner atom is 04 + 02 = 06. Here a is the lattice parameter.

1.5.2 Face centred cubic (FCC) lattice

The unit cell of an FCC lattice contains atoms at the corners as well as at the centre of each face. In this case, the nearest neighbours of corner atoms are the face centred atoms and distance between them is $a/\sqrt{2}$. For a corner atom, the number of nearest atoms from one unit cell is 03. To each corner, eight unit cells have been attached and each face centred atom is common to two unit cells. Hence, the coordination number of a corner atom will be (03 × 8)/2 = 12.

1.5.3 Body centred cubic (BCC) lattice

The unit cell of a BCC lattice contains atoms at the corners as well as at its centre. In this case, the nearest neighbours of corner atoms is the body centred atoms and the distance between them is $\sqrt{3}a/2$. For a corner atom, the number of nearest atoms from one unit cell is 01. To each corner, eight unit cells have been attached. Hence, the coordination number of a corner atom will be 01 × 8 = 08.

1.5.4 Hexagonal closed packed (HCP) lattice

The unit cell of an HCP crystal structure is shown in Fig. 1.9. The top and bottom faces of the unit cell are regular hexagons (inner angles 120°) with an atom at each corner and at the centre. Another plane called the mid-plane that provides three additional internal atoms to the unit cell is situated at $c/2$ from the orthocentre of alternate equilateral triangles at the top or basal hexagonal face. If a and c represent, respectively, the edge and height of the unit cell, then ideally, the c/a ratio should be $\sqrt{8/3}$. In this case, the nearest neighbours of the central atom are the corner atoms as well as the atoms on the mid-plane and the distance between them is a. The coordination number of the central atom on the hexagonal face is 6 atoms on the hexagonal plane + 3 atoms on the bottom mid-plane + 3 atoms on the top mid-plane = 12.

1.6 Atomic Packing Factor (APF)

Atomic packing factor or packing density is a measure of the density of crystalline solids. The more is the atomic packing factor, the more closely the atoms are packed and larger is the density of the solid. It is defined as the ratio of the volume of all the atoms in the unit cell to its total volume, i.e.,

$$\text{APF} = \frac{\text{Volume of all the atoms in the unit cell}}{\text{Volume of the unit cell}}$$

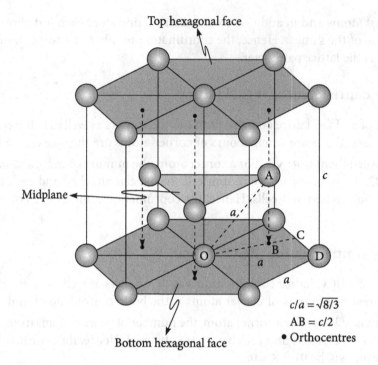

Top hexagonal face

Midplane

Bottom hexagonal face

$c/a = \sqrt{8/3}$
$AB = c/2$
• Orthocentres

Figure 1.9 | Unit cell of an HCP crystal structure

In the calculation of the atomic packing factor, it is assumed that the atoms have spherical shapes and that atoms in a unit cell are in contact with each other. In the following, we will calculate the atomic packing factor in SC, FCC, BCC and HCP.

1.6.1 Simple cubic (SC) lattice

The unit cell of a simple cubic lattice contains atoms only at the corners. To each corner, eight unit cells have been attached and hence, each corner atom is shared by eight unit cells, i.e., each corner atom contributes only 1/8 of its volume. In other words, the eight corner atoms contribute as a whole only one atom to an unit cell. If a is the lattice parameter and r is the atomic radius then we have

$2r = a$ or $r = a/2$.

The volume of one atom is

$$\frac{4\pi r^3}{3} = \frac{4\pi(a/2)^3}{3} = \frac{\pi a^3}{6}$$

Hence, $\text{APF} = \dfrac{\text{Volume of all the atoms}}{\text{Volume of the unit cell}} = \dfrac{\pi a^3}{6} \times \dfrac{1}{a^3} = \dfrac{\pi}{6} = 0.524$

1.6.2 Face centred cubic (FCC) lattice

The unit cell of an FCC lattice contains atoms at the corners as well as at the centre of each face. To each corner, eight unit cells have been attached and hence, each corner atom is shared by eight unit cells, i.e., each corner atom contributes only 1/8 of its volume. In other words, the eight corner atoms contribute as a whole only one atom to an unit cell. Each facial atom is shared by two unit cells, i.e., each facial atom contributes only 1/2 of its volume. Since there are six faces, the facial atoms contribute as a whole only three atoms. Thus, in an FCC, there are 03 atoms per unit cell. If a is the lattice parameter and r is the atomic radius, then we have

$$(4r)^2 = a^2 + a^2 \text{ or } r = \frac{a}{2\sqrt{2}}$$

The total volume of three atoms is

$$3 \times \frac{4\pi}{3} r^3 = 3 \times \frac{4\pi}{3} \left(\frac{a}{2\sqrt{2}} \right)^3 = \frac{\pi a^3}{3\sqrt{2}}$$

Hence, $APF = \dfrac{\text{Volume of all the atoms}}{\text{Volume of the unit cell}}$

$$= \frac{\pi a^3}{3\sqrt{2}} \times \frac{1}{a^3} = \frac{\pi}{3\sqrt{2}} = 0.740$$

1.6.3 Body centred cubic (BCC) lattice

The unit cell of a BCC lattice contains atoms at the corners as well as at its centre. To each corner, eight unit cells have been attached and hence, each corner atom is shared by eight unit cells. Eight corner atoms contribute as a whole only one atom. Thus, in BCC, there are 02 atoms in a unit cell. If a is the lattice parameter and r is the atomic radius, then we have

$$(4r)^2 = a^2 + a^2 + a^2 \text{ or } r = \frac{\sqrt{3}a}{4}$$

The total volume of two atoms is

$$2 \times \frac{4\pi}{3} r^3 = 2 \times \frac{4\pi}{3} \left(\frac{\sqrt{3}a}{4} \right)^3 = \frac{\pi\sqrt{3}a^3}{8}$$

Hence, $APF = \dfrac{\text{Volume of all the atoms}}{\text{Volume of the unit cell}}$

$$= \frac{\pi\sqrt{3}a^3}{8} \times \frac{1}{a^3} = \frac{\pi\sqrt{3}}{8} = 0.680$$

1.6.4 Hexagonal closed packed (HCP) lattice

The number of hexagonal unit cells attached to a corner point on a plane is 3 ($360° \div 120° = 3$) and thus, each corner atom is shared by 6 (3 + 3) unit cells. The central atom on the hexagonal face is shared by 2 unit cells. Thus, altogether, the number of atoms per unit cell is

$6 \times \dfrac{1}{6} + \dfrac{1}{2}$ (contribution from top face) + 3 (internal atoms on the mid-plane) + $6 \times \dfrac{1}{6} + \dfrac{1}{2}$ (contribution from bottom face) = 06.

From Fig. 1.9, we have

$$OA = a$$

$$OD = a$$

$$CD = a/2$$

$$OC = \sqrt{OD^2 - CD^2} = \sqrt{a^2 - (a/2)^2} = \frac{\sqrt{3}}{2}a$$

Since B is the orthocenter, we know from geometry that

$$OB = \frac{2}{3} \times OC$$

or $$OB = \frac{a}{\sqrt{3}}$$

From Fig. 1.9, we have

$$OA^2 = OB^2 + BA^2$$

or $$a^2 = (a/\sqrt{3})^2 + (c/2)^2$$

or $$\frac{c}{a} = \sqrt{\frac{8}{3}} = 1.633$$

However, for some HCP metals, this ratio deviates from this ideal value. It varies from 1.58 (Be) to 1.89 (Cd). As there is no reason to suppose that the atoms in these crystals are not in contact, it follows that they must be ellipsoidal in shape rather than spherical.

As evident in Fig. 1.9, 6 triangular prisms constitute a unit cell of HCP. Hence, the volume of the unit cell is

$$6 \times \text{area of triangular base} \times \text{height}$$

or $$6 \times \frac{\sqrt{3}\, a^2}{4} \times c = 6 \times \frac{\sqrt{3}\, a^3}{4} \times \frac{c}{a}$$

or $$= 6 \times \frac{\sqrt{3}\, a^3}{4} \times \sqrt{8/3} = 3\sqrt{2}\, a^3$$

All the atoms on the top hexagonal face touch each other and also touch the atoms on the mid-plane. The same is the case with the atoms on the bottom face. If r is the atomic radius, we have

$$r = a/2$$

Hence, the volume of all the atoms in the unit cell is

$$= 6 \times \frac{4\pi}{3} r^3 = 6 \times \frac{4\pi}{3} (a/2)^3 = \pi a^3$$

Taking the ratio of the volume of all the atoms in the unit cell to the volume of the unit cell, we have

$$\text{APF} = \frac{\pi a^3}{3\sqrt{2}a^3} = \frac{\pi}{3\sqrt{2}} = 0.74$$

From this, we conclude that FCC and HCP crystalline solids have more density than that of SC and BCC solids.

1.7 Structures of Typical Crystals

1.7.1 Diamond structure

Diamond is a metastable carbon polymorph (identical chemical composition but different crystalline structure) at room temperature and atmospheric pressure. Each carbon bonds to four other carbons in a tetrahedral configuration, and these bonds are totally covalent. Thus, the coordination number is four. The space lattice of a diamond is FCC. The diamond structure can be visualized as an inter-penetration of two FCC structures along the main body diagonal by a distance of $a/4$. The position of the origin of the second FCC unit cell is $(a/4, a/4, a/4)$. Hence, the nearest neighbour distance will be

$$\sqrt{(0-a/4)^2 + (0-a/4)^2 + (0-a/4)^2} = \sqrt{3}\frac{a}{4}$$

where a is the lattice parameter. This length is the bond length and also defines the diameter of the carbon atom in the structure. The coordinates of the four atoms within the unit cell are shown in Fig. 1.10. The two atoms at $(0, 0, 0)$ and $(a/4, a/4, a/4)$ constitute the basis.

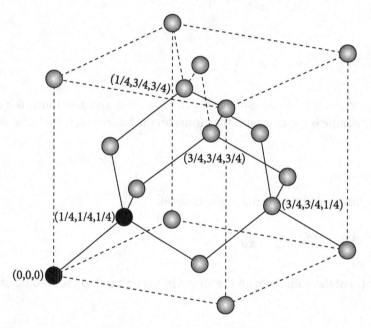

Figure 1.10 Crystal structure of a diamond showing the tetrahedral bond arrangements. The two atoms shown as blackened circles constitute the basis

APF of the diamond cubic structure

The unit cell as shown in Fig. 1.10, has 8 corner atoms, 6 face centred atoms and 4 atoms within the body. The contribution of all the atoms to the unit cell is

$$8 \times \frac{1}{8} + 6 \times \frac{1}{2} + 4 = 8$$

The diameters of the carbon atom as calculated earlier is

$$2r = \sqrt{3}\frac{a}{4}$$

The volume of 8 carbon atoms is

$$= 8 \times \frac{4\pi}{3} r^3 = \frac{4\pi}{3} \left(\sqrt{3} \frac{a}{8} \right)^3 = \frac{\pi \sqrt{3}}{16} a^3$$

Thus, the APF of the diamond cubic structure is obtained as

$$APF = \frac{\dfrac{\pi \sqrt{3}}{16} a^3}{a^3} = \frac{\pi \sqrt{3}}{16} = 0.34$$

Due to low APF, the diamond structure is relatively empty. The silicon, germanium and *a–tin* crystallizes in diamond cubic structure.

1.7.2 Cubic ZnS structure

The mineralogical terms for zinc sulfide (or zinc sulphide) are zinc blende, or sphalerite. The unit cell is shown in Fig. 1.11. The structure is very similar to the diamond cubic structure except that here the two atoms in the basis are of different kinds. The diamond belongs to the centrosymmetric crystal system having a centre of inversion whereas cubic ZnS belongs to

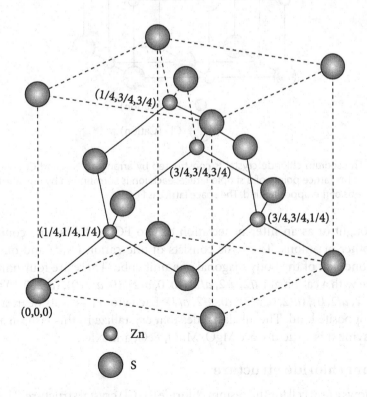

Figure 1.11 Crystal structure of cubic zinc sulphide, ZnS

the non-centrosymmetric crystal system having no centre of inversion. The coordinates of the four Zn atoms within the unit cell are $(a/4, a/4, a/4)$, $(a/4, 3a/4, 3a/4)$, $(3a/4, 3a/4, 3a/4)$, $(3a/4, 3a/4, a/4)$ and the coordinates of the sulphur atoms are $(0, 0, 0)$, $(0, a/2, a/2)$, $(a/2, 0, a/2)$, $(a/2, a/2, 0)$. There are four molecules of ZnS per unit cell. Each atom is surrounded by four atoms of the other type in a regular tetrahedral configuration. The compounds having cubic zinc sulphide structure are CuCl, InSb, SiC and CdS.

1.7.3 Sodium chloride structure

Perhaps the most common crystal structure is the sodium chloride (NaCl) structure. The coordination number for both cations (Na$^+$) and anions (Cl$^-$) is 6. The Na$^+$ and Cl$^-$ radii ratio lies approximately between 0.414 and 0.732. A unit cell for this crystal structure is shown in Fig. 1.12.

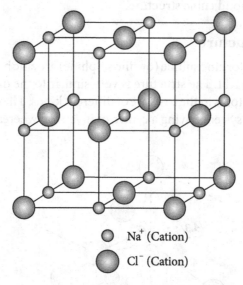

$\bigcirc$ Na$^+$ (Cation)

$\bigcirc$ Cl$^-$ (Cation)

Figure 1.12 | The sodium chloride crystal is constructed by arranging Na$^+$ and Cl$^-$ ions alternately at the lattice points of a simple lattice; each ion is surrounded by 6 nearest neighbour ions of the opposite kind. The space lattice is FCC

It may be thought of as an inter-penetration of two FCC lattices, one composed of the cations, the other of anions. The basis consists of one cation (Na$^+$) and one anion (Cl$^-$) separated by one-half of the body diagonal of a unit cube. There are four units of NaCl in each unit cube with a cation at $(a/2, a/2, a/2)$, $(0, 0, a/2)$, $(0, a/2, 0)$, $(a/2, 0, 0)$ and an anion at $(0, 0, 0)$, $(a/2, a/2, 0)$, $(a/2, 0, a/2)$, $(0, a/2, a/2)$. Each atom has as its nearest neighbour 6 atoms of the opposite kind. The alkali halides that crystallize in this fashion are KCl, KBr, and oxides having this structure are MgO, MnO, FeO, and NiO.

1.7.4 Cesium chloride structure

Figure 1.13 shows a unit cell for the cesium chloride (CsCl) crystal structure. The space lattice is of the simple cubic type. The cation Cs$^+$ at $(0,0,0)$ and anion Cl$^-$ at $(a/2, a/2, a/2)$ constitute a basis. Since each ion is at the centre of a cube of ions of the other kind, the coordination number is 8. In our case as shown in the figure, the anions Cl⁻ are located at each of the corners of a cube, whereas the cube centre is a single cation Cs⁺. Interchange of anions with cations,

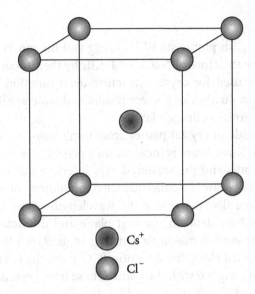

Figure 1.13 | Crystal structure of cesium chloride CsCl

and vice versa, produces the same crystal structure. It may be thought of as the inter-penetration of two simple cubic lattices with the corner of one sub-lattice as the body centre of the other. This is not a BCC crystal structure because ions of two different kinds are involved.

1.8 Diffraction by Crystal

Interplanar spacing of crystal planes in crystals are of the order of a few angstroms. If a suitable radiation of wavelength of the order of interplanar spacing is incident on a crystal, it will be scattered by the atoms of the crystals. The scattered rays will form a diffracted beam that contains all the information about the crystals. Thus, crystal structures along with their properties can be determined by analyzing the diffracted beam. The diffracted beam is defined as a beam composed of a large number of scattered rays mutually reinforcing one another.

The de Broglie wavelengths $[\lambda = (h/p)]$ of accelerated electrons, neutrons and X-rays are of the same order as that of interplanar spacing of crystals. Therefore, crystals act as three-dimensional diffraction gratings for X-rays, electrons, protons, and neutrons. When these radiations are incident on a crystal, maximum diffraction occurs when certain conditions relating to interplanar spacing, wavelength of the incident radiation and angle of diffraction are satisfied. These conditions at which maximum diffraction occur are determined by Bragg's law and Laue's conditions. In general, for crystal structure determination by diffraction method, X-rays are used. Visible light cannot be used for crystal diffraction because the wavelength of visible light is thousands times larger than the interplanar spacing of crystal planes.

1.8.1 Bragg's law

In the year 1913, two English physicists W.H. Bragg and his son W.L. Bragg, particularly the latter, determined the structure of NaCl, KCl, KBr by their theory of X-ray diffraction. The basic equation they used for crystal structure determination became very popular due to its simplicity, reproduction of correct results and wide applications. Later on, the equation became to be known as Bragg's law.

A perfect crystal is made of crystal planes containing basis in periodic arrangements. When a monochromatic X-ray beam is incident on a crystal, it is elastically scattered from periodically arranged atoms and the scattered rays interfere due to the phase difference between them. Thus, Bragg's law is a consequence of periodicity of atoms in crystals.

Let d be the interplanar distance and λ the wavelength of the monochromatic X-ray beam. Consider a ray AB incident on the first plane and diffracted along the direction BC from the plane by the atom B making a glancing angle θ with the plane. Similarly, the parallel ray $A'B'$ is diffracted along the direction $B'C'$ from the next plane by the atom B' making the same glancing angle θ with the plane. These have been depicted in Fig. 1.14.

The diffracted beams BC and $B'C'$ will interfere constructively or destructively depending on their path difference. To determine the path difference between the two rays ABC and $A'B'C'$, normals BD and BE are drawn from the point B to the ray $A'B'$ and $B'C'$. The path difference δ between the diffracted beams BC and $B'C'$ as seen from Fig. 1.14 is given by

$$\delta = DB'E = DB' + B'E$$

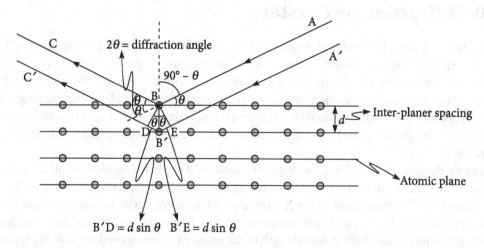

Figure 1.14 | Diffraction of monochromatic X-ray by crystal planes

In the triangle BDB',

$$\sin\theta = \frac{DB'}{BB'} = \frac{DB'}{d}, \text{ since } BB' = d$$

or $DB' = d\sin\theta$

Similarly, $B'E = d\sin\theta$

Addition of these two equations gives

$$DB' + B'E = 2d\sin\theta$$

or $\delta = 2d\sin\theta$ 　　　　　　　　　　　　　　　　　　(1.11)

The diffracted beams BC and $B'C'$ will interfere constructively and produce maximum intensity if the path difference δ between them is an integral multiple of the wavelength, i.e., $\delta = n\lambda$. Thus, we have

$$2d\sin\theta = n\lambda$$ 　　　　　　　　　　　　　　　　　　(1.12)

$n = 1, 2, 3, 4 \ldots$ is called the order of diffraction. Equation (1.12) is the famous Bragg's law.

In deriving this equation, we have not taken into consideration the effect of the refractive index of the medium on the direction of propagation of X-rays. According to Snell's law, refractive index $\mu\left(=\dfrac{c}{v} = \dfrac{\lambda v}{\lambda' v} = \dfrac{\lambda}{\lambda'}\right)$ is given as

$$\frac{\lambda}{\lambda'} = \mu = \frac{\sin i}{\sin r} = \frac{\sin(90-\theta)}{\sin(90-\theta')} = \frac{\cos\theta}{\cos\theta'}$$

$$n\frac{\lambda}{\mu} = 2d\sin\theta = 2d\sqrt{1-\cos^2\theta'}$$

$$n\lambda = 2d\sqrt{\mu^2 - \cos^2\theta}$$

$$= 2d\sqrt{\mu^2 - 1 + \sin^2\theta}$$

$$= 2d\sin\theta\sqrt{1 - \frac{1-\mu^2}{\sin^2\theta}}$$

$$= 2d \sin\theta \sqrt{1 - \frac{(1-\mu)(1+\mu)}{\sin^2\theta}}$$

$$= 2d \sin\theta \left(1 - \frac{2(1-\mu)}{\sin^2\theta}\right)^{1/2} \quad (\text{since } \mu + 1 \approx 2)$$

$$= 2d \sin\theta \left(1 - \frac{1}{2}\frac{2(1-\mu)}{\sin^2\theta}\right)$$

$$= 2d \sin\theta \left(1 - \frac{1-\mu}{\frac{(n\lambda)^2}{(2d)^2}}\right), \; (\text{since } 2d \sin\theta = n\lambda)$$

$$n\lambda = 2d \sin\theta \left(1 - \frac{4(1-\mu)d^2}{n^2\lambda^2}\right)$$

This equation is the modified Bragg's law with a small correction term $\frac{4(1-\mu)d^2}{n^2\lambda^2}$ which is much less than 1. For higher order diffraction, this equation boils down to $2d\sin\theta = n\lambda$.

Discussions

The direction of the diffracted beam shown in Fig is considered as a negative edge. 1.14 is the direction along which constructive interference occurs. For a first order diffraction, n = 1 and the path difference is λ since $\delta = n\lambda$. The diffraction is due to the rays diffracted from the first and second plane. For second order diffraction, $n = 2$ and the path difference is 2λ. The diffraction is due to the rays diffracted from the first and third plane. Similarly, for the 3rd, 4th, ... order diffraction. The rays scattered by all the atoms in all the planes are therefore in phase and interfere constructively to form a diffracted beam in the direction shown in Fig. 1.14.

The diffraction of X-rays by crystals and the reflection of visible light by mirrors appear very similar since in both cases, the angle of incidence is equal to the angle of reflection as shown in Fig. 1.14. However, fundamentally, they differ in the following ways.

Reflection of visible light	Diffraction of X-rays
The reflection of visible light takes place on a thin surface layer only	The diffracted beam from a crystal is built up of rays scattered by all the atoms which lie in the path of the incident beam.

The reflection of visible light takes place for any angle of incidence	The diffraction of monochromatic X-rays by a crystal occurs for those particular angles of incidence satisfying Bragg's law.
The reflection of visible light by a good mirror is almost 100%.	The intensity of a diffracted X-ray beam is very low as compared to that of the incident beam.

Experimental aspects of Bragg's law

The angle between the direction of incident beam and the diffracted beam as can be seen in Fig. 1.14 is 2θ and is called the diffraction angle. This angle 2θ is measured experimentally. Since $\sin\theta$ cannot exceed unity, we have

$$\frac{n\lambda}{2d} = \sin\theta < 1 \quad \text{or} \quad \frac{n\lambda}{2d} < 1 \tag{1.13}$$

Equation (1.13) shows that $n\lambda$ must be less than two times the interplanar spacing for the diffraction to occur. The smallest value of $n = 1$ which implies that the condition for diffraction at any observable angle 2θ is

$$\lambda < 2d$$

For most crystals, $d \approx 3$ Å or less which means that the wavelength should not exceed 6 Å for the diffraction phenomena to occur in crystals.

To facilitate the experimental aspects, Bragg's law can be modified in such a way that any order of diffraction can be considered as the first order diffraction. Bragg's law can be written as

$$2\frac{d}{n}\sin\theta = 1 \times \lambda$$

Now since the coefficient of λ is unity, we can always consider diffraction of any order as a first order from the planes, real or virtual, having interplanar spacing (d/n). Thus, we can express Bragg's law (1.12) in a more useful form as

$$2d\sin\theta = \lambda \tag{1.14}$$

Characteristic features of Bragg's law

i. Bragg's law is a consequence of the periodicity of the space lattice.

ii. The law does not refer to the arrangement or basis of the atoms associated with its lattice point.

iii. The composition of the basis determines the relative intensity of various orders of diffraction from a given set of parallel planes.

iv. Bragg's diffraction can occur only for wavelength $\lambda < 2d$.

v. For the same order and spacing, the angle θ decreases as the wavelength decreases.

Example 1.7

An X-ray tube operates at a potential difference of 40 kV with a copper target. For a certain crystal, the first order diffraction is observed at a diffraction angle 31.6° for the wavelength 1.54 Å. Calculate the interplanar spacing of the crystal.

Solution

The data given in the question are $n = 1$, $2\theta = 31.6°$, $\lambda = 1.54$ Å.

$$2d \sin\theta = n\lambda \implies d = \frac{n\lambda}{2\sin\theta} = \frac{1 \times 1.54}{2 \times \sin 15.8}\overset{\text{o}}{\text{A}} = 2.83 \overset{\text{o}}{\text{A}}$$

Example 1.8

A beam of X-rays of wavelength 1.54 Å is incident on a crystal at a glancing angle 13°40′ when the first order Bragg's reflection occurs. Calculate the glancing angle for the third order reflection.

Solution

The data given in the question are $n_1 = 1$, $n_2 = 3$, $\theta_1 = 13°40′ = 13.67°$, $\lambda = 1.54$ Å.

$$\frac{2d \sin\theta_2}{2d \sin\theta_1} = \frac{n_2 \lambda}{n_1 \lambda}$$

or $\quad \sin\theta_2 = \frac{n_2}{n_1} \times \sin\theta_1 = \frac{3}{1} \times \sin 13.67 = 0.71$

or $\quad \theta_2 = 45.2°$

Hence, the glancing angle for the third order reflection is 45.2°.

Example 1.9

The spacing of {100} planes in an NaCl crystal is 2.820 Å. The X-rays incident on the surface of this crystal is found to give rise to first order reflection at a grazing angle of 15.8°. Calculate the wavelength of the X-rays used.

Solution

The data given in the question are $d = 2.820$ Å, $n = 1$, $\theta = 15.8°$.

From Bragg's law, we have

$$\lambda = \frac{2d\sin\theta}{n} = \frac{2\times 2.820\times\sin 15.8}{1}\overset{o}{A} = 1.54\overset{o}{A}$$

Example 1.10

A beam of X-rays of wavelength 1.54 Å is incident on a cubic crystal at 13°40′ when the first order Bragg's reflection occurs from {112} planes. Calculate the interatomic spacing.

Solution

The data given in the question are $\lambda = 1.54$Å, $\theta = 13°40′ = 13.67°$, $\{hk\ell\} = \{112\}$
From Bragg's law, we get

$$d = \frac{n\lambda}{2\sin\theta} = \frac{1\times 1.54}{2\times\sin 13.67}\overset{o}{A} = 3.26\overset{o}{A}.$$

The relation between interatomic distance of a cubic lattice a and interplanar spacing d is given by

$$d = \frac{a}{\sqrt{h^2 + k^2 + \ell^2}}$$

or $a = d\sqrt{h^2 + k^2 + \ell^2} = 3.26\overset{o}{A}\sqrt{1^2 + 1^2 + 2^2} = 7.98\overset{o}{A}$

1.8.2 Diffraction directions

The directions along which diffracted beam has the maximum intensity are called diffraction directions. These are the directions which satisfy Bragg's law. The diffraction angle 2θ gives the diffraction direction. The interplanar spacing in ($hk\ell$) planes in an orthogonal ($\alpha = \beta = \gamma = 90°$) crystal system is given from Eq. (1.6) as

$$\frac{1}{d_{hk\ell}^2} = \frac{h^2}{a_1^2} + \frac{k^2}{a_2^2} + \frac{\ell^2}{a_3^2} \tag{1.15}$$

For ($hk\ell$) diffraction from Bragg's law, we can have

$$d_{hk\ell}^2 = \frac{\lambda^2}{4\sin^2\theta} \tag{1.16}$$

Putting Eq. (1.16) into (1.15), we have

$$\sin^2 \theta = \frac{\lambda^2}{4}\left(\frac{h^2}{a_1^2}+\frac{k^2}{a_2^2}+\frac{\ell^2}{a_3^2}\right) \tag{1.17}$$

Equation (1.17) gives the direction along which the diffracted beam has maximum intensity in orthogonal crystal systems.

i. Cubic system $\sin^2 \theta = \dfrac{\lambda^2}{4}\left(\dfrac{h^2+k^2+\ell^2}{a^2}\right)$ $\tag{1.18}$

ii. Tetragonal system $\sin^2 \theta = \dfrac{\lambda^2}{4}\left(\dfrac{h^2+k^2}{a_1^2}+\dfrac{\ell^2}{a_3^2}\right)$ $\tag{1.19}$

iii. Orthorhombic system $\sin^2 \theta = \dfrac{\lambda^2}{4}\left(\dfrac{h^2}{a_1^2}+\dfrac{k^2}{a_2^2}+\dfrac{\ell^2}{a_3^2}\right)$ $\tag{1.20}$

iv. Hexagonal system $\sin^2 \theta = \dfrac{\lambda^2}{3}\left(\dfrac{h^2+hk+k^2}{a_1^2}\right)+\dfrac{\lambda^2\ell^2}{4a_3^2}$ $\tag{1.21}$

Combining Bragg's law with the expression for interplanar spacing of $(hk\ell)$ planes, we conclude that diffraction directions are solely determined by the shape and size of the unit cell of a crystal system. Conversely from the diffraction directions, we can possibly determine the shape and size of the unit cell of a crystal system.

1.9 Reciprocal Lattice

In the previous sections, we have seen that X-ray diffraction is ideally equivalent to reflections by a set of crystal planes and are describable by Bragg's law. Yet there are diffraction effects which Bragg's law is totally unable to explain, particularly those involving diffuse scattering. The reciprocal lattice concept first given by German physicist Ewald in 1921 provides a general theory of X-ray diffraction. Bragg's law of X-ray diffraction can be expressed in more simple form in terms of the reciprocal lattice. The wave–mechanical behaviour of the electrons in the periodic crystal lattice is also more readily understood by the concept of reciprocal lattice.

In real crystals, we have to deal with many sets of crystal planes, with a variety of orientations and spacings that diffract a beam of X-rays. They become too difficult to be visualized. However, we know that the orientation of a plane is determined by a normal drawn on the plane. For our convenience, a normal is one-dimensional, whereas a plane is two-dimensional. Therefore, wherever the question of multitude of planes arises, we can always think of them in terms of their normals. Now, like direct lattices, we will introduce the concept of reciprocal lattices geometrically in the following way.

If the length assigned to each normal is proportional to the reciprocal of the interplanar spacing of that plane, the points at the end of the normals drawn from a common origin form a lattice called the reciprocal lattice. From this definition of reciprocal lattice, the following steps are followed in constructing a reciprocal lattice.

i.   Select a point as the origin.

ii.   From this origin, draw normals to every set of planes in the direct lattice.

iii.   Set the length of each normal equal to the reciprocal of the interplanar spacing for this particular set of planes.

iv.   Put a point at the end of each normal.

The set of points so obtained is the reciprocal lattice. Thus, the points in the reciprocal lattice represent a tabulation of

i.   The normals to all the direct lattice planes.

ii.   The interplanar spacings of the direct lattice planes.

Therefore, each point in the reciprocal lattice preserves all the characteristics of each direct lattice plane, the point represents. The direction of a point from the origin specifies the orientation of the planes it represents and the distance of the point from the origin represents the interplanar spacing of the set of planes. Thus, each point in the reciprocal lattice contains all the information about the set of plane it represents.

1.9.1 Reciprocal lattice vector

If $\vec{a}_1$, $\vec{a}_2$, and $\vec{a}_3$ are fundamental translation vectors in the direct lattice, then the fundamental translation vectors $\vec{b}_1$, $\vec{b}_2$ and $\vec{b}_3$ in reciprocal lattice are defined by

$$\vec{b}_1 = 2\pi \frac{\vec{a}_2 \times \vec{a}_3}{\vec{a}_1 \cdot \vec{a}_2 \times \vec{a}_3}, \ \vec{b}_2 = 2\pi \frac{\vec{a}_3 \times \vec{a}_1}{\vec{a}_1 \cdot \vec{a}_2 \times \vec{a}_3}, \ \vec{b}_3 = 2\pi \frac{\vec{a}_1 \times \vec{a}_2}{\vec{a}_1 \cdot \vec{a}_2 \times \vec{a}_3} \tag{1.22}$$

The factors 2π are not used by crystallographers in defining the fundamental translation vectors in the reciprocal lattice.

From the process of a construction of a reciprocal lattice, it is pretty clear that $\vec{b}_1$ is perpendicular to $\vec{a}_2$ and $\vec{a}_3$, $\vec{b}_2$ is perpendicular to $\vec{a}_3$ and $\vec{a}_1$ and $\vec{b}_3$ is perpendicular to $\vec{a}_1$ and $\vec{a}_2$. These can be inferred from Eq. (1.22). All these relations are expressed compactly by

$$\vec{b}_i \cdot \vec{a}_j = 2\pi \delta_{ij} \tag{1.23}$$

where $\delta_{ij} = 1$ if $i = j$ and $\delta_{ij} = 0$ if $i \neq j$

Like the lattice translation vector $\vec{T} = n_1\vec{a}_1 + n_2\vec{a}_2 + n_3\vec{a}_3$, the reciprocal lattice vector $\vec{G}$ is defined as

$$\vec{G} = v_1\vec{b}_1 + v_2\vec{b}_2 + v_3\vec{b}_3 \tag{1.24}$$

where v_1, v_2, and v_3 are integers. The reciprocal lattice vector $\vec{G}$ defines all the points in the reciprocal lattice.

Thus, we have associated two lattices, one a direct lattice and the other a reciprocal lattice to every crystal structure. A diffraction pattern of a crystal is a map of the reciprocal lattice of the crystal, whereas a microscopic image, which can be resolved on a fine enough scale is a map of the direct lattice. The reciprocal lattice is a lattice in Fourier space or reciprocal space, whereas the direct lattice is a lattice in real space or ordinary space, the space in which we are living. When we rotate a crystal, we rotate both the reciprocal lattice and the direct lattice. The unit of fundamental translation vectors in the direct lattice is the unit of length, i.e., meter but the unit of fundamental translation vectors in reciprocal lattice is inverse of unit of length, i.e., meter^{-1}. As n_1, n_2 and n_3 of $\vec{T} = n_1\vec{a}_1 + n_2\vec{a}_2 + n_3\vec{a}_3$ are the coordinates of a point in real space, v_1, v_2 and v_3 of $\vec{G} = v_1\vec{b}_1 + v_2\vec{b}_2 + v_3\vec{b}_3$ are the coordinates of a point in reciprocal space or Fourier space. The wavevectors $\vec{k}$s are always drawn in reciprocal space. The crystal planes in the real space are represented by points in reciprocal space.

An interesting relation is obtained by taking the dot product of the lattice translation vector $\vec{T}$ and the reciprocal lattice vector $\vec{G}$. $\vec{G} \cdot \vec{T}$ is the projection of vector $\vec{T}$ in the direction of $\vec{G}$. Thus, we have

$$\vec{G} \cdot \vec{T} = (v_1\vec{b}_1 + v_2\vec{b}_2 + v_3\vec{b}_3) \cdot (n_1\vec{a}_1 + n_2\vec{a}_2 + n_3\vec{a}_3)$$

Taking the help of Eq. (1.23) from this equation, we get

$$\vec{G} \cdot \vec{T} = 2\pi(v_1 n_1 + v_2 n_2 + v_3 n_3)$$

Since n_1, n_2, n_3 v_1, v_2, and v_3 are integers, $v_1 n_1 + v_2 n_2 + v_3 n_3$ should be an integer say equal to N. Therefore this equation becomes

$$\vec{G} \cdot \vec{T} = 2\pi \times \text{int}\,eger = 2\pi N \tag{1.25}$$

or $e^{i(\vec{G} \cdot \vec{T})} = 1 \tag{1.26}$

Example 1.11

The basis vectors of a hexagonal lattice may be defined as

$$\vec{a}_1 = \frac{\sqrt{3}}{2}a\hat{x} + \frac{1}{2}a\hat{y}, \vec{a}_2 = -\frac{\sqrt{3}}{2}a\hat{x} + \frac{1}{2}a\hat{y}, \vec{a}_3 = c\hat{z}.$$

Calculate the reciprocal lattice translation vectors.

Solution

We have already calculated

$$V = \vec{a}_1.\vec{a}_2 \times \vec{a}_3 = \frac{\sqrt{3}}{2}a^2c \text{ in Example 1.1.}$$

$$\vec{a}_1 \times \vec{a}_2 = \left(\frac{\sqrt{3}}{2}a\hat{x} + \frac{1}{2}a\hat{y}\right) \times \left(-\frac{\sqrt{3}}{2}a\hat{x} + \frac{1}{2}a\hat{y}\right) = \hat{z}\frac{\sqrt{3}}{2}a^2$$

or $$\vec{b}_3 = 2\pi \frac{\vec{a}_1 \times \vec{a}_2}{\vec{a}_1 \cdot \vec{a}_2 \times \vec{a}_3} = \frac{2\pi}{c}\hat{z}$$

$$\vec{a}_2 \times \vec{a}_3 = \left(-\frac{\sqrt{3}}{2}a\hat{x} + \frac{1}{2}a\hat{y}\right) \times c\hat{z} = \frac{\sqrt{3}}{2}ac\hat{y} + \frac{ac}{2}\hat{x}$$

or $$\vec{b}_1 = 2\pi \frac{\vec{a}_2 \times \vec{a}_3}{\vec{a}_1 \cdot \vec{a}_2 \times \vec{a}_3} = \frac{2\pi}{\sqrt{3}a}\hat{x} + \frac{2\pi}{a}\hat{y}$$

$$\vec{a}_3 \times \vec{a}_1 = c\hat{z} \times \left(\frac{\sqrt{3}}{2}a\hat{x} + \frac{1}{2}a\hat{y}\right) = \frac{\sqrt{3}}{2}ac\hat{y} - \frac{ac}{2}\hat{x}$$

or $$\vec{b}_2 = 2\pi \frac{\vec{a}_3 \times \vec{a}_1}{\vec{a}_1 \cdot \vec{a}_2 \times \vec{a}_3} = -\frac{2\pi}{\sqrt{3}a}\hat{x} + \frac{2\pi}{a}\hat{y}$$

1.9.2 Properties of reciprocal lattices

A reciprocal lattice is an invariant geometrical object, whose properties are fundamental in the theory of solids. Some geometrical properties of a reciprocal lattice are as follows:

i.   Every reciprocal lattice vector is normal to a certain crystal plane of the direct lattice. Let $\vec{G} = h\vec{b}_1 + k\vec{b}_2 + \ell\vec{b}_3$ be a reciprocal lattice vector. We shall show that the vector $\vec{G}$ is perpendicular to the $(hk\ell)$ plane in the direct lattice.

In Fig. 1.15, the $(hk\ell)$ plane is shown. From the definition of Miller indices, the intercepts which the $(hk\ell)$ plane makes with $\vec{a}_1$, $\vec{a}_2$, and $\vec{a}_3$ axes are $\frac{a_1}{h}$, $\frac{a_2}{k}$, and $\frac{a_3}{\ell}$

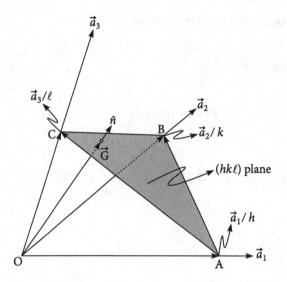

Figure 1.15 Relation between $\vec{G} = h\vec{b}_1 + k\vec{b}_2 + \ell\vec{b}_3$ and $(hk\ell)$ plane in the direct lattice

respectively. This has been shown in Fig. 1.7. As shown in this figure, the vector $\overrightarrow{AB}$ lies on the $(hk\ell)$ plane. Applying the rules of vectors to the vector triangle OAB, we have

$$\overrightarrow{AB} = \frac{\vec{a}_2}{k} - \frac{\vec{a}_1}{h}$$

Taking the dot product of $\vec{G}$ and $\overrightarrow{AB}$, we get

$$\vec{G} \cdot \overrightarrow{AB} = \left(h\vec{b}_1 + k\vec{b}_2 + \ell\vec{b}_3 \right) \cdot \left(\frac{\vec{a}_2}{k} - \frac{\vec{a}_1}{h} \right) = 2\pi - 2\pi = 0$$

This equation shows that $\vec{G}$ is perpendicular to $\overrightarrow{AB}$.
Similarly, from the vector triangle OAC, we have

$$\overrightarrow{AC} = \frac{\vec{a}_3}{\ell} - \frac{\vec{a}_1}{h}$$

Taking the dot product of $\vec{G}$ and $\overrightarrow{AC}$, we get

$$\vec{G} \cdot \overrightarrow{AC} = \left(h\vec{b}_1 + k\vec{b}_2 + \ell\vec{b}_3 \right) \cdot \left(\frac{\vec{a}_3}{\ell} - \frac{\vec{a}_1}{h} \right) = 2\pi - 2\pi = 0$$

This equation shows that $\vec{G}$ is perpendicular to $\overline{AC}$.

Since the vector $\vec{G} = h\vec{b}_1 + k\vec{b}_2 + \ell\vec{b}_3$ is normal to two vectors $\overline{AB}$ and $\overline{AC}$ on a single plane, the vector itself is normal to the plane containing the two vectors.

ii.   If the components of $\vec{G}$ have no common factor, then G is inversely proportional to the interplanar spacing of the lattice planes normal to $\vec{G}$.

Let $\hat{n}$ be a unit vector normal to the plane ABC as shown in Fig 1.15. Hence, we define the unit vector $\hat{n}$ as

$$\hat{n} = \frac{\vec{G}}{G} = \frac{h\vec{b}_1 + k\vec{b}_2 + \ell\vec{b}_3}{G}$$

As can be seen from Fig. 1.15, the interplanar spacing $d_{hk\ell}$ is given by

$$d_{hk\ell} = \hat{n} \cdot \frac{\vec{a}_1}{h} = \frac{h\vec{b}_1 + k\vec{b}_2 + \ell\vec{b}_3}{G} \cdot \frac{\vec{a}_1}{h} = \frac{h\vec{b}_1 \cdot \vec{a}_1 + k\vec{b}_2 \cdot \vec{a}_1 + \ell\vec{b}_3 \cdot \vec{a}_1}{hG}$$

$$= \frac{h2\pi + 0 + 0}{hG}$$

or $d_{hk\ell} = \dfrac{2\pi}{G}$ (1.27)

Thus $d_{hk\ell} \propto \dfrac{1}{G}$

iii. The volume of the unit cell of the reciprocal lattice $\vec{b}_1 \cdot \vec{b}_2 \times \vec{b}_3$ is inversely proportional to the volume of a unit cell $\vec{a}_1 \cdot \vec{a}_2 \times \vec{a}_3$ of the direct lattice.

The volume of the unit cell of the reciprocal lattice is given by

$$\vec{b}_1 \cdot \vec{b}_2 \times \vec{b}_3 = 2\pi \frac{\vec{a}_2 \times \vec{a}_3}{\vec{a}_1 \cdot \vec{a}_2 \times \vec{a}_3} \cdot 2\pi \frac{\vec{a}_3 \times \vec{a}_1}{\vec{a}_1 \cdot \vec{a}_2 \times \vec{a}_3} \times 2\pi \frac{\vec{a}_1 \times \vec{a}_2}{\vec{a}_1 \cdot \vec{a}_2 \times \vec{a}_3}$$

$$= \frac{8\pi^3}{\left(\vec{a}_1 \cdot \vec{a}_2 \times \vec{a}_3\right)^3} \left(\vec{a}_2 \times \vec{a}_3\right) \cdot \left\{\left(\vec{a}_3 \times \vec{a}_1\right) \times \left(\vec{a}_1 \times \vec{a}_2\right)\right\}$$

$$= \frac{8\pi^3}{\left(\vec{a}_1 \cdot \vec{a}_2 \times \vec{a}_3\right)^3} \left(\vec{a}_2 \times \vec{a}_3\right) \cdot \left[\left\{\vec{a}_3 \cdot \left(\vec{a}_1 \times \vec{a}_2\right)\right\} \vec{a}_1 - \left\{\vec{a}_1 \cdot \left(\vec{a}_1 \times \vec{a}_2\right)\right\} \vec{a}_3\right]$$

$$= \frac{8\pi^3}{\left(\vec{a}_1 \cdot \vec{a}_2 \times \vec{a}_3\right)^3} \left(\vec{a}_1 \cdot \vec{a}_2 \times \vec{a}_3\right)^2$$

$$\text{or } \vec{b}_1 \cdot \vec{b}_2 \times \vec{b}_3 = \frac{8\pi^3}{\vec{a}_1 \cdot \vec{a}_2 \times \vec{a}_3} \tag{1.28}$$

Thus, we have $\vec{b}_1 \cdot \vec{b}_2 \times \vec{b}_3 \propto \dfrac{1}{\vec{a}_1 \cdot \vec{a}_2 \times \vec{a}_3}$

iv.   The direct lattice is the reciprocal of its own reciprocal lattice.

We know from Eq. (1.23), that

$$\vec{a}_i \cdot \vec{b}_i = 2\pi \tag{1.29}$$

v.   This equation proves that the direct lattice is the reciprocal of its own reciprocal lattice.

vi.   The unit cell of a reciprocal lattice need not be a parallelepiped. In fact, we almost always deal with the Wigner–Seitz cell of the reciprocal lattice.

1.10 Diffraction and Reciprocal Lattices

Now we take a more general approach in which X-rays scattered from different individual atoms can recombine to give a diffracted beam. The great utility of a reciprocal lattice however lies in its connection with diffraction problems. We shall consider how X-rays scattered by the atom O at the origin of the direct lattice are affected by those scattered by any other atom at A. Let the coordinates of A with respect to the origin O be $n_1\vec{a}_1, n_2\vec{a}_2$, and $n_3\vec{a}_3$, where $n_1, n_2,$ and n_3 are integers. Hence, we can represent $\overline{OA} = \vec{r}$ by

$$\vec{r} = \overline{OA} = n_1\vec{a}_1 + n_2\vec{a}_2 + n_3\vec{a}_3 \tag{1.30}$$

Let us consider the parallel rays scattered by two identical atoms at O and A in the crystal separated by a distance $\vec{r} = \overline{OA}$.

This phenomenon is illustrated in Fig. 1.16. The diffracted rays are parallel. To determine the conditions under which diffraction will occur, we have to determine the phase difference between the rays scattered by the atoms at O and A. The phase difference is obtained from the path difference by multiplying the latter with $(2\pi/\lambda)$. The lines OB and OC in Fig. 1.16 are the incident and diffracted wave fronts respectively. Let

$\vec{k}_O$ = wave vector or propagation vector in the direction of incidence X-ray.

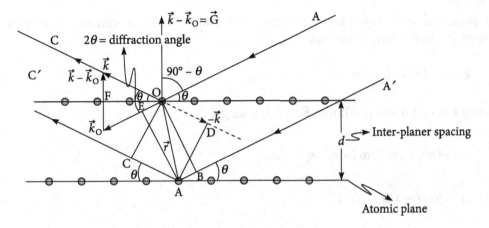

Figure 1.16 | Scattering of X-rays by atoms at O and A

$k_O = \dfrac{2\pi}{\lambda}$ = magnitude of wave vector or propagation vector $\vec{k}_O$.

$\hat{k}_O$ = unit vector in the direction of $\vec{k}_O$.

$\vec{k}$ = wave vector or propagation vector in the diffraction direction of X-ray.

$k = \dfrac{2\pi}{\lambda}$ = magnitude of wave vector or propagation vector $\vec{k}$.

$\hat{k}$ = unit vector in the direction of $\vec{k}$.

The path difference δ between the rays scattered by the atoms at O and A is given as

$\delta = BA + AC = OE + OD$ = Component of OA along $\hat{k}_O$ + Component of OA along $\hat{k}$

or $\qquad \delta = \hat{k}_O \cdot \vec{r} + \left(-\hat{k}\right)\cdot \vec{r} = \left(\hat{k}_O - \hat{k}\right)\cdot \vec{r}$ 　　　　　　　　　　　　(1.31)

Therefore, the phase difference φ between the rays scattered by the atoms at O and A will be given by

$\varphi = \dfrac{2\pi}{\lambda}\delta = -\dfrac{2\pi}{\lambda}\left(\hat{k} - \hat{k}_O\right)\cdot \vec{r} = -\left(\vec{k} - \vec{k}_O\right)\cdot \vec{r}$ 　　　　　　　　(1.32)

Now we know that $(\vec{k} - \vec{k}_O)$ is a vector in the reciprocal lattice normal to the crystal plane in the direct lattice. Therefore, the vector $(\vec{k} - \vec{k}_O)$ is called a scattering normal. By Eq. (1.32), uation diffraction is related to the reciprocal lattice. The vector in the reciprocal lattice is represented by $\vec{G} = h\vec{b}_1 + k\vec{b}_2 + \ell\vec{b}_3$, where h, k and ℓ are the coordinates of a point in reciprocal space. Thus, we can have

$$\vec{k} - \vec{k}_0 = h\vec{b}_1 + k\vec{b}_2 + \ell\vec{b}_3 \tag{1.33}$$

Putting Eqs (1.33) and (1.30) into Eq. (1.32), we get

$$\varphi = -\left(h\vec{b}_1 + k\vec{b}_2 + \ell\vec{b}_3\right).n_1\vec{a}_1 + n_2\vec{a}_2 + n_3\vec{a}_3$$

or $\quad \varphi = -2\pi\left(hn_1 + kn_2 + \ell n_3\right) \tag{1.34}$

We know pretty well that for the diffraction maxima to occur $\varphi = 2\pi n$ with $n = 1, 2, 3, \ldots$, i.e., n is an integer. Comparing Eq. (1.34) with $\varphi = 2\pi n$, we conclude that $hn_1 + kn_2 + \ell n_3$ must be an integer. According to Eq. (1.30), n_1, n_2 and n_3 are integers and so h, k, and ℓ has to be integers with the result that $hn_1 + kn_2 + \ell n_3$ is an integers. Thus, $hn_1 + kn_2 + \ell n_3$ is an integer when h, k and ℓ are integers. Therefore, we conclude that in the expression $\vec{k} - \vec{k}_0 = h\vec{b}_1 + k\vec{b}_2 + \ell\vec{b}_3$, h, k and ℓ all are integers. Thus, we arrive at one of most important expressions of X-ray diffraction

$$\vec{k} - \vec{k}_0 = h\vec{b}_1 + k\vec{b}_2 + \ell\vec{b}_3 = \vec{G} \tag{1.35}$$

where h, k and ℓ are integers. From Eq. (1.25), we have

$$\Delta\vec{k} = \vec{G}, \text{ where } \Delta\vec{k} = \vec{k} - \vec{k}_0 \tag{1.36}$$

or   Scattering normal = Reciprocal lattice vector.

In the elastic scattering of X-ray photons, its energy $h\nu$ is conserved, i.e., the energy of the incident photon and the scattered photon are equal. In this case, the magnitudes of $\vec{k}$ and $\vec{k}_0$ are equal and are equal to $(2\pi/\lambda)$. From Eq. (1.35), we have

$$\vec{k} + \vec{G} = \vec{k}_0 \quad \text{or} \quad (\vec{k} + \vec{G})\cdot(\vec{k} + \vec{G}) = k_0^2 \tag{1.37}$$

Since $k = k_0 = \dfrac{2\pi}{\lambda}$, we have from Eq. (1.37),

$$2\vec{k}\cdot\vec{G} + G^2 = 0 \tag{1.38}$$

This is the central result of the theory of elastic scattering of waves in a periodic lattice. If $\vec{G}$ is a reciprocal lattice vector, then $-\vec{G}$ is also a reciprocal lattice vector and with this substitution, Eq. (1.38) becomes

$$2\vec{k}\cdot\vec{G}=G^2 \tag{1.39}$$

It is to be noted that $\Delta\vec{k}=\vec{G}$ and $2\vec{k}\cdot\vec{G}=G^2$ are Bragg's equation in terms of the reciprocal vector. Bragg's law, $2d\sin\theta=\lambda$ can be derived from $\Delta\vec{k}=\vec{G}$ and $2\vec{k}\cdot\vec{G}=G^2$ easily.

1.10.1 Evaluation of scattering normal $\Delta\vec{k}=\vec{k}-\vec{k}_0$

The angle between $\vec{k}_o$ and $\vec{k}$ as shown in Fig. 1.16 is 2θ, which is called diffraction angle. Therefore, we have

$$\left|\vec{k}-\vec{k}_0\right|=\sqrt{k^2+k_0^2-2kk_0\cos2\theta}=k\sqrt{2}\sqrt{1-\cos2\theta}=2k\sin\theta$$

or $\quad \left|\vec{k}-\vec{k}_0\right|=\dfrac{4\pi}{\lambda}\sin\theta$ \hfill (1.40)

1.10.2 Laue's conditions

Vector form

Laue's conditions for X-ray diffraction can be derived from equation $\vec{k}-\vec{k}_0=h\vec{b}_1+k\vec{b}_2+\ell\vec{b}_3$ by dot product of this equation by $\vec{a}_1,\vec{a}_2,$ and $\vec{a}_3$ separately.

$$\vec{a}_1\cdot(\vec{k}-\vec{k}_0)=\vec{a}_1\cdot(h\vec{b}_1+k\vec{b}_2+\ell\vec{b}_3)$$

or $\quad \vec{a}_1\cdot(\vec{k}-\vec{k}_0)=2\pi h$ \hfill (1.41)

Similarly, $\vec{a}_2\cdot(\vec{k}-\vec{k}_0)=2\pi k$ \hfill (1.42)

and $\quad \vec{a}_3\cdot(\vec{k}-\vec{k}_0)=2\pi\ell$ \hfill (1.43)

The Eqs (1.41) to (1.43) are Laue's conditions in vector form which must be satisfied simultaneously for X-ray diffraction to occur.

Scalar form

If α, β and γ are the angles the vector $(\vec{k} - \vec{k}_0)$ makes with the crystallographic axes $\vec{a}_1, \vec{a}_2$, and $\vec{a}_3$ respectively, Eq. (1.41) becomes

$$a_1 \left| \vec{k} - \vec{k}_0 \right| \cos\alpha = 2\pi h$$

or $\quad a_1 \dfrac{4\pi}{\lambda} \sin\theta \cos\alpha = 2\pi h$

or $\quad 2a_1 \sin\theta \cos\alpha = \lambda h$ $\hfill$ (1.44)

Similarly, $\quad 2a_2 \sin\theta \cos\beta = \lambda k$ $\hfill$ (1.45)

and $\quad 2a_3 \sin\theta \cos\gamma = \lambda \ell$ $\hfill$ (1.46)

Equations (1.44) to (1.46) are called Laue's conditions or Laue's equations of X-ray diffraction in scalar form.

Bragg's law from Laue's conditions

The three Eqs (1.44) to (1.46) serve to determine a unique value for θ and scattering normal, thus defining a scattering direction. The direction cosines $\cos\alpha$, $\cos\beta$ and $\cos\gamma$ of the scattering normal for a particular wavelength are proportional to (h/a_1), (k/a_2) and $\dfrac{\ell}{a_3}$ respectively as known from Eqs (1.44) to (1.46). The neighbouring planes whose Miller indices are $(hk\ell)$ intersect crystallographic axes $\vec{a}_1, \vec{a}_2$, and $\vec{a}_3$ at (a_1/h), (a_2/k), and $\dfrac{a_3}{\ell}$. Therefore, the direction cosines of the normal to the $(hk\ell)$ planes, are also proportional to (h/a_1), (k/a_2), and $\dfrac{\ell}{a_3}$ respectively. Thus, the scattering normal $(\vec{k} - \vec{k}_0)$ is identical to the normal to the $(hk\ell)$ planes. Therefore, we conclude that the $(hk\ell)$ planes may be regarded as the diffracting planes Bragg conceived.

Now putting

$$h = \frac{a_1 \cos\alpha}{d}, \quad k = \frac{a_2 \cos\beta}{d} \quad \text{and} \quad \ell = \frac{a_3 \cos\gamma}{d}$$

into Eqs (1.44) to (1.46), Laue's conditions boils down to Bragg's law as

$$2d \sin\theta = \lambda$$

Thus, Bragg's law is derived from the equation of Laue's conditions.

1.10.3 Bragg's law from $\Delta \vec{k} = \vec{G}$

Given that $\Delta \vec{k} = \vec{G}$

or $\quad \left| \vec{k} - \vec{k}_0 \right| = \left| \vec{G} \right|$

or $\quad \dfrac{4\pi}{\lambda} \sin \theta = \dfrac{2\pi}{d}$ [by Eqs (1.40) and (1.27)]

or $\quad 2d \sin \theta = \lambda$

Thus, Bragg's law is proved from the equation $\Delta \vec{k} = \vec{G}$.

1.11 Intensity of Diffracted Beam

The intensity of the diffracted beam depends on the positions of the atoms in the unit cell. For some positions of the atoms in the unit cell, the diffracted beam may be extinct! In a reverse way, we can determine the positions of the atoms in the unit cell (hence the type of unit cell) by observing the diffracted intensities. This can be accomplished by establishing a perfect relationship between intensity and position of atoms in the unit cell. Since this is a complicated method, we shall proceed step by step; first scattering by an electron, then by an atom and finally by the unit cell.

1.11.1 Scattering by an electron

An X-ray is an electromagnetic radiation. When X-ray is incident on an electron at rest, sinusoidally varying electromagnetic force sets the electron into oscillation as a result of which the electron emits electromagnetic radiation. We call this scattering of X-rays by electrons. It is assumed that this scattering is coherent, i.e., incident wavelength and scattered wavelength are equal (reverse is the incoherent scattering). This coherent scattering is called Thomson scattering and is depicted in Fig. 1.17.

The intensity at point P is given by

$$I_p = I_0 \frac{K}{r^2} \times \frac{1 + \cos^2 2\theta}{2} \quad \text{with } K = \left(\frac{\mu_0 e^2}{4\pi m_e} \right)^2 \tag{1.47}$$

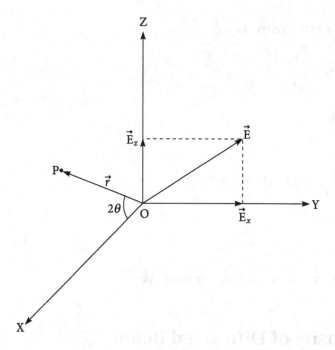

Figure 1.17 | Thomson scattering of X-rays by an electron. The incident X-ray beam is travelling in the direction of OX and encounters an electron at O. P is a point on the XZ plane. $\dfrac{1+\cos^2 2\theta}{2}$ is called the polarization factor

This equation shows that the intensity decreases inversely with the square of the distance. It also shows that the scattered beam is stronger in the forward direction ($2\theta = 0°$) or the backward direction ($2\theta = 180°$) than in a direction at right angles to the incident beam.

Apart from Thomson scattering, an electron can scatter X-ray by the Compton effect. This Compton scattering is an incoherent scattering. It cannot take part in diffraction because its phase is randomly related to that of the incident beam. Therefore, Compton scattering contributes nothing to the diffracted beam intensity.

1.11.2 Scattering by an atom

Bragg's law gives the condition for diffraction from point scattering centers in which electrons have been associated with geometrical points of the space lattice. However, in reality, X-rays are scattered by a cloud of electrons surrounding the nuclei rather than by point particles. Since the size of the atom is of the order of the X-ray wavelength, all parts of the atom do not scatter in phase and the interference of waves scattered from different parts of the electron distribution has to be considered. Due to the heavy mass of the nucleus of the atom, the nucleus scatters X-rays too weakly to be taken into account.

The atomic scattering factor or atomic form factor f of an atom is defined as the ratio of the radiation amplitude scattered by the electron distribution in an atom to the radiation amplitude scattered by an electron. Mathematically, it is given as

$$f = \frac{\text{Amplitude of the wave scattered by an atom}}{\text{Amplitude of the wave scattered by an electron}} \qquad (1.48)$$

The atomic scattering factor is a measure of the efficiency of the given atom to scatter incident X-rays in a given direction. Schematically and ideally, scattering of X-rays by an atom is shown in Fig. 1.18.

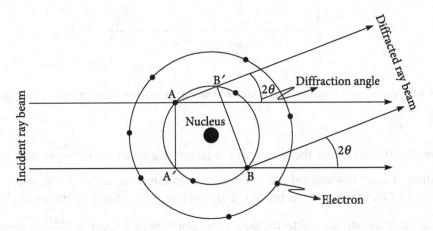

Figure 1.18 | Scattering of beam of X-rays by an isolated atom. To simplify the problem, electrons have been shown as points. $2\theta = 0°$ defines the forward direction and $2\theta = 180°$ defines the backward direction

As shown in Fig. 1.18, the path difference between the scattered waves scattered by the electrons at A and B is zero in the forward direction, resulting in zero phase difference. Hence, the scattered waves produce a diffraction maximum in the forward direction. The amplitude of the wave in the forward direction is equal to the addition of amplitudes of each wave scattered by each electrons of the atom. If there are Z electrons scattering waves each of amplitude a, then the amplitude of the wave scattered by an atom will be Za in the forward direction. Putting this in Eq. (1.48), we get $f = Z$.

In general, the atomic scattering factor f is given by

$$f = \int_{r=0}^{\infty} 4\pi r^2 \rho(r) \frac{\sin \mu r}{\mu r} dr \quad \text{with} \quad \mu = \frac{4\pi}{\lambda} \sin \theta \qquad (1.49)$$

where $\rho(r)$ represents the electron number density around the nucleus.

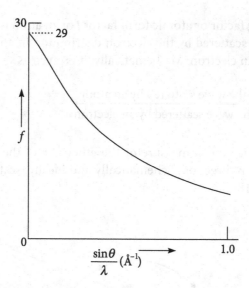

Figure 1.19 Variation of the atomic form factor or atomic scattering factor of copper with respect to $\frac{\sin\theta}{\lambda}$

Equation (1.49) shows that the atomic form factor is a function of $\frac{\sin\theta}{\lambda}$ and $\rho(r)$, the distribution of electrons around the nucleus. Due to the presence of the term $\rho(r)$ in the expression (1.49), the atomic scattering factor will vary for atoms of different elements.

The variation of the atomic scattering factor f of copper with respect to $\frac{\sin\theta}{\lambda}$ is depicted in Fig. 1.19. Since the intensity of the scattered rays is proportional to the square of the resultant amplitude, the intensity of the scattered rays scattered by an atom is obtained by taking the square of the ordinate of Fig. 1.18.

1.11.3 Scattering by a unit cell

In the previous sections, we have discussed the scattering of X-rays by an electron and by an atom. The scattering of X-rays by an electron and by an atom can be extended to include scattering by a unit cell. To calculate the intensity of the diffracted beam from real crystals, we have to discuss the scattering of X-rays by the unit cell because the entire crystal is produced by the repetition of the fundamental unit cell. Since the atoms are arranged regularly in a unit cell, the scattered radiation is limited to certain definite directions. The arrangement of atoms in a unit cell affects the intensity of the diffracted beam. This is a condition over Bragg's law. In certain directions, there will be no diffracted beam even though Bragg's law is satisfied for that direction!

The phase difference φ between the rays scattered by two atoms according to Eq. (1.34) will be given by

$$\varphi = 2\pi\left(hu + kv + \ell w\right) \tag{1.50}$$

where u, v and w are the fractional coordinates of one atom and the other atom is situated at the origin on the ($hk\ell$) plane. As we know, any wave can be expressed in the exponential form by $y = re^{i\varphi}$. Similarly, we can express the wave scattered by a given atom in the complex exponential form by

$$\psi = re^{i\varphi} = fe^{i2\pi(hu+kv+\ell w)} \tag{1.51}$$

If there are different atoms of different elements in the unit cell, then

$$f_1 e^{i2\pi(hu_1+kv_1+\ell w_1)}, f_2 e^{i2\pi(hu_2+kv_2+\ell w_2)}, f_3 e^{i2\pi(hu_3+kv_3+\ell w_3)}, \dots \text{ and so on} \tag{1.52}$$

are the scattered waves scattered by each atom. The resultant wave F of all the waves scattered by each atom will be given according to the superposition principle by

$$F = f_1 e^{i2\pi(hu_1+kv_1+\ell w_1)} + f_2 e^{i2\pi(hu_2+kv_2+\ell w_2)} + f_3 e^{i2\pi(hu_3+kv_3+\ell w_3)} + \cdots \tag{1.53}$$

The geometrical structure factor or simply structure factor is defined as the resultant wave of all the waves scattered by all the atoms in the unit cell. According to this definition of geometrical structure factor, Eq. (1.53) gives the expression for geometrical structure factor. This equation may be written more compactly by

$$F_{hk\ell} = \sum_{1}^{N} f_n e^{2\pi i(hu_n+kv_n+\ell w_n)} \tag{1.54}$$

The summation is taken over all the atoms in the unit cell. u_n, v_n and w_n are the fractional coordinates of the nth atom and f_n is the atomic scattering factor of the nth atom. Eq. (1.54) shows that the structure factor in general is complex. Since the structure factor is the resultant wave, this equation expresses both amplitude and phase of the resultant wave. We know that the magnitude of the resultant wave gives the amplitude of the resultant wave, i.e., $|\psi|$ = amplitude. Therefore, $|F|$ gives the amplitude of the resultant wave in terms of amplitude of the wave scattered by an electron. Mathematically,

$$|F| = \frac{\text{Amplitude of the wave scattered by all the atoms in a unit cell}}{\text{Amplitude of the wave scattered by an electron}} \tag{1.55}$$

We know that the intensity of a wave is proportional to the square of the amplitude of the wave. Therefore, the intensity of the resultant wave scattered by all the atoms in the unit cell is proportional to $|F|^2$. Thus, Eq. (1.54) turns out to be a very important relation in X-ray crystallography since it gives the intensity of any ($hk\ell$) reflection from a knowledge of the atomic positions in the unit cell.

1.11.4 Structure factor calculations

For structure factor calculation, we will apply the formulae $e^{n\pi i} = (-1)^n$ and $e^{n\pi i} = e^{-n\pi i}$ regularly. Here n is any integer, odd or even.

Primitive unit cell

The simplest unit cell is the primitive cell of the simple cubic unit cell of simple cubic crystals containing one atom which may be assumed to be at the origin. Since the atom is at the origin, its fractional coordinates are given by

i. $u_1 = 0$

ii. $v_1 = 0$

iii. $w_1 = 0$

From Eq. (1.54) for one atom, we have

$$F = f_1 e^{2\pi i(h.0 + k.0 + \ell.0)} = f_1 e^{2\pi i \times 0} = f_1$$

In this case, $|F|^2 = f_1^2$. Thus, $|F|^2$ is independent of h, k and ℓ and is the same for all reflections.

Base-centered unit cell

A base-centered cell has two atoms of the same type per unit cell; one atom is located at the origin with fractional coordinates (0,0,0) and the other atom is located at the center of the basal plane with fractional coordinates $\left(\dfrac{1}{2}, \dfrac{1}{2}, 0\right)$. From Eq. (1.54) for two atoms, we have

$$F = f_1 e^{2\pi i(hu_1 + kv_1 + \ell w_1)} + f_2 e^{2\pi i(hu_2 + kv_2 + \ell w_2)} = f_1 e^{2\pi i(h.0 + k.0 + \ell.0)} + f_2 e^{2\pi i\left(\frac{h}{2} + \frac{k}{2} + \ell.0\right)}$$

Since the two atoms are of the same type, we set $f_1 = f_2 = f$ and obtain

$$F = f(1 + e^{\pi i(h+k)})$$

If $h + k$ is even, we have $F = f(1+1) = 2f$

If $h + k$ is odd, we have $F = f(1-1) = 0$

In either case, index ℓ has no effect on the structure factor. Reflections from (111), (113), (023), (205) planes all have the same value of the structure factor, i.e., 2f. Similarly reflections from (011), (103), (123), (235), and so on all have a zero structure factor as a result of which there will be no diffracted beams from these planes.

Body-centered unit cell

A body-centered cell has two atoms of the same type per unit cell – one atom is located at the origin with fractional coordinates $(0,0,0)$ and other atom is located at the center of the unit cell with fractional coordinates $\left(\dfrac{1}{2},\dfrac{1}{2},\dfrac{1}{2}\right)$. From Eq. (1.54) for two atoms, we have

$$F = f_1 e^{2\pi i(hu_1+kv_1+\ell w_1)} + f_2 e^{2\pi i(hu_2+kv_2+\ell w_2)} = f_1 e^{2\pi i(h.0+k.0+\ell.0)} + f_2 e^{2\pi i\left(\frac{h}{2}+\frac{k}{2}+\frac{\ell}{2}\right)}$$

Since the two atoms are of the same type, we set $f_1 = f_2 = f$ and obtain

$$F = f\left(1 + e^{\pi i(h+k+\ell)}\right)$$

If $h + k + \ell$ is even, we have $F = f(1 + 1) = 2f$

If $h + k + \ell$ is odd, we have $F = f(1 - 1) = 0$

In this case, reflections from the (011), (103), (123), (235), (352), and so on planes, all have the same value of the structure factor, i.e., 2f. Similarly, reflections from the (111), (113), (023), (225), and so on planes all have a zero structure factor as a result of which there will be no diffracted beams from these planes.

FCC unit cell

An FCC cell has four atoms of the same type per unit cell – one atom is located at the origin with fractional coordinates $(0,0,0)$ and the other three atoms are located at the center of the facial planes with fractional coordinates

$$\left(\frac{1}{2},\frac{1}{2},0\right), \quad \left(\frac{1}{2},0,\frac{1}{2}\right) \text{,and } \left(0,\frac{1}{2},\frac{1}{2}\right).$$

From Eq. (1.54) for four atoms, we have

$$F = f_1 e^{2\pi i(hu_1+kv_1+\ell w_1)} + f_2 e^{2\pi i(hu_2+kv_2+\ell w_2)} + f_3 e^{2\pi i(hu_3+kv_3+\ell w_3)} + f_4 e^{2\pi i(hu_4+kv_4+\ell w_4)}$$

$$= f_1 e^{2\pi i(h.0+k.0+\ell.0)} + f_2 e^{2\pi i\left(\frac{h}{2}+\frac{k}{2}+\ell.0\right)} + f_3 e^{2\pi i\left(\frac{h}{2}+k.0+\frac{\ell}{2}\right)} + f_4 e^{2\pi i\left(h.0+\frac{k}{2}+\frac{\ell}{2}\right)}$$

Since the four atoms are of the same type, we set $f_1 = f_2 = f_3 = f_4 = f$ and obtain

$$F = f + f e^{\pi i(h+k)} + f e^{\pi i(h+\ell)} + f e^{\pi i(k+\ell)}$$

or $\quad F = f\left[1 + e^{\pi i(h+k)} + e^{\pi i(h+\ell)} + e^{\pi i(k+\ell)}\right]$

If $h+k, h+\ell$ and $k+\ell$ all are even, we have $F = f(1+1+1+1) = 4f$　$F = f(1+1+1+1) = 4f$.

If any two of $h + k$, $h + \ell$ and $k + \ell$ are odd, we have $F = 0$.

In this case, diffraction can not occur from (100), (112), (123), (235), (205), and so on planes since $F = 0$ for all these planes and diffraction occurs from (111), (200), (220), (115), and so on planes since $F \neq 0$ for all these planes.

From these calculations, we can conclude that the structure factor is independent of shape and size of the unit cell and depends only on the ($hk\ell$) planes.

Questions

1.1　What do you mean by crystalline solids?

1.2　What do you mean by amorphous solids?

1.3　What crystals are called binary crystals?

1.4　Give few examples of binary crystals

1.5　What do you mean by point lattice?

1.6　What is a unit cell?

1.7　Define fundamental translation vectors.

1.8　What are called lattice parameters?

1.9　Define lattice translation vector.

1.10　What do you mean by Bravais lattice?

1.11　What do you mean by basis?

1.12　Define primitive unit cell.

1.13　What is the necessity of defining lattice directions and lattice directions?

1.14　What do you mean by direction indices in crystallography?

1.15　What are the steps required to find out direction indices in crystal?

1.16　The unit cell has four body diagonals. How, then, is the number of members in <111> eight?

1.17　The unit cell has twelve edges. How, then, is the number of members in <100 > six?

1.18　The unit cell of an orthogonal crystal has six faces. How, then, is the number of members in <110> twelve?

1.19　What do you mean by a family of lattice directions?

1.20　What are Miller indices?

1.21　What are the steps required to find out the Miller indices of crystal planes?

1.22　How are the Miller indices of a crystal plane found out in the case of hexagonal crystals?

1.23   What do you mean by a family of crystal planes?

1.24   What are the salient features of Miller indices?

1.25   Show that in a simple cubic crystal, the direction [$hk\ell$] is perpendicular to the plane ($hk\ell$).

1.26   What is linear density of atoms?

1.27   What is planar density of atoms?

1.28   Distinguish between interplanar spacing and interatomic spacing.

1.29   How can you conclude that FCC and HCP crystalline solids have more density than that of SC and BCC solids?

1.30   What is coordination number? Calculate the coordination number of FCC, BCC, HCP structures.

1.31   What is atomic packing factor? Calculate it for an HCP structure.

1.32   Calculate the atomic packing factor of a diamond cubic structure.

1.33   Why cannot X-rays be diffracted by optical grating?

1.34   Define diffracted beam.

1.35   Derive Bragg's law of X-ray diffraction.

1.36   Explain Bragg's law of X-ray diffraction.

1.37   What are the characteristic features of Bragg's law?

1.38   Explain how Bragg's law is used to determine the wavelength of X-rays.

1.39   Distinguish between visible light reflection and X-ray diffraction.

1.40   Why cannot visible light be diffracted by crystals?

1.41   Define diffraction angle.

1.42   Define diffraction directions.

1.43   Explain how any order X-ray diffraction can be considered as a first order diffraction.

1.44   Explain why Compton scattering is neglected in X-ray diffraction.

1.45   Why dose a nucleus scatter X-rays negligibly?

1.46   The wavelength of radiation should not exceed two times the interplanar spacings for the diffraction phenomena to occur in crystals. Explain

1.47   How are diffraction directions calculated from Miller indices?

1.48   What is the importance of the reciprocal lattice concept?

1.49   Define reciprocal lattice

1.50   Explain how a reciprocal lattice is constructed.

1.51   Enumerate the properties of reciprocal lattice.

1.52   Distinguish between reciprocal lattice and real lattice.

1.53   Define reciprocal lattice vector.

1.54   Find the reciprocal lattice vectors of a simple cubic lattice.

1.55   Give a comparison between real space and reciprocal space.

1.56 Evaluate the dot product of the lattice translation vector $\vec{T}$ and the reciprocal lattice vector $\vec{G}$. What are the properties of reciprocal lattice?

1.57 Show that every reciprocal lattice vector is normal to a certain crystal plane of the direct lattice.

1.58 Show that the magnitude of the reciprocal lattice vector is inversely proportional to the interplanar spacing of the lattice planes.

1.59 Show that the volume of the unit cell of the reciprocal lattice is inversely proportional to the volume of a unit cell in the direct lattice.

1.60 The direct lattice is the reciprocal of its own reciprocal lattice. Prove.

1.61 What is the magnitude of $\vec{k}\cdot\vec{G}$?

1.62 What is the magnitude of $\vec{k}_0\cdot\vec{G}$?

1.63 Derive Bragg's law in terms of reciprocal lattice vectors.

1.64 Calculate the magnitude of the scattering normal.

1.65 Derive Laue's conditions.

1.66 State Laue's conditions in vector form

1.67 State Laue's conditions in scalar form

1.68 Derive Bragg's law from Laue's conditions.

1.69 What is the angle between $\vec{G}$ and $\vec{k}$? What is the angle between $\vec{G}$ and $\vec{k}_0$?

1.70 Derive Bragg's law from $\Delta\vec{k}=\vec{G}$

1.71 What are the factors on which the intensity of the diffracted beam depend?

1.72 What do you mean by the scattering of X-rays by electrons?

1.73 Distinguish between Thomson and Compton scattering.

1.74 Define atomic scattering factor. What does it represent?

1.75 Write an expression for the atomic form factor.

1.76 Prove from Eq. (1.49), $f = Z$ in the forward direction.

Problems

1.1 If 0.25, 0.5, and 0.75 are the coordinates of a point on a line, determine the direction indices of the line. [Ans [123]]

1.2 The density of NaCl is $2.167\dfrac{gm}{cm^3}$ and its molecular mass is 58.46. Find the lattice parameter. [Ans 5.64 Å]

1.3 In a certain crystal, a plane makes intercepts 15 mm, 20 mm, and 8 mm along three crystallographic axes having lengths 3 Å, 5 Å and 2 Å respectively. Find the Miller indices of the plane. [Ans (455)]

1.4 Find the Miller indices of a plane making intercepts of 1, 2, and ∞ with crystallographic axes. [Ans (210)]

1.5   Determine the Miller indices of a plane that makes intercepts of 2 Å, 3 Å, and 4Å on the crystallographic axes of an orthorhombic crystal with $a_1 : a_2 : a_3 = 4:3:2$.

[Ans (421)]

1.6   The lattice constant of a cubic lattice is a. Calculate the spacing between {120} planes.   [Ans $\dfrac{a}{\sqrt{5}}$]

1.7   An X-ray tube operates at a potential difference 40 kV with a copper target. For certain crystals, the first order diffraction is observed at a glancing angle 137° for the wavelength 1.54 Å. Calculate the interplanar spacing of the crystal.   [Ans 3.25 Å]

1.8   The spacing of {100} planes in NaCl crystal is 2.820 Å. X-rays incident on the surface of this crystal is found to give rise to first order reflection at a grazing angle of 15.8°. Calculate the wavelength of X-rays used.   [Ans 15.4 Å]

1.9   A beam of X-rays of wavelength 15.4 Å is incident on a crystal at 4.5° when the first order Bragg's reflection occurs. Calculate the glancing angle for the third order reflection.   [Ans 13.6°]

1.10   Determine the glancing angle on the cube face {100} of a rock salt crystal corresponding to second order reflection. Given $a = 2.814$ Å and $\lambda = 0.710$ Å.   [Ans 14.6Å]

1.11   X-rays of wavelength 1.5418 Å are diffracted by planes (111) in a cubic crystal at an angle 30° in first order reflection. Calculate the interatomic distance.   [Ans 2.67 Å]

1.12   Show that in a simple cubic lattice, interplanar spacings of {101}, {110} and {011} planes are in the ratio 1:1:1.

1.13   The distance between (111) planes in a face-centered cubic crystal is 2 Å. Determine the lattice parameter.   [Ans $2\sqrt{3}$Å]

1.14   Determine the highest order diffraction order that can be observed with an X-ray radiation of wavelength 1.2 Å from a quartz crystal.   [Ans 7]

1.15   Copper is an FCC crystal and its atomic radius is 0.1278 nm. Calculate the interplanar spacing of (321) atomic plane.   [Ans 0.0966 nm]

Multiple Choice Questions

1.   Which of the following is not a crystal?
    (i)   alumina   (ii)   glass
    (iii)   table salt   (iv)   ferrites

2.   Which of the following does not give volume of a unit cell?
    (i) $\left|\vec{a_1}.\vec{a_2} \times \vec{a_3}\right|$ (ii) $\left|\vec{a_2}.\vec{a_3} \times \vec{a_1}\right|$
    (iii) $\left|\vec{a_3}.\vec{a_1} \times \vec{a_2}\right|$ (iv)   none of the above

3.   How many Bravais lattices are there?
    (i)   13   (ii)   14
    (iii)   15   (iv)   none of the above

4.   How many crystal systems are there?
      (i)   5   (ii) 6
      (iii) 7   (iv) 8

5.   The direction along the face diagonal of a unit cell of a cubic crystal is denoted by
      (i)   [111]   (ii) [123]
      (iii) [100]   (iv) [110]

6.   In a BCC, which of the following directions has the maximum linear density?
      (i)   [111]   (ii) [123]
      (iii) [100]   (iv) [110]

7.   In an FCC, which of the following directions has the maximum planar density?
      (i)   (110)   (ii) (100)
      (iii) (211)   (iv) (111)

8.   Does a diffracted beam exist for any direction satisfying Bragg's law?
      (i)   Yes   (ii) No

9.   The coordination number of the central atom on the hexagonal face (HCP lattice) is
      (i)   12   (ii) 8
      (iii) 6   (iv) 4

10.   The coordination number of a corner atom in a BCC lattice is
      (i)   12   (ii) 8
      (iii) 6   (iv) 4

11.   Which of the following statements is correct? Atomic packing factor of
      (i)   SC and FCC are equal   (ii) FCC and BCC are equal
      (iii) BCC and HCP are equal   (iv) FCC and HCP are equal

12.   The space lattice of diamond is
      (i)   SC   (ii) BCC
      (iii) FCC   (iv) HCP

13.   The space lattice of sodium chloride is
      (i)   SC   (ii) BCC
      (iii) FCC   (iv) HCP

14.   Which of the following statements is correct?
      (i)   FCC and HCP crystalline solids have more density than that of SC and BCC solids
      (ii) FCC and HCP crystalline solids have less density than that of SC and BCC solids
      (iii) SC and BCC crystalline solids have more density than that of FCC and HCP
      (iv) SC and BCC crystalline solids have less density than that of FCC and HCP

15.   The reciprocal lattice concepts was first given by
    (i)   Newton   (ii)   Ewald
    (iii) Einstein   (iv) Maxwell

16.   When we rotate a crystal, we rotate
    (i)   only the reciprocal lattice   (ii)   only the direct lattice
    (iii) both reciprocal and direct lattices   (iv) lattice can not be rotated.

17.   The points in reciprocal space correspond to what thing in real space?
    (i)   points   (ii)   lines
    (iii) planes   (iv) nothing

18.   In a BCC lattice, in which of the following planes is there no Bragg reflection?
    (i)   (110)   (ii)   (221)
    (iii) (211)   (iv) (112)

19.   In an FCC lattice, in which of the following planes is there no Bragg reflection?
    (i)   (112)   (ii)   (220)
    (iii) (115)   (iv) (111)

20.   Structure factor of a unit cell depends upon
    (i)   Shape of the unit cell   (ii)   Volume of the unit cell
    (iii) Shape and volume of the unit cell   (iv) Only on $(hk\ell)$ planes

Answers

1 (ii)	2 (iv)	3 (ii)	4 (iii)	5 (iv)	6 (i)	7 (i)	8 (ii)
9 (i)	10 (ii)	11 (iv)	12 (iii)	13 (iii)	14 (i)	15 (ii)	16 (iii)
17 (iii)	18 (ii)	19 (i)	20 (iv)				

2 Defects in Crystals

2.1 Introduction

Imperfections are inherent in nature. Except God nothing is perfect. In science, many ideas are made ideal for the sake of simple interpretations so that beginners can comprehend concepts. In the previous chapter, we have defined an ideal crystal as a solid in which atoms, molecules or ions are arranged in a periodic pattern in three dimensions in such a manner that each atom has identical surroundings. However, in real crystalline materials the arrangement of atom or molecules is not so perfect. The regular patterns are interrupted by crystallographic defects. Any deviation from the perfect atomic arrangement in a crystal is said to contain imperfections or defects.

While it is perhaps intuitive to think of defects as bad things, they are in fact necessary, even crucial, to the behaviour of materials. The presence of a relatively small number of defects has a profound impact on the structure-sensitive properties of materials. Intentional introduction of defects is important in many kinds of material processing. Production of advanced semiconductor devices requires not only a rather perfect Si crystal as starting material, but also involves introduction of specific imperfections in that perfect material. Forging a metal tool introduces defects and increases strength and elasticity of the tool. Thus, imperfections in solids are necessary evils! Almost all technology involving materials depends on the existence of some kind of defects. Together with bonding, crystal structure and defects decide the properties of any material.

Defects exist in all solid materials. For ease of their characterization, defects may be classified into the following categories on the basis of their geometry or dimensionality.

i.   Point defects (zero-dimensional)

ii.   Line defects (one-dimensional)

iii.   Surface defects (two-dimensional)

iv.   Volume defects (three-dimensional)

2.2 Point Defects

Point defects are said to be there where an atom is missing or is in an irregular place in the lattice structure. Typically, these defects involve at most a few extra or missing atoms. Various point defects are enlisted here.

i.   Vacancy

ii.   Interstitials

iii.   Frenkel and Schottky defects

iv.   Colour centres

v.   Polarons

vi.   Excitons

vii.   Antisite defects

viii. Topological defects

2.2.1 Vacancy

The simplest of a point defect is a vacancy or a vacant lattice point as shown in Fig. 2.1.

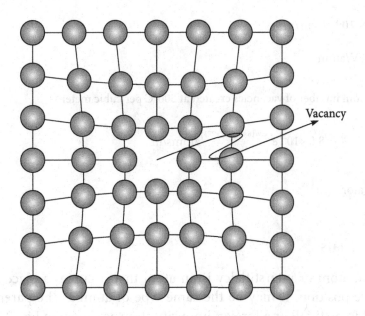

Figure 2.1 | Two-dimensional representation of a vacancy

Vacancies are formed during solidification and also as a result of atomic vibrations which in turn depend upon temperatures. The higher is the temperature, the more often atoms jump from one equilibrium position to another and the larger number of vacancies can be found in a crystal. At equilibrium, the number of vacancies, N_v, increases with the absolute temperature, T, and can be estimated using the equation

$$N_v = Ne^{-Q_v/kT}$$

where N is the number of regular lattice sites, k is the Boltzmann constant (8.62×10^{-5} eV/K), and Q_v is the energy needed to form a vacancy in a perfect crystal, i.e., is the energy required to move an atom from the interior of a crystal to its surface. When the density of vacancies becomes relatively large, there is a possibility for them to cluster together and form voids.

Example 2.1

Calculate the equilibrium number of vacancies per cubic meter for copper at 500° C. The energy for vacancy formation in copper is 0.9 eV/atom and the number of atoms per unit volume is 8.0×10^{28} atoms/m³ at 500° C.

Solution

Data given

$$T = (273 + 500)K = 773 \text{ K}$$

$$N = 8.0 \times 10^{28} \text{ atoms/m}^3$$

$$Q_v = 0.9 \text{ eV/atom}$$

The equilibrium number of vacancies created at 500°C per cubic meter is

$$N_v = Ne^{-Q_v/kT} = 8.0 \times 10^{28} e^{-0.9/\left(8.62 \times 10^{-5} \times 773\right)} \text{atoms/m}^3$$

$$1.1 \times 10^{23} \text{ atoms/m}^3$$

2.2.2 Interstitials

An interstitial atom or interstitialcy is an atom that occupies a place outside the normal lattice position. It may be the same type of atom as the parent crystal is made of (self interstitial) or a foreign impurity atom as shown in Fig. 2.2(a).

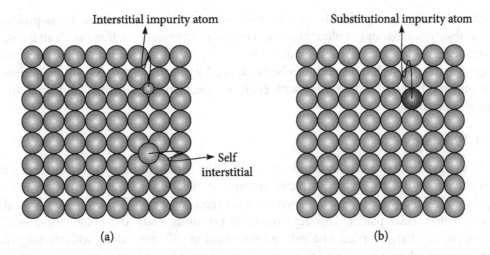

Figure 2.2 | (a) Two-dimensional representation of interstitials. (b) Substitutional impurity atom

Interstitialcy is most probable if the atomic packing factor is low. Creation of a self-interstitial causes substantial distortions in the surrounding lattice and costs more energy when compared to the energy needed for creating a vacancy. Under equilibrium conditions, self-interstitials are present in lower concentrations than vacancies. Foreign, usually smaller atoms (carbon, nitrogen, hydrogen, oxygen) are called interstitial impurities. They introduce less distortion to the lattice and are more common in real materials. They are also more mobile. If the foreign atom replaces or substitutes a lattice atom, it is called a substitutional impurity (Fig. 2.2(b)). An example of interstitial impurity atoms is the carbon atoms that are added to iron to make steel.

2.2.3 Frenkel and Schottky defects

In ionic crystals, the bonding is provided by coulombic forces between positively and negatively charged ions. Point defects in ionic crystals are charged as well. The coulombic forces are very large and any charge imbalance has a very strong tendency to balance itself. To maintain charge neutrality, several point defects are created. Thus, in an ionic crystal, if there is a vacancy in a positive ion site, a vacancy is created at a neighbouring negative ion site to maintain the charge neutrality condition. This pair of vacancies is called a Schottky defect.

A Frenkel defect is either a pair of cation (positive ion) vacancy and a cation interstitial or it may be an anion (negative ion) vacancy and anion interstitial. However, anions are much larger than cations and it is not easy for an anion interstitial to form. In both Frenkel and Schottky defects, the pair of point defects stays close to each other because of the strong coulombic attraction of their opposite charges.

2.2.4 Colour centres

The colouration of crystals is due to defects in the crystals. It is possible to colour the crystals in a number of different ways by adding selected impurities to create colour centres in selected transparent materials. Atomic and electronic defects of various types produce optical absorption bands or colour centres in transparent crystals such as the alkali halides, alkaline earth fluorides, or metal oxides. These are general phenomena found in a wide range of materials. Colour centres can be produced by gamma rays or X-rays, by addition of impurities, by bombarding them with energetic electrons or neutrons, and sometimes through electrolysis.

2.2.5 Polarons

When a negative charge carrier is put into a solid, the surrounding ions will interact with it. Positive ions will be slightly attracted towards it. The ions can adjust their positions slightly, balancing their interactions with the charge carrier and the forces that hold the ions in their regular places. This adjustment of positions leads to a polarization locally centred on the charge carrier. The induced polarization will move along with the negative charge carrier when it is moving through the medium. After it passes, the region returns to its normal configuration. The combination of the charge carrier and the surrounding polarization is called polaron. The effect is most pronounced in ionic solids consisting of positive and negative ions. A polaron interacts with the surrounding atoms in the solids as if it is a negatively charged particle. Its effective mass is greater than that of a free electron and in NaCl it is more than double.

2.2.6 Excitons

An exciton is a bound state of an electron and a hole, attracted to each other by the electrostatic coulombic force. It is an electrically neutral quasi-particle that exists in insulators, semiconductors and some liquids. Due to electrical neutrality, it is very difficult to detect. The exciton is regarded as an elementary excitation in condensed matter that can transport energy without transporting net electric charge. Exciton vanishes when its electron–hole pair recombines. When the exciton vanishes, its energy content may be converted to photons, or may be transferred to neighbouring electrons, creating new excitons. Due to energy transportation or energy conversion, there is some localized lattice distortion.

2.2.7 Antisite defects

Antisite defects occur in an ordered alloy or compound when atoms of different type exchange positions. For example, some alloys have a regular structure in which every other atom is a different species. For illustration, assume that type A atoms sit on the corners of a cubic lattice, and type B atoms sit in the centre of the cubes. If one cube has an A atom at its centre, the atom is on a site usually occupied by a B atom, and is thus an antisite defect. This is neither a vacancy nor an interstitial, nor an impurity.

2.2.8 Topological defects

Topological defects are regions in a crystal where the normal chemical bonding environment is topologically different from the surroundings. For instance, in a perfect sheet of graphene, all atoms are in rings containing six atoms. If the sheet contains regions where the number of atoms in a ring is different from six, while the total number of atoms remains the same, a topological defect has formed. An example is the Stone Wales defect in nanotubes, which consists of two adjacent 5-membered and two 7-membered atom rings.

2.3 Line Defects (One-Dimensional)

Any deviation from a perfectly periodic arrangement of atoms along a line is called a line defect or line imperfection. Line imperfections (one-dimensional defects) are commonly also called dislocations. The line along which dislocation occurs is called a dislocation line in the solid. Dislocations strongly influence the mechanical properties of material if they occur in high densities. These are created during the solidification process, plastic deformation, vacancy condensation or atomic mismatch in solid solutions.

Dislocations can be best understood by referring to the two limiting cases:

i.   Edge dislocation
ii.   Screw dislocation

2.3.1 Edge dislocation

Edge dislocation or Taylor–Orowan dislocations are caused by the termination of atomic planes inside a crystal. The dislocation line is the edge of this incomplete atomic plane. It has been marked in Fig. 2.3. The analogy with a stack of paper is apt: if half a piece of paper is inserted in a stack of paper, the defect in the stack is only noticeable at the edge of the half sheet. Within the region around the dislocation line, there is some localized lattice distortion. Because of an incomplete plane of atoms, the atoms above the dislocation line are squeezed together and are in a state of compression whereas atoms below are pulled apart and experience tensile stresses. Thus, regions of compression and extension are associated with an edge dislocation. The magnitude of this distortion decreases with distance away from the dislocation line; at positions far away, the crystal lattice is virtually perfect.

Edge dislocation is considered positive when compressive stresses are present above the dislocation line it is represented by ⊥. A unit vector $\hat{t}$ along the dislocation line may be assigned. For positive dislocation, the unit vector $\hat{t}$ points into the plane of the paper along the dislocation line. If the compressive stresses exist below the dislocation line, it is considered as a negative edge dislocation and is represented by T and the unit vector $\hat{t}$ points out of the plane of paper along the dislocation line. A schematic view of edge dislocations are shown in Fig. 2.3.

Dislocation is characterized by a vector $(\vec{b})$ called Burger's vector. The direction and magnitude of Burger's vector can be determined by constructing a loop around the disrupted region and noticing the extra inter-atomic spacing needed to close the loop. It can be best explained by considering two-dimensional crystals as shown in Fig. 2.3.

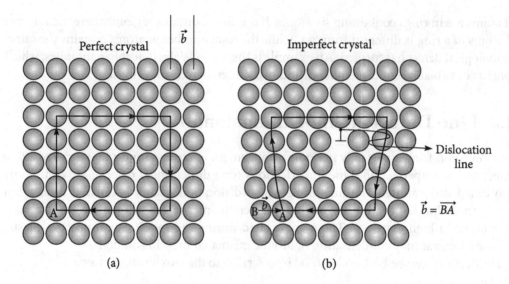

Figure 2.3 (a) Burger's circuit in perfect and imperfect crystals. (b) Burger's circuit in imperfect crystals and definition of Burger's vector

In perfect crystals, if we start from point A and move 4 steps upward, 5 steps right, 4 steps downward and 5 steps left, we reach the starting point A again. The directional path thus traced is called Burger's circuit and in a perfect crystal, it is closed or complete. If we do the same operation in an imperfect crystal containing edge dislocation, we arrive at point B and we need an extra backward step to reach A. The direction and magnitude of this extra step defines Burger's vector $\vec{b}$, i.e.,

$$\vec{b} = \overrightarrow{BA}$$

For positive dislocation, Burger's circuit is clockwise and for negative dislocation, it is anti-clockwise. Its magnitude is equal to the inter-atomic spacing. A pure edge dislocation can glide or slip in a direction perpendicular to its length, i.e., along its Burger's vector in the slip plane defined by $\vec{b}$ and $\hat{t}$ vectors. Dislocation moves by slip, conserving the number of atoms in the incomplete plane. It may move vertically by a process known as climb, if diffusion of atoms or vacancies can take place at an appropriate rate. Atoms are to be added to the incomplete plane for negative climb, i.e., the incomplete plane increases in extent downwards, and vice versa. Thus, climb motion is considered as non-conservative. The movement by climb is controlled by diffusion process.

2.3.2 Screw dislocations

The screw dislocation or Burger's dislocation is slightly more difficult to visualize. Basically, it comprises a structure in which a helical path is traced around the dislocation line by the atomic planes in the crystal lattice. The motion of a screw dislocation is also a result of shear stress, but the defect line movement is perpendicular to the direction of the stress and the atomic displacement. To visualize a screw dislocation, imagine a block of metal with a shear stress applied across one end so that the metal begins to rip. This is shown in Fig. 2.4(a). Figure 2.4(b) shows the plane of atoms just above the rip. The atoms represented by 1–5 rows of circles have not yet moved from their original position. The atoms represented by 9–13 rows of circles have moved to their new position in the lattice and have re-established metallic bonds. The atoms represented by 6–8 rows of circles are in the process of moving. It can be seen that only a portion of the bonds are broken at any given time. As was the case with the edge dislocation, movement in this manner requires a much smaller force than breaking all the bonds across the middle plane simultaneously. If the shear force is increased, the atoms will continue to slip to the right; 6–8 rows of atoms will find their way back into a proper spot in the lattice and 1–5 rows of atoms will slip out of position. In this way, the screw dislocation will move upward in the image, which is perpendicular to the direction of the stress.

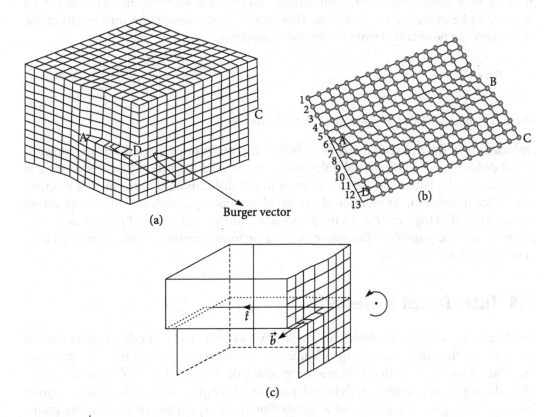

Figure 2.4 (a) A metal cuboid begins to rip under shear stress, (b) The plane of atoms just above the rip, (c) Schematic view of a negative screw dislocation

Screw dislocation has its dislocation line parallel to Burger's vector. A screw dislocation is like a spiral ramp with an imperfection line down its axis. Screw dislocations result when planes are displaced relative to each other through shear. Shear stresses are associated with the atoms adjacent to the screw dislocation; therefore, extra energy is involved. Screw dislocation is considered positive if Burger's vector and unit vector $\hat{t}$ are parallel, and vice versa. A positive screw dislocation is represented by ⟳ a dot surrounded by a circular direction in the clockwise direction, whereas the negative screw dislocation is represented by ⟲ a dot surrounded by a circular direction in the anti-clockwise direction. A schematic view of a negative screw dislocation is shown in Fig. 2.4(c).

Most dislocations found in crystalline materials are probably neither pure edge nor pure screw, but exhibit components of both types; these are termed mixed dislocations. The nature of a dislocation (i.e., edge, screw, or mixed) is defined by the relative orientations of the dislocation line and Burgers vector. Burger's vector is perpendicular to the dislocation line in case of edge dislocation and parallel to the dislocation line in case of screw dislocation. It is neither perpendicular nor parallel for a mixed dislocation. It is unique to a dislocation.

Dislocations have distortional energy associated with them as is evident from the presence of tensile/compressive/shear stresses around a dislocation line. Strains can be expected to be in the elastic range, and thus, stored elastic energy per unit length of the dislocation can be obtained from the following equation

$$E \approx \frac{1}{2}Gb^2$$

where G = shear modulus, b^2 = square of Burger's vector.

Dislocations in the real crystal can be classified into two groups based on their geometry; (i) full dislocations and (ii) partial dislocations. In partial dislocations, Burger's vector will be a fraction of a lattice translation, whereas in full dislocation, it is an integral multiple of a lattice translation. As mentioned earlier, elastic energy associated with a dislocation is proportional to the square of its Burger's vector; dislocation will tend to have as small a Burger's vector as possible. This explains the reason for separation of dislocations that tend to stay away from each other.

2.4 Interfacial Defects

Until now, we considered defects of single crystals that have periodic, regular atomic arrangement throughout the sample. Single crystals, however, can be rarely found in real materials unless the growth conditions are specially designed and controlled as in the case of producing silicon single crystals for microelectronic chips or blades for turbine engines. Instead, solids generally consist of a number of small crystallites or grains. The grains range from nanometres to millimetres in size. These materials are called poly-crystals. The individual grains are separated by grain boundaries.

Surface or interfacial defects are boundaries that have two dimensions and normally separate regions of the materials that have different crystal structures and/or crystallographic orientations. These imperfections include external surfaces, grain boundaries, twin boundaries, stacking faults and phase boundaries. They refer to the regions of distortions that lie about a surface having thickness of a few atomic diameters. These imperfections are not thermodynamically stable, rather they are meta-stable imperfections. They arise from the clustering of line defects into a plane.

2.4.1 External surface

The crystal structure terminates at the external surface. Surface atoms are not bonded to the maximum number of nearest neighbours, and are therefore in a higher energy state than the atoms at interior positions. These unsaturated bonds of these surface atoms give rise to a surface energy, expressed in units of energy per unit area. To reduce this energy, materials tend to minimize the total surface area as in case of liquid drop.

2.4.2 Grain boundaries

Grain boundaries are important in several ways. They present paths for atoms to diffuse into the material and scatter light passing through transparent materials to make them opaque. They also affect mechanical properties. The boundaries limit the lengths and motions of dislocations that can move. This means that smaller grains (more grain boundary surface area) strengthens materials at ambient temperature.

Crystalline solids are usually made of a number of grains separated by grain boundaries. Grain boundaries are several atoms distance wide, and there is mismatch of orientation of grains on either side of the boundary as shown in Fig. 2.5.

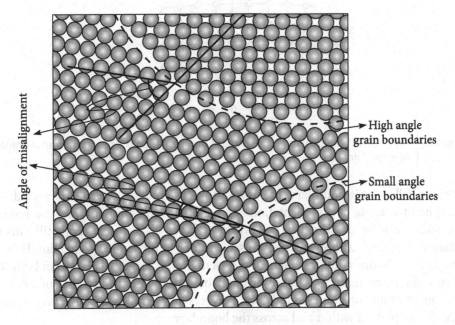

Figure 2.5 | Schematic illustration of grain boundaries showing high and small angle grain boundaries

When this misalignment is small, of the order of a few degrees (<10°), it is called low angle grain boundary. These boundaries can be described in terms of aligned dislocation arrays. If the low grain boundary is formed by edge dislocations, it is called a tilt boundary; if it is formed due to screw dislocations, it is called a twist boundary. A small angle grain boundary is formed when edge dislocations are aligned in the manner of Fig. 2.6.

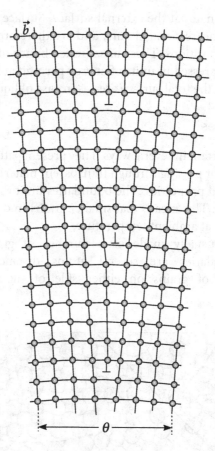

Figure 2.6 | Origin of tilt boundary having an angle of misorientation θ. It results from an alignment of edge dislocations

This type is called a tilt boundary; the angle of misorientation θ, is also depicted in the figure. When the angle of misorientation θ is parallel to the boundary, a twist boundary results, which can be described by an array of screw dislocations. Both tilt and twist boundaries are planar surface imperfections in contrast to high angle grain boundaries. For high angle grain boundaries, the angle of disorientation is large (>15°). Grain boundaries are chemically more reactive because of grain boundary energy. In spite of disordered orientation of atoms at grain boundaries, polycrystalline solids are still very strong as cohesive forces present within and across the boundary.

At ambient temperatures, grain boundaries give strength to a material. Therefore, in general, fine grained materials are stronger than coarse grained ones because they have more grain boundaries per unit volume. However, at higher temperatures, grain boundaries act to weaken a material due to corrosion and other factors.

2.4.3 Twin boundaries

It is a special type of grain boundary across which there is specific mirror lattice symmetry. Twin boundaries occur in pairs such that the orientation change introduced by one boundary is restored by the other as depicted in Fig. 2.7.

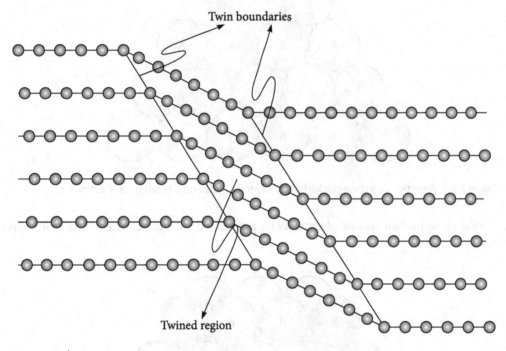

Figure 2.7 | Schematic diagram showing a pair of twin boundaries

The region between the pair of boundaries is called the twinned region. Twins which form during the process of recrystallization are called annealing twins, whereas deformation twins form during plastic deformation as in the case of shape memory alloys. Twinning occurs on a definite crystallographic plane and in a specific direction, both of which depend on the crystal structure. Annealing twins are typically found in metals that have an FCC crystal structure and low stacking fault energy, while mechanical/deformation twins are observed in BCC and HCP metals. Annealing twins are usually broader and with straighter sides than mechanical twins. Twins do not extend beyond a grain boundary.

2.5 Stacking Faults

Both face-centred cubic and hexagonal close-packed crystal structures are generated by the stacking of the close-packed atomic planes on top of one another. The difference between the two structures lies in the stacking sequence.

Let the centres of all the atoms in one close-packed plane be labelled A. As observed in Fig. 2.8, two sets of equivalent triangular depressions (marked B and C) are formed by three adjacent atoms, into which the next close-packed plane of atoms may fit.

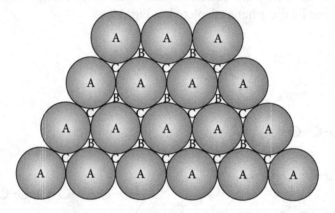

Figure 2.8 | A portion of a close-packed plane of atoms; A, B, and C positions are indicated

A second close-packed atomic plane may be positioned with the centres of its atoms over either B or C sites.

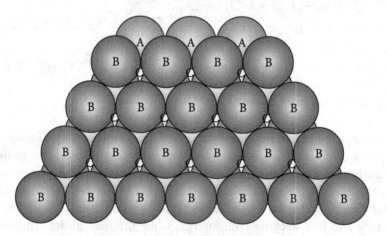

Figure 2.9 | The AB stacking sequence for close-packed atomic planes

As shown in Fig. 2.9, if B positions are arbitrarily chosen, the stacking sequence is termed as AB. The real distinction between FCC and HCP lies in where the third close-packed layer is positioned. For HCP, the centres of this layer are aligned directly above the original A positions. This stacking sequence, AB AB AB ... is repeated over and over. For the face-centred crystal structure, the centres of the third plane are situated over the C sites of the first plane. This gives an ABC ABC ABC ... stacking sequence.

Stacking faults are defects in the stacking sequence of atomic planes. Stacking sequence in an FCC crystal is ABC ABC ABC ..., and the sequence for HCP crystals is AB AB AB.... When there is disturbance in these stacking sequences, formation of stacking faults takes place.

2.6 Bulk or Volume Defects

Volume defects as the name suggests are defects in three-dimensions. These include pores, cracks, foreign particle inclusions and other phases. These defects are normally introduced either intentionally or non-intentionally during processing and fabrication steps. All these defects are capable of acting as stress raisers, and thus, are deleterious to the parent metal's mechanical behaviour. However, in some cases, foreign particles are added purposefully to strengthen the parent material. The procedure is called dispersion hardening where foreign particles act as obstacles to the movement of dislocations, which facilitates plastic deformation. The second-phase particles act in two distinct ways; particles are either cut by the dislocations or they resist cutting and dislocations are forced to bypass them. Strengthening due to ordered particles is responsible for the good high-temperature strength on many super-alloys. However, pores are detrimental because they reduce effective load-bearing area and act as stress concentration sites.

2.7 Atomic Vibrations

Atomic vibrations occur, even at absolute zero temperature (a quantum mechanical effect) and increase in amplitude with temperature. In fact, the temperature of a solid is really just a measure of the average vibrational activity of atoms and molecules. Vibrations displace transient atoms from their regular lattice site, which destroys the perfect periodicity. In a sense, these atomic vibrations may be thought of as imperfections or defects. At room temperature, a typical vibrational frequency of atoms is of the order of 10 vibrations per second, whereas the amplitude is a few thousandths of a nanometer. Many properties and processes in solids are manifestations of this vibrational atomic motion. For example melting occurs once the atomic bonds are overcome by vigorous vibrations.

Questions

2.1   Explain how imperfections in solids are necessary evils.

2.2   Classify defects based on their geometry

2.3   What are different types of point defects?

2.4 Show graphically the variation of equilibrium number of vacancies with temperature.

2.5 Explain how vacancy defects are created in crystals.

2.6 What is interstitialcy? How is it useful?

2.7 Give a comparative account of Frenkel defect and Schottky defect.

2.8 Write short notes on colour centres, polarons, excitons.

2.9 Explain the antisite defect

2.10 Explain the topological defect.

2.11 Differentiate between edge dislocation and screw dislocation.

2.12 What is Burger's vector? Explain how it is helpful to describe edge dislocation. Explain how Burger's vector is helpful to describe screw dislocation.

2.13 Cite the relative Burger's vector–dislocation line orientations for edge, screw, and mixed dislocations.

2.14 Dislocation will tend to have as small a Burger's vector as possible. Explain

2.15 Explain the phenomenon of screw dislocation in crystalline solids.

2.16 Explain the phenomenon of edge dislocation in crystalline solids.

2.17 Discus different types of interfacial defects in materials.

2.18 Give a comparative account of different types of grain boundaries.

2.19 Explain how fine grained materials are stronger than coarse grained materials.

2.20 Differentiate between stacking sequence of FCC and HCP structures.

2.21 Explain how volume defects in materials are useful.

2.22 Why are atomic vibrations considered imperfections or defects in materials.

Multiple Choice Questions

1. Defects in crystals
   (i) Are never beneficial
   (ii) are sometimes beneficial
   (iii) are always beneficial
   (iv) always weaken materials

2. Introduction of defects during forging of a metal tool
   (i) increases strength and elasticity of the tool
   (ii) decreases strength and elasticity of the tool
   (iii) increases strength and decreases elasticity of the tool
   (iv) decreases strength and increases elasticity of the tool

3. In highly sophisticated electronic industries to manufacture BJT
   (i) there should not be any defects in the materials
   (ii) there should be large defects in the materials
   (iii) there should be controlled amount of defects in the materials
   (iv) none of the above

4. Which of the following is not a point defect?

    (i)   Vacancy   (ii)   interstitials

    (iii) edge dislocations   (iv)   interstitials

5. Point defects in ionic crystals are

    (i)   negatively charged   (ii)   positively charged

    (iii) neutral   (iv)   charged

6. Which of the following is not a line defect?

    (i)   Vacancy   (ii)   screw dislocation

    (iii) edge dislocations   (iv)   none of the above

7. Which of the following is a point defect?

    (i)   Screw dislocation   (ii)   edge dislocations

    (iii) polarons   (iv)   stacking faults

8. Which of the following is a line defect?

    (i)   Frenkel defects   (ii)   Screw dislocation

    (iii) Schottky defects   (iv)   excitons

9. The dislocation line in a crystal is the

    (i)   Line passing through a vacancy

    (ii) line passing through an impurity atom

    (iii) line perpendicular to the incomplete atomic plane

    (iv) edge of the incomplete atomic plane

10. Edge dislocation is considered positive when

    (i)   compressive strains present above the dislocation line

    (ii) extensive stresses present above the dislocation line

    (iii) compressive stresses present below the dislocation line

    (iv) compressive stresses present above the dislocation line

Answers

1 (ii)	2 (i)	3 (iii)	4 (iii)	5 (iv)	6 (i)	7 (iii)	8 (ii)

9 (iv)   10 (iv)

3   X-rays

3.1   Introduction

X-rays were accidentally discovered in 1895 by the physicist Roentgen and were so named because their nature was unknown at that time. After performing a series of experiments, Roentgen concluded that when a beam of fast moving electrons strikes a solid target, an invisible high penetrating radiation is produced. It took almost two decades to establish its exact nature. Actually, X-rays are electromagnetic waves of very short wavelength. The X-rays used in diffraction have wavelengths ranging from 0.5 Å – 2.5 Å. Till today, X-rays are an invaluable tool in the field of material characterization and medical sciences.

3.2   Production of X-rays by a Coolidge Tube

According to the electromagnetic theory, accelerating or decelerating charged particles emits radiation. X-rays are produced when fast moving electrons are decelerated rapidly; deceleration occurs when they strike a target of suitable material. Thus, the basic requirements for the production of X-rays are

i.   a source of electrons

ii.   effective means of accelerating the electrons

iii.   a target of suitable materials of high atomic weight

The production of X-rays by a Coolidge tube is described briefly here. The modern Coolidge tube designed by Coolidge is shown in Fig. 3.1.

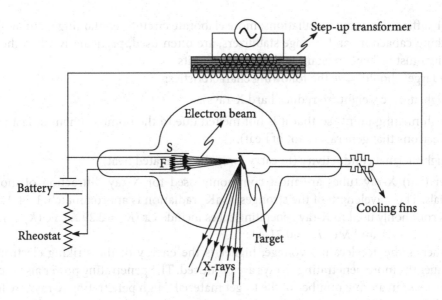

Figure 3.1 │ The schematic diagram of a self-rectifying filament Coolidge tube

It consists of a highly evacuated hard glass bulb containing a cathode and an anode. The cathode consists of a tungsten filament F and is heated by passing a current through it. The electrons are emitted by the process of thermionic emission from the cathode. The filament is surrounded by a molybdenum cylinder S kept at a negative potential so as to focus the electrons emitted from the filament into a small area on the target.

The target consists of a copper block in which a piece of tungsten or molybdenum is embedded. The target is cut at an angle of 45° with the horizontal. A high D.C. potential of about more than 30,000 volts is applied between filament F and the target T. Due to this high potential difference, the electrons emitted from the filament are accelerated to a high speed and high energy. When these accelerated electrons strike the target, they give up their kinetic energy and thereby, produce X-rays. Most of the kinetic energy of the electrons striking the target is converted into heat, less than 1% being transformed into X-rays. In order to save the target from heat, it is constantly cooled by a cooling arrangement. Generally, the target is cooled by flowing cooled water into it. The intensity of X-rays depends upon the number of electrons striking the target, i.e., the rate of emission of electrons from the filament. Thus, in a modern Coolidge tube, the intensity of X-rays can be controlled by regulating the current flowing through the filament. Although one might think that an X-ray tube can be operated only from a DC source, since the electron flow must occur only in one direction, it is actually possible to operate a tube from an A.C. source such as a transformer because of the rectifying properties of the tube itself. Current exists during the half-cycle in which the filament is negative with respect to the target; during the reverse half-cycle, the filament is positive, but no electrons can flow since only the filament is hot enough to emit electrons. Thus, a simple circuit such as shown in

Fig. 3.1 suffices for many installations; more elaborate circuits, containing rectifying tubes, smoothing capacitors and voltage stabilizers, are often used, particularly when the X-ray intensity must be kept constant within narrow limits.

The target should have the following characteristics:

i.   high atomic weight (to reduce hard X-rays).

ii.   high melting point (so that it is not melted due to the bombardment of fast moving electrons that generates a lot of heat).

iii. high thermal conductivity (to carry away the generated heat).

Copper (Cu) X-ray tubes are most commonly used for X-ray diffraction of inorganic materials. The wavelength of the strongest CuK$_\alpha$ radiation is approximately 1.54 Å. Other targets commonly used in X-ray generating tubes include Cr (K_α = 2.29 Å), Fe (K_α = 1.94 Å), Co (K_α = 1.79 Å), and Mo (K_α = 0.71 Å).

Higher is the accelerating voltage, higher is the energy of the striking electrons and consequently, more penetrating X-rays are produced. The penetrating power also increases with increase in atomic number of the target material. High penetrating X-rays are termed as hard X-rays while low penetrating X-rays are termed as soft X-rays. Thus, the quality of X-rays, i.e., its penetrating power depends on the cathode–target potential difference and the atomic number of the target.

Example 3.1

An X-ray tube is operating at 100 kV and 10 mA. If only 1% of the electric power supplied is converted into X-rays, at what rate per second is the heat produced in the target?

Solution

The data given are

V = 100 kV = 10^5 V, i = 10 mA = 10 × 10^{-3} A

The percentage of electric power converted into heat = (100 – 1)% = 99%
The rate of heat produced in the target per second

$= V \times i \times 99\% = 10^5 \times 10 \times 10^{-3} \times 0.99$ J/s = 900 J/s

3.3 Origin of X-rays

We know that X-rays are produced when high speed electrons strike the target material. At a tube voltage of 30,000 volts, this speed is about one-third that of light. Not all electrons are decelerated in the same way. Some electrons are stopped in one impact and produce X-ray photons of minimum wavelength. Other electrons lose fractions of their total kinetic energy continuously due to successive collisions with the atoms of the target. Thus, loss of kinetic energy of electrons is in two ways giving rise to a continuous spectrum and characteristic spectrum.

3.3.1 Origin of continuous X-ray spectrum

The continuous spectrum of an X-ray is due to the rapid and continuous deceleration of the electrons on hitting the target. A few fast moving electrons penetrate deep into the interior of the atoms of the target material and collide with its atoms. Due to this, the electrons are deflected from their original paths. In this way, the electrons are decelerated continuously and there is continuous loss of energy. This continuous loss of energy during deceleration is given off in the form of radiation called X-rays. The X-rays consist of a continuous range of frequencies up to maximum frequency v_{max} or minimum wavelength λ_{min}. This is called a continuous spectrum.

If an electron moving at speed v comes to rest suddenly by losing all its kinetic energy $\left(\dfrac{1}{2}mv^2\right)$, a photon containing equal amount of energy, hv will be created. Thus, we have

$$hv_{max} = \frac{1}{2}mv^2 \tag{3.1}$$

If electron of charge q gains $\dfrac{1}{2}mv^2$ energy by moving under a potential difference of V volt, Eq. (3.1) becomes

$$hv_{max} = \frac{1}{2}mv^2 = qV \tag{3.2}$$

or $\qquad \dfrac{hc}{\lambda_{min}} = qV$

or $\qquad \lambda_{min} = \dfrac{hc}{qV} = \dfrac{6.62\times10^{-34}\times3\times10^{8}}{1.6\times10^{-19}V}m = \dfrac{12400}{V}\overset{\circ}{A} \tag{3.3}$

Example 3.2

The operating voltage of an X-ray tube is 40 kV. Find the maximum speed of the electrons striking the surface of the anode.

Solution

$$\frac{1}{2}mv_{m}^2 = qV$$

or $\qquad v_{m} = \sqrt{\dfrac{2qV}{m}} = \sqrt{\dfrac{2\times1.6\times10^{-19}\times40000}{9.11\times10^{-31}}}\,\text{m/s} = 1.18\times10^{8}\,\text{m/s}$

Example 3.3

An X-ray tube operates at 45 kV. Calculate the shortest wavelength, of the X-rays produced.

Solution

The X-ray produced will be the shortest wavelength if the whole of the energy acquired by the electron is converted into X-rays.

$$\lambda_{min} = \frac{12400}{V}\,\overset{o}{A} = \frac{12400}{45000}\,\overset{o}{A} = 0.28\,\overset{o}{A}$$

Example 3.4

If the potential difference between the anode and the cathode is 45 kV, what is the cut-off frequency of the emitted X-rays?

Solution

The short wavelength limit λ_{swl} of X-rays is given as

$$\lambda_{swl} = \frac{12400}{V}\,\overset{o}{A} = \frac{12400}{45000}\,\overset{o}{A} = 0.28\,\overset{o}{A}$$

The cut-off frequency v_0 of X-rays is given as

$$v_O = \frac{c}{\lambda_{swl}} = \frac{3\times10^8}{0.28\times10^{-10}}\,Hz = 10.71\times10^{18}$$

Example 3.5

Find the short wavelength limit of a continuous X-ray spectrum when it shifts by 0.4 Å, if the voltage applied to the X-ray tube is doubled.

Solution

Let the short wavelength limit be λ_{1swl} when tube voltage is V_1 and λ_{2swl} when tube voltage is V_2. Thus, according to the question, we have

$$\lambda_{2swl} = \lambda_{1swl} - 0.4\,\overset{o}{A} \quad \text{and} \quad \frac{V_2}{V_1} = 2$$

Since, $\lambda_{swl} = \frac{12400}{V}\,\overset{o}{A}$, we get $\frac{\lambda_{1swl}}{\lambda_{2swl}} = \frac{V_2}{V_1} = 2$

or $\lambda_{1swl} = 2\lambda_{2swl} = 2\left(\lambda_{1swl} - 0.4\,\overset{o}{A}\right)$

or $\lambda_{1swl} = 0.8 \overset{\text{o}}{A}$

Equation (3.3) gives the short wavelength limit λ_{min} as a function of the applied voltage V (in volt). It is independent of the target material. If an electron is not completely stopped in one encounter but undergoes a glancing impact which only partially decreases its speed, then only a fraction of its energy qV is emitted as radiation and the photons produced has energy less than hv_{max}. In terms of wave motion, the corresponding X-ray has a frequency lower than v_{max} and a wavelength longer than λ_{min}. The totality of these wavelengths, ranging upward from λ_{min}, constitutes the continuous spectrum as shown in Fig. 3.2.

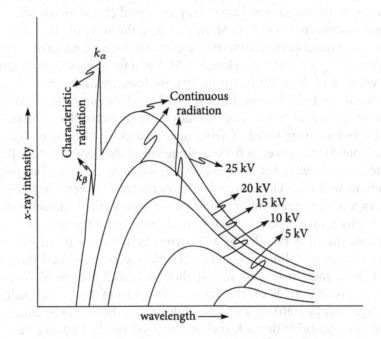

Figure 3.2. The X-ray spectrum of molybdenum as a function of applied voltage ranging from 5 kV to 25 kV. In addition to the applied voltage, the appearance of the characteristic spectra depends upon the target material

As shown in the figure, as the applied voltage increases, the curves become higher and shift to the left. This is because, as applied voltage increases, the number of photons produced per second and the average energy per photon both increases. The total X-ray energy emitted per second, which is proportional to the area under one of the curves of Fig. 3.2, also depends on the atomic number of the target and on the tube current i, which is a measure of the number of electrons per second striking the target. This total X-ray intensity of continuous spectrum is given by

$$I_C = AiZV^m \tag{3.4}$$

where A is a proportionality constant and m is a constant with a value of about 2. Therefore, where large amounts of continuous radiation are desired, heavy metals like tungsten ($Z = 74$) are used as a target and as high a voltage as possible is applied. It is important to remember that the target material affects the intensity but not the wavelength distribution of the continuous spectrum.

3.3.2 Origin of the characteristic spectrum

When the voltage on an X-ray tube is raised above a certain critical value, the characteristic of the target metal, sharp intensity maxima appear at certain wavelengths, superimposed on the continuous spectrum as shown in Fig. 3.2. Since they are narrow and their wavelengths are characteristic of the target metal used, they are called characteristic lines. These lines fall into several sets, referred to as K, L, M, and so on, in the order of increasing wavelength, all the lines together forming the characteristic spectrum of the metal used as the target. For a copper target, the K_α-line has a wavelength 1.54 Å and for a molybdenum target, the K_α-line has a wavelength of about 0.7 Å. Normally these lines are useful in X-ray diffraction.

The continuous spectrum is caused by the rapid deceleration of electrons by the target, but the origin of the characteristic spectrum lies in the atoms of the material itself. A few fast moving electrons having speed of about one-third of the velocity of light may knock out the tightly bound electrons even from the inner-most shells (like K, L shells, and so on) leaving the atoms in excited states. Immediately, the vacancies so created are filled up by the electrons from higher shells. The energy difference radiated in these electronic transitions appears as X-rays of very small but of definite wavelengths called characteristic lines. The wavelengths of this X-rays depend upon the atomic number of the target material.

The K shell vacancy may be filled by an electron from anyone of the outer shells, thus giving rise to a series of K-lines. The characteristic spectra produced due to transition of electrons from L shell to K shell and M shell to K shell are the K_α-line and K_β-line respectively. It is possible to fill a K shell vacancy either from the L or M shell, so that one atom of the target may be emitting a K_α-line while its neighbour may be emitting a K_β-line. However, it is more probable that a K shell vacancy will be filled by an L electron than by an M electron, and the result is that the K_α-line is stronger than the K_β-line. The K_α-line has longer wavelength than that of the K_β-line. It also follows that it is impossible to excite one K-line without exciting all the others. L-characteristic lines originate in a similar way. An electron knocked out of the L shell may be filled by an electron from some outer shell, thus producing L_α-lines, L_β-lines and so on.

Example 3.6

The energy of a K-electron, in a certain metal is –20 keV and that of an L-electron is –2 keV. Find the wavelength of the X-rays emitted when these electrons jump from the L shell to the K shell.

Solution

$$E_L = -2 \text{ keV} = -2 \times 10^3 \times 1.6 \times 10^{-19} \text{ J}$$

$E_K = -20 \text{ keV} = -20 \times 10^3 \times 1.6 \times 10^{-19} \text{ J}$

The energy of the X-ray photon is given as

$$E = E_L - E_K = h\nu = \frac{hc}{\lambda}$$

or $\lambda = \dfrac{hc}{E_L - E_K} = \dfrac{6.62 \times 10^{-34} \times 3 \times 10^8}{18 \times 10^3 \times 1.6 \times 10^{-19}} \text{ m} = 0.69 \overset{\text{o}}{\text{A}}$

There should be a critical excitation voltage for characteristic radiation. K-lines can be produced if the tube voltage is such that the bombarding electrons have enough energy to knock an electron out of the K shell of a target atom. If W_K is the work required to remove a K-electron, then the bombarding electrons must have this much kinetic energy, i.e.,

$$\frac{1}{2}mv^2 = W_K \tag{3.5}$$

Less energy is required to remove an L-electron than a K-electron, since the former is farther from the nucleus. Therefore, it follows that the L-excitation voltage is less than that of K and that K-lines cannot be produced without L, M, and so on, lines accompanying it.

These characteristic lines may be seen in the uppermost curve of Fig. 3.2 for voltage 25 kV (for molybdenum target). Since the critical K-excitation voltage, i.e., the voltage necessary to excite K-characteristic radiation, is 20.01 kV for molybdenum, the lines do not appear in the lower curves of the same figure. An increase in voltage above the critical voltage shifts the continuous spectrum to still shorter wavelengths and increases the intensities of the characteristic lines relative to the continuous spectrum but does not change their wavelengths. The intensity of any characteristic line, measured above the continuous spectrum, depends both on the tube current I and the amount by which the applied voltage V exceeds the critical excitation voltage for that line. For a K-line, the intensity I_K is given by

$$I_K = Bi\left(V - V_K\right)^n \tag{3.6}$$

Here

B = proportionality constant

V_K = critical K-excitation voltage

$n \approx 1.5$

The intensity of a characteristic line can be quite large. For example, in the radiation from a copper target operated at 30 kV, the K_α-line has an intensity about 90 times that of the wavelengths immediately adjacent to it in the continuous spectrum. Besides being very intense, characteristic lines are also very narrow, having half-width of about 0.001 Å. The existence of this strong and sharp K_α-line makes X-ray diffraction possible, since most diffraction experiments require the use of intense monochromatic K_α radiation. The K_β radiation is usually removed by use of a filter, a monochromator or an energy-selective detector.

3.4 Absorption of X-rays

The most spectacular properties of X-rays are its ability to penetrate materials that are opaque to less energetic radiation. When X-rays encounter any form of matter, they are partly transmitted and partly absorbed. High-energy X-ray photons are more likely to interact with most tightly electrons, i.e., with electrons in the K or L shells. Because of this, X-rays have more penetrating power. Experiment shows that the intensity I of an X-ray beam decreases exponentially as it passes through any homogeneous substance. Thus, in other words

$$\frac{dI}{I} \propto -dx$$

or $$\frac{dI}{I} = -\mu dx \qquad (3.7)$$

Here,

$$\frac{dI}{I} = \text{fractional decrease in X-ray intensity}$$

$$\mu = \text{proportionality constant.}$$

The proportionality constant μ is called linear absorption coefficient or macroscopic absorption coefficient or linear attenuation coefficient. From Eq. (3.7), it is defined as the fractional decrease in X-ray intensity after passing through unit distance. Its value depends upon the X-rays wavelength and the material used.

From Eq. (3.7), the intensity of the transmitted X-rays I is obtained as

$$\int_{I_0}^{I} \frac{dI}{I} = -\int_{0}^{x} \mu dx$$

or $$I(x) = I_0 e^{-\mu x} \qquad (3.8)$$

Here,

I_0 = intensity of the incident X-rays.

$I(x)$ = intensity of the X-rays after passing through a distance x.

The linear absorption coefficient μ is proportional to density ρ. Hence, for a particular material, $\dfrac{\mu}{\rho}$ is constant. $\dfrac{\mu}{\rho}$ is called the mass absorption coefficient and its value for different materials is tabulated in scientific data books. Hence, Eq. (3.8) can be modified to

$$I = I_0 e^{-\frac{\mu}{\rho}\rho x} \qquad (3.9)$$

The variation of mass absorption coefficient $\dfrac{\mu}{\rho}$ with wavelength is shown in Fig. 3.3.

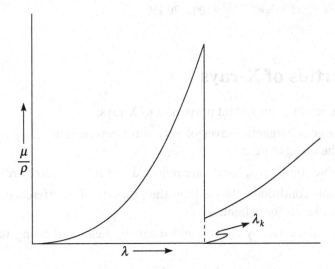

Figure 3.3 | The variation of mass absorption coefficient of a typical material with wavelength

Abruptly $\dfrac{\mu}{\rho}$ decreases to a minimum value for certain wavelength called K-absorption edge λ_K. For nickel, it is just around 1.54 Å. It corresponds to the critical energy for ejection of K-shell electrons of the atom. The curve consists of two similar branches separated by a sharp discontinuity at the absorption edge. Along each branch, the absorption coefficient varies with wavelength approximately according to a relation of the form

$$\frac{\mu}{\rho} = k\lambda^3 Z^3 \qquad (3.10)$$

Matter absorbs X-rays in two distinct ways, (i) by scattering and (ii) by true absorption, and these two processes together make up the total absorption measured by the quantity $\frac{\mu}{\rho}$. Short wavelength (high energy) X-rays are therefore highly penetrating and are termed hard as mentioned earlier, while long-wavelength X-rays are easily absorbed and are said to be soft.

Example 3.7

The linear absorption coefficient of aluminium is 0.693/cm for the K_α line. What percentage of the intensity of this line will pass through a 5 mm plate of aluminium?

Solution

The data given are $\mu = 0.693$/cm, $x = 5$ mm $= 0.5$ cm

The percentage of the intensity

$$= \frac{I}{I_0} \times 100 = e^{-\mu x} \times 100 = e^{-0.693 \times 0.5} \times 100 = 70.7\%$$

3.5 Properties of X-rays

Following are some of the important properties of X-rays:

i.   X-rays are electromagnetic waves of very short-wavelength. They travel in straight lines with the speed of light.

ii.   Under suitable conditions, X-rays are reflected and refracted like ordinary light.

iii.   Under suitable conditions, they exhibit the property of interference, diffraction and polarization like ordinary light.

iv.   They are not deflected by electric and magnetic fields indicating that the rays are uncharged.

v.   X-rays can penetrate through substances that are opaque to ordinary light, for example, wood, flesh, thick paper, thin sheets of metals. Lead offers maximum resistance to X-rays.

vi.   They cause fluorescence in many substances like barium, cadmium, tungstate, zinc sulphide, and so on.

vii.   They are capable of causing ionization in a gas through which they pass.

viii.   They can affect a photographic plate.

ix.   X-rays have a destructive effect on living tissue. When they are exposed to the human body, they cause reddening of the skin, sores and serious injuries to the tissues and glands. They can destroy white corpuscles of the blood.

x. They are capable of ejecting electrons from surfaces of metals, that is, they causes photoelectric effect.

xi. They produce secondary X-rays on striking certain heavy metals.

3.6 Moseley's Law

In 1913, Moseley studied the characteristic X-ray spectrum of a number of elements by using them as targets in an X-ray tube. When the rays were analyzed by means of a spectrometer, two series, namely K-series and L-series, were observed. He drew the following conclusions from his observations:

i.   There was a clear regularity in the lines of a series.

ii. Spectra produced in case of various targets were similar to each other except that the lines had different wavelengths.

iii. For a particular line, the frequency v varied in a regular manner from one element to the next in the periodic table of elements.

A graph between $\sqrt{v}$ (along the Y-axis) and the atomic number Z (along the X-axis) of the target was found to be a straight line as shown in Fig. 3.4. The result is popularly known as Moseley's law. It states that the square root of the frequency of any particular spectral line is proportional to the atomic number of the target element. Mathematically, Moseley's law can be expressed as

$$v = a(Z - \sigma)^2 \tag{3.11}$$

where a and σ are constants.

The energy of an electron E_n moving in a shell having principal quantum number n as calculated from Bohr's theory of hydrogen atom is given by

$$E_n = -\frac{m_0 q^4 Z^2}{8\varepsilon_0^2 h^2 n^2} \tag{3.12}$$

m_0 = rest mass of electron

$q = 1.6 \times 10^{-19}$ C = charge of electron

Z = atomic number

$\varepsilon_0 = 8.85 \times 10^{-12}$ C^2/Nm^2 = permittivity of free space

$h = 6.63 \times 10^{-34}$ J.s = Planck's constant

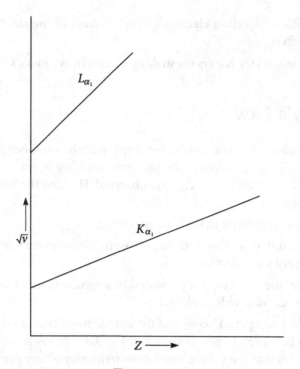

Figure 3.4 | Moseley's relation between $\sqrt{v}$ and Z and for two characteristic lines

If an electron jumps from a higher energy state n_2 of energy E_{n_2} to a lower energy state n_1 of energy E_{n_1}, its excess energy is emitted in the form of a photon of energy hv. Thus, we have

$$hv = E_{n_2} - E_{n_1} = \frac{m_0 q^4 Z^2}{8\varepsilon_0^2 h^2 n_1^2} - \frac{m_0 q^4 Z^2}{8\varepsilon_0 h^2 n_2^2}$$

or $$v = \frac{m_0 q^4 Z^2}{8\varepsilon_0^2 h^3}\left(\frac{1}{n_1^2} - \frac{1}{n_2^2}\right) = Rc\left(\frac{1}{n_1^2} - \frac{1}{n_2^2}\right)Z^2 \qquad (3.13)$$

Here, $R = \dfrac{m_0 q^4}{8\varepsilon_0 h^3 c}$ = Rydberg's constant = $1.0973731 \times 10^7\,\mathrm{m^{-1}}$.

c = speed of light in vacuum.

The resemblance of this equation

$$v = Rc\left(\frac{1}{n_1^2} - \frac{1}{n_2^2}\right)Z^2$$

to Moseley's law encourages us to analyze it a bit more. If one electron is removed from the K-shell, the remaining outer electrons will be attracted to the nucleolus not by the nuclear charge Ze, but by a charge $(Ze-e)$, the remaining K-shell electrons screen the outer electrons from the full nuclear attractions. Thus, the σ term in Moseley's law is a nuclear screening constant. By combining this discussion with Eq. (3.13), we can have

$$v = Rc\left(\frac{1}{n_1^2} - \frac{1}{n_2^2}\right)(Z-\sigma)^2 \tag{3.14}$$

The wavelength of the characteristic line is given as

$$\lambda = \frac{c}{v} = \frac{1}{R\left(\dfrac{1}{n_1^2} - \dfrac{1}{n_2^2}\right)(Z-\sigma)^2} \tag{3.15}$$

For K-series lines, $\sigma = 1$. The wavelength of the K_α-line ($n_1 = 1$, $n_2 = 2$) as obtained from this formula is

$$\lambda = \frac{4}{3R(Z-1)^2} \tag{3.16}$$

Choosing copper as the target ($Z = 29$) and using this formula, we get the wavelength for K_α-line $\lambda = 1.55$ Å which is in close agreement with the experimental value 1.54 Å!

This law paved the way for the present adjustment of elements in the periodic table. Previously, elements were arranged in the order of increasing atomic mass. A straight line graph between $\sqrt{v}$ and Z, as explained earlier, indicated that the fundamental quantity which increases regularly from one element to the next was its atomic number and not the atomic mass as thought previously. The origin of different characteristic lines is illustrated in Fig. 3.5.

Example 3.8

X-rays from a cobalt target tube consists of one strong K-line of cobalt ($Z = 27$) and two weak K-lines for impurities. They are $K_{\alpha(\text{Cobalt})} = 0.1785$ nm $K_{\alpha(\text{impurity1})} = 0.2285$ nm and $K_{\alpha(\text{impurity2})} = 0.1537$ nm. Determine the atomic number of these impurities and hence, identify the elements. (For K-lines, take $\sigma = 1$ in Moseley's law)

Solution

For K_α-lines, $n_1 = 1$, $n_2 = 2$, and $\sigma = 1$. From Moseley's law, we have

$$\lambda = \frac{4}{3R(Z-1)^2}$$

or $Z = 1 + \sqrt{\dfrac{4}{3R\lambda}}$ with $R = 1.0973731 \times 10^7 \, \text{m}^{-1}$

Data given are $K_{\alpha(\text{impurity1})} = 0.2285$ nm and $K_{\alpha(\text{impurity2})} = 0.1537$ nm. We have

$$Z = 1 + \frac{3.48571 \times 10^{-4}}{\sqrt{\lambda}} = 1 + \frac{3.48571 \times 10^{-4}}{\sqrt{0.2285 \times 10^{-9}}} = 24$$

and $$Z = 1 + \frac{3.48571 \times 10^{-4}}{\sqrt{\lambda}} = 1 + \frac{3.48571 \times 10^{-4}}{\sqrt{0.1537 \times 10^{-9}}} = 29$$

Thus, the atomic number of the impurities are 24 and 29 which correspond to chromium and copper respectively.

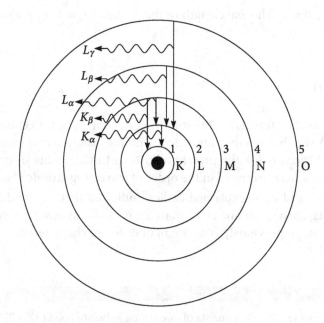

Figure 3.5 | Electronic transitions in an atom producing characteristic spectra (schematic). Emission processes are indicated by arrows

3.7 Bragg's X-rays Spectrometer

The gaps in the atomic planes of crystals are of the order of X-ray wavelength. Therefore, crystals can play the role of gratings for X-rays in the same manner as that of optical grating for visible light. Experimentally, Bragg's law ($2d \sin \theta = n\lambda$) can be utilized in two ways. By using X-rays of known wavelength λ and measuring θ, we can determine the spacing d of various planes in a crystal; this is structure analysis. Alternatively, we can use a crystal with planes of known spacing d, measure θ, and thus, determine the wavelength λ of the radiation used; this is X-ray spectroscopy. The essential features of an X-ray spectrometer or Bragg's spectrometer are shown in Fig. 3.6.

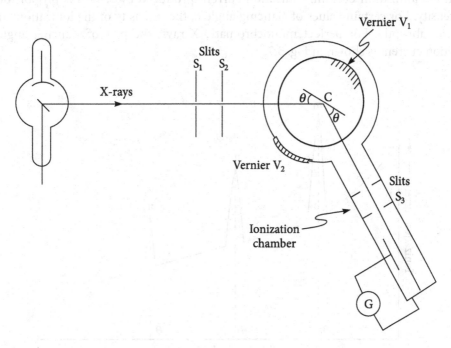

Figure 3.6 | Schematic diagram of Bragg's spectrometer

3.7.1 Determination of crystal structure by Bragg's X-ray spectrometer

The structure determination of a crystal in solid form is done in the following way by using Bragg's X-ray spectrometer. The crystal mounted on the turn-table acts like a reflection grating. The X-rays from X-ray tube are narrowed to obtain a fine pencil of beam by passing them through lead slits S_1 and S_2. The beam is now allowed to fall on a crystal C mounted on a circular turn-table of the spectrometer. This turn-table is capable of rotating about a vertical axis passing through its centre. The rotation can be read on a circular graduated scale S fitted with the vernier V_1. The reflected beam for a particular glancing angle then passes through slits S_3 and enters the ionization chamber. The ionization chamber is simply a gas container with two electrodes. It is capable of rotating about a vertical axis passing through the centre of the spectrometer table. The position of the ionization chamber can be read by a second vernier V_2. The turn-table and ionization chamber are linked together in such a way that when the turn-table rotates through an angle θ, the ionization chamber turns through 2θ. In this way, the beam is always reflected into the ionization chamber whatever may be the glancing angle at the surface of the crystal.

The X-rays entering the ionization chamber ionize the gas which causes a current to flow between the two electrodes that can be measured by galvanometer G. The ionization current is measured for different values of glancing angle θ. A plot is then obtained between θ and the ionization current. Ionization current produced by X-rays is proportional to its intensity. For certain values of glancing angle θ, the intensity of the ionization current increases abruptly. For perfect monochromatic X-rays, the plot of glancing angle and ionization current is shown in Fig. 3.7.

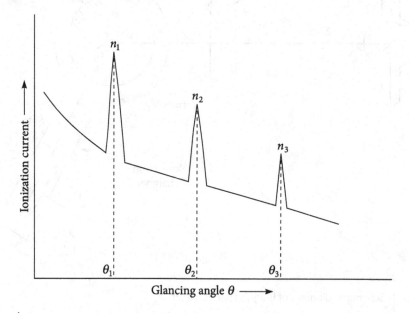

Figure 3.7 Variation of ionization current with glancing angle for monochromatic X-rays

For polychromatic X-rays, the plot of the glancing angle and ionization current is shown in Fig. 3.8. In both these figures, it is found that the ionization current does not fall to zero for any value of glancing angle θ; it is maximum for certain angles. This shows the existence of a continuous spectrum over which the characteristic line spectrum is superposed. With the help of Bragg's law ($2d \sin \theta = n\lambda$) and by measuring the glancing angle at which reflection occurs (the angle where intensity becomes maximum), we can determine the interplanar spacing of crystals by taking X-rays of known wavelength.

3.7.2 Powder method

The method of structure determination mentioned in the previous section is possible if a single crystal of reasonable size is available. However, practically, this is not always possible. Therefore, the Debye–Scherrer technique is often employed. The solid crystal is ground into a very fine homogeneous powder with a crystallite size less than 10 μm

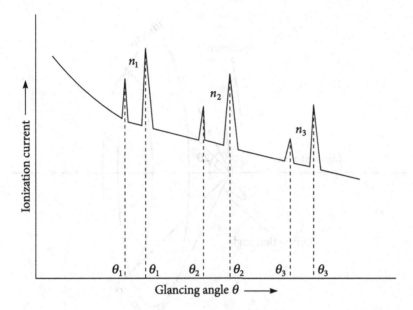

Figure 3.8 | Variation of ionization current with glancing angle for X-rays containing two wavelengths

and inserted into a thin glass capsule. The powdered sample contains an enormous number of fine crystallites with random orientations of crystal planes. Each set of planes in a crystal will give rise to a cone of diffraction. Each cone consists of a set of closely spaced dots each one of which represents a diffraction from a single crystallite. This has been illustrated in Fig. 3.9.

The basics of the experimental set up of the powder method are shown in Fig. 3.10. The X-rays from the X-ray tube are allowed to pass through a monochromator which ideally absorbs all wavelengths except the K_α-line. The beam is collimated by passing through a collimator. This fine pencil of X-rays is made to fall on the powdered specimen, which is placed at the centre of a drum-shaped cassette. A photographic film in the shape of a strip of 2.5 cm width is attached to the inner circumference of the drum.

The basic principle underlying this powder technique is that in the powder, the millions of micro-crystals have all possible random orientations. Among these very large number of micro-crystals, there will always exist some crystals whose lattice planes are so oriented that they satisfy Bragg's relation $n\lambda = 2d \sin \theta$. Diffraction therefore takes place from these planes.

As the parallel lattice planes with a given spacing d and the same value of n and θ occur in all positions around the axis of the incident beam, the reflected rays produce a cone with a semi-vertical angle 2θ. For various sets of d and n, various cones of rays are obtained. This has been depicted in Fig. 3.10. The scattered X-rays are incident on the photographic film. The intersections of the different cones of the photographic film are a series of concentric circular rings. Radii of these rings recorded on the film can be used to find the glancing angle. Now, the interplanar spacing of crystalline substances d can be calculated. The pattern recorded on the photographic film is shown in Fig. 3.11 when the film is laid flat.

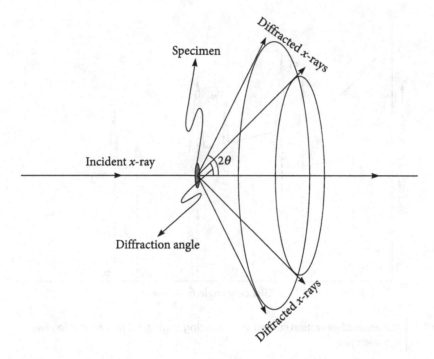

Figure 3.9 | X-rays are scattered in a sphere (called the Ewald sphere) around the sample

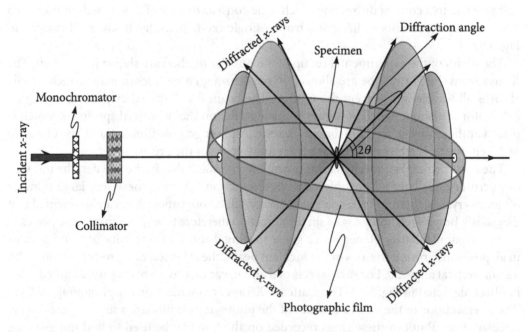

Figure 3.10 | Experimental setup of the powder method

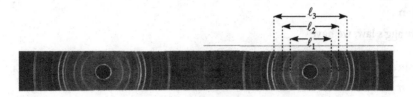

Figure 3.11 | Powder X-ray diffraction pattern on a photographic film

Due to the narrow width of film, only parts of circular rings are registered on it. The curvature of arcs reverses when the angle of diffraction exceeds 90°.

Let ℓ_1, ℓ_2, ℓ_3 and so on, be the distances between symmetrical lines on stretched photographic film and D, the diameter of cylindrical film, i.e., inner diameter of the drum. Then,

$$\frac{\ell_1}{\pi D} = \frac{4\theta_1}{360°}$$

or $$\theta_1 = \frac{90°}{\pi D}\ell_1$$

Similarly,

$$\theta_2 = \frac{90°}{\pi D}\ell_2 \text{ and } \theta_3 = \frac{90°}{\pi D}\ell_3$$

Using these values of θ in Bragg's formula, interplanar spacing d of crystals can be calculated.

Applications of the powder method

Powder diffraction methods are used to study many characteristics of materials. A few of them include

i.   phase composition
ii.   quantitative phase analysis
iii. unit cell lattice parameters
iv.   average crystallite size of nanocrystalline samples
v.   crystallite microstrain
vi.   texture
vii. residual strain

Example 3.9

Calculate the longest wavelength that can be analyzed by a rock salt crystal of spacing $d = 2.82$ Å in the first order.

Solution

From Bragg's law, we have

$$\frac{2d \sin \theta_{max}}{n} = \lambda_{max}$$

or $\lambda_{max} = \dfrac{2d \times 1}{n} = \dfrac{2 \times 2.82 \,\overset{O}{A} \times 1}{1} = 5.64 \,\overset{O}{A}$

Example 3.10

The spacing of planes in a crystal is 1.8 Å and the angle for the first order Bragg's reflection is 30°. Determine the energy of the X-ray photon in keV.

Solution

From Bragg's law, we have

$$\lambda = \frac{2d \sin \theta}{n} = \frac{2 \times 1.8 \,\overset{O}{A} \sin 30}{1} = 1.8 \,\overset{O}{A}$$

The energy of the X-ray photon

$$E = \frac{hc}{\lambda} = \frac{6.62 \times 10^{-34} \times 3 \times 10^{8}}{1.8 \times 10^{-10}} J = 11.03 \times 10^{-16} J = \frac{11.03 \times 10^{-16}}{1.6 \times 10^{-19}} eV = 6.90 \,keV$$

3.8 Uses of X-rays

Due to the property of X-rays that they are highly penetrating and have small wavelength, the rays have been enormously useful in various fields such as industry, engineering, medicine and scientific research work.

i.   It is a versatile tool in NDT methods of materials.

ii.   X-rays are used to detect any defect in radio valves, tennis balls, rubber tyres and the presence of pearls in oysters.

iii. X-rays can be used to test the homogeneity of welded joints, insulating materials, and so on.

iv.   X-rays are used to detect cracks in structures. They are very helpful in detecting any cracks in the body of the aeroplane and motor cars.

v.   X-rays can be used to analyze the structure of alloys and other composite bodies by determining the crystal form in an ingot with the help of diffraction of X-rays. In this way, alloys like cobalt–nickel steels, bronzes, duraluminium, porcelain insulators, and the like have been analyzed.

vi.   X-rays are used in determining the atomic number and identification of various chemical elements.

vii. Diffraction of X-rays, both wide angle and small angle, provide information regarding the molecular structure of crystalline, nano-structured materials, polymers and proteins in solid as well as in liquid state.

viii. The most common use of X-rays is to get a photograph of the interior of the human body. The photograph is known as a radiograph. Radiographs can be used to detect fractures, diseased organs, foreign matter like bullets and formation of stones in human body.

ix.   Hard X-rays are used to destroy tumours very deep inside the body.

x.   The property of X-rays that they are harmful to living tissues enables us to cure certain skin diseases by subjecting them to an exposure of X-rays. The X-rays, while passing through the diseased portion, kill the germs, thus curing it. It is called radiotherapy.

xi.   X-rays are also employed by custom/intelligence officials to screen the bags and luggage of persons crossing the border, so that they may not be carrying any objectionable materials with them.

Questions

3.1   Describe with a diagram, the production of X-rays by a modern Coolidge tube.

3.2   Explain the origin of the continuous X-ray spectrum.

3.3   Derive an expression for short wavelength limit of the continuous X-ray spectrum.

3.4   Explain the origin of the characteristic X-ray spectrum.

3.5   Give a comparative account of the characteristic X-ray spectrum and the continuous X-ray spectrum.

3.6   Explain why K-lines cannot be produced without producing L-lines.

3.7   Describe the phenomenon of interaction of X-rays with matter.

3.8   Distinguish between hard X-rays and soft X-rays.

3.9   Derive an expression for the intensity of X-rays as it transmits through a material in terms of material constants.

3.10 How does the linear absorption coefficient vary with atomic number of the absorber? Show it graphically.

3.11 What do you mean by critical excitation voltage in X-ray production?

3.12 Discuss some of the important properties of X-rays.

3.13 Derive Moseley's law in the X-ray spectrum with explanation of symbols. How does it change the arrangement of elements in the periodic table?

3.14 Describe the construction and working of Bragg's X-ray spectrometer.

3.15   Describe structure determination by powder diffraction method. What are its advantages over other methods?

3.16   Describe the uses of X-rays in different fields of science and technology.

Problems

3.1   The energy of an L-electron in a certain metal is −18 keV and that of an M-electron is −2 keV. Find the wavelength of X-rays emitted when these electrons jump from the M-shell to the L-shell.   [Ans   0.78 Å]

3.2   The operating voltage of an X-ray tube is 50 kV. Find the maximum speed of the electrons striking the surface of the anode and the shortest wavelength of the X-rays produced.   [Ans 1.32×10^8 m/s, 0.25 Å]

3.3   How much voltage must be applied between the anode and the cathode of an X-ray tube to produce X-rays of the shortest wavelength, 0.5 Å.   [Ans   24.8 kV]

3.4   Find the short wavelength limit of the continuous X-ray spectrum, when it shifts by 0.2 Å, if the voltage applied to the X-ray tube is tripled.   [Ans   0.3 Å]

3.5   An X-ray tube is operating at 80 kV and 10 mA. If only 1% of the electric power supplied is converted into X-rays, at what rate is the target being heated per second?
                                                                                 [Ans   792 J]

3.6   Calculate the shortest wavelength of X-ray when an X-ray tube with copper target operates at 40 kV.   [Ans   0.31 Å]

3.7   The linear absorption coefficient of copper is 13.9/cm for the K_α-line. What percentage of the intensity of this line will pass through a 5 mm plate of copper?   [Ans   0.096%]

3.8   X-rays from a certain copper target tube consists of one strong K-line of copper (Z = 29) and two weak K-lines for impurities. They are $K_{\alpha(Cu)}$ = 0.1542 nm, $K_{\alpha(imp1)}$ = 0.1441 nm, and $K_{\alpha(imp2)}$ = 0.1666 nm. Determine the atomic number of these impurities and hence, identify the elements. (For K-lines, take σ = 1 in Moseley's law)   [Ans   30, 28, Zn, Ni]

Multiple Choice Questions

1.   Which of the following is the correct justification for the name X-rays?
     (i)   rays are variable
     (ii) nature of the rays was unknown initially
     (iii) rays are due to unknown material particles
     (iv) rays are not visible

2.   From the energy point of view, X-rays are
     (i)   above gamma ray   (ii) below gamma ray
     (iii) above UV-ray   (iv) below UV-ray

3. Which of the following is correct?
   (i) small fraction of kinetic energy of striking electrons are converted into X-rays
   (ii) large fraction of kinetic energy of striking electrons are converted into X-rays
   (iii) large fraction of potential energy of striking electrons are converted into X-rays
   (iv) small fraction of potential energy of striking electrons are converted into X-rays

4. Which of the following is not a characteristics of a target?
   (i) high atomic weight
   (ii) high melting point
   (iii) high thermal conductivity
   (iv) high density

5. The continuous X-ray spectrum is produced due to
   (i) Continuous deceleration of striking electrons inside the target
   (ii) Continuous deceleration of knocked out electrons inside the target
   (iii) Continuous acceleration of striking electrons inside the target
   (iv) Continuous acceleration of knocked out electrons inside the target

6. The characteristic X-ray spectrum is produced due to
   (i) sudden deceleration of striking electrons by the target
   (ii) sudden deceleration of knocked out electrons from the inner-most shells
   (iii) filling up of vacancies in the inner-most shells by the outer shell electrons
   (iv) filling up of vacancies in the outer shells by the inner shell electrons

7. Which of the following makes X-ray diffraction possible?
   (i) existence of continuous X-ray spectrum
   (ii) existence of K_β-line in the X-ray spectrum
   (iii) existence of K_γ-line in the X-ray spectrum
   (iv) existence of K_α-line in the X-ray spectrum

8. In case of X-ray absorption, mass absorption coefficient
   (i) increases with increase of density of materials
   (ii) decreases with increase of density of materials
   (iii) is independent of material density
   (iv) none of the above

9. Which of the following is not a property of X-rays?
   (i) They are electromagnetic waves of very short-wavelength
   (ii) They are electromagnetic waves of very long wavelength
   (iii) They travel in straight lines with the speed of light
   (iv) X-rays are electromagnetic waves of very high frequencies

10. In accordance with Moseley's law if v is the frequency of the characteristic spectrum of different targets, then

    (i) graph between $\sqrt{v}$ and the atomic number Z of the target material is a parabola

    (ii) A graph between v and the atomic number Z of the target material is a straight line

    (iii) A graph between v and the atomic number Z of the target material is a parabola

    (iv) graph between $\sqrt{v}$ and the atomic number Z of the target material is a straight line

Answers

1 (ii)	2 (ii, iii)	3 (i)	4 (iv)	5 (i)	6 (iii)	7 (iv)	8 (iii)
9 (ii)	10 (iv)						

4 Bonding in Solids

4.1 Introduction

The forces which bind together the atoms or molecules of a substance are called bonds or bonding force. Depending upon the magnitude of this bonding force, the substance remains in different states like solid, liquid or gaseous state. The root cause for which carbon exists as both graphite and diamond is bonding. The magnitude of this force decreases from solid to gas. Many physical properties of materials like melting point, boiling point, elasticity, thermal expansion, electrical and thermal conductivity are possible to predict with knowledge of the bonding forces or bonding energy.

4.2 Bonding Forces

Two different forces must be present to establish bonding in a substance. An attractive force is necessary for any bonding; simultaneously, a repulsive force is required to keep the atoms from coalescing into a "point particle". As the two atoms approach each other from a larger distance, they exert attractive and repulsive forces as shown in Fig. 4.1; the magnitude of each is a function of the inter-atomic separation. The attractive forces, which keep the atoms together forcing them to form a solid is given by

$$F_A = \frac{A}{r^i} \tag{4.1}$$

The repulsive forces which keep the atoms apart forcing them not to coalesce into a point is given by

$$F_R = -\frac{B}{r^j} \tag{4.2}$$

In these equations, A and B are proportionality constants and i, j are real numbers. The nature of these forces for two isolated atoms is illustrated in Fig. 4.1.

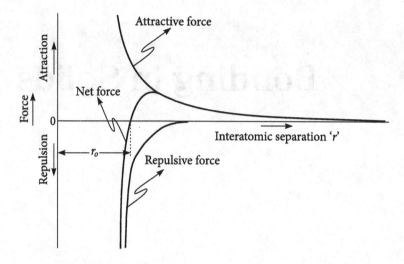

Figure 4.1 | The dependence of repulsive, attractive, and net forces on inter-atomic separation for two isolated atoms

When the outer electron shells of the two approaching atoms begin to overlap, a strong repulsive force comes into play. Further decrease of distance is not possible and atoms are at equilibrium. The reason for the strong repulsion at very short distances is Pauli's exclusion principle. At equilibrium, the attractive and repulsive forces balance each other. Now the atoms will oppose any attempt to disturb them. The centres of the two atoms will remain separated by the equilibrium distance r_0 as indicated in the figure.

4.3 Bonding Energies

Many times, it is more convenient to work with the potential energies between two atoms instead of forces. Mathematically potential energy U and force F are related by

$$F = -\frac{\partial U}{\partial r} \tag{4.3}$$

The nature of these potential energies of two isolated atoms is illustrated in Fig. 4.2.

The total potential energy U is the sum of the repulsive energy U_R and the attractive energy U_A, i.e.,

$$U = U_R + U_A$$

or $\quad U = \dfrac{C}{r^m} - \dfrac{D}{r^n} \tag{4.4}$

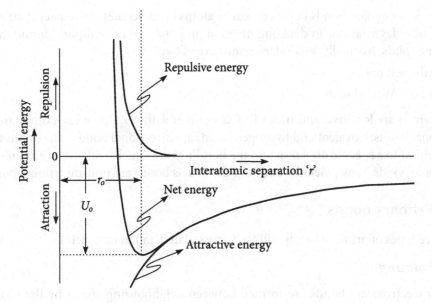

Figure 4.2 | The dependence of repulsive, attractive, and net potential energies on inter-atomic separation for two isolated atoms

Here C and D are constants and m and n the repulsive and attractive exponents. Every system attains a stable state by acquiring minimum potential energy. When atoms come closer to form bonds, their valence electrons re-arrange themselves so as to reach a stable state by acquiring minimum potential energy. Thus, equilibrium separation r_0, corresponds to the separation distance at the minimum of the total potential energy curve. This inter-atomic equilibrium separation r_0 is called bond length. The bonding energy U_0 is the minimum potential energy. The bonding energy is defined as the energy required to separate the atoms to an infinite separation. The magnitude of this bonding energy and the shape of the energy-versus inter-atomic separation curve vary from material to material, and they both depend on the type of atomic bonding.

4.4 Classification of Bonds

Broadly speaking, there are two types of bonds; one is primary bonds or inter-atomic bonds, and the other is secondary bonds or inter-molecular bonds. Primary bonds are strong bonds which hold the atoms together. There are three types of primary bonds:

i.   Ionic bonds

ii.   Covalent bonds

iii. Metallic bonds.

The three types of primary bonding reflect the ways in which atoms can group together by gaining or losing or sharing electrons, so that they can get inert gas electron configurations.

Though, secondary bonds between nearby atoms or molecules are weaker than primary bonds, they play vital role in deciding material properties of most liquids, liquid mixtures and some solids. Normally, secondary bonds are of two types.

i.   Hydrogen bonds
ii.   Van der Waals bonds

Bonds which are localized and occur in fixed angles with respect to each other are called directional bonds. Covalent and hydrogen bonds are directional bonds. The bonds that are not localized to a specific direction and the bondings are equal at all angles are called non-directional bonds. Ionic, metallic and Van der Waals bonds are non-directional bonds.

4.4.1 Primary bonds

The three types of primary bonds will be discussed in detail in this section.

Ionic bonding

Ionic or electrovalent bonds are formed between neighbouring atoms by the transfer of electrons from one atom to other atoms. Solids with more than one type of atom often possess ionic bonds. This includes ceramic materials, such as oxides and silicates, as well as salts. It is always found in compounds that are composed of both metallic and non-metallic elements. Ionic bonding involves the transfer of electrons from an electropositive atom to an electronegative atom. Atoms of a metallic element easily give up their valence electrons to the non-metallic atoms. In the process, all the atoms acquire stable or inert gas configurations and in addition, they become ions. The bonding force is the Coulomb attraction between the two resulting ions creating a very strong bond. Thus, none of the atoms in an ionic solid are neutral; all atoms in the crystal are ions. Ionizing both atoms usually costs some energy and is called ionization potential/ionization energy. It is defined as the energy required to remove an electron from the atom or molecule in the gaseous state.

NaCl is the most common example of ionic bonding. When sodium and chlorine atoms are placed together, there is a transfer of electrons from the sodium Na to the chlorine Cl atom as shown in Fig. 4.3.

When Na ($11 = 1s^2, 2s^2\ 2p^6, 3s^1$) gives up one electron, it becomes Na^+ ($1s^2, 2s^2\ 2p^6$). It becomes more stable by having completely filled 2p sub-shell. When Cl^- ($17 = 1s^2, 2s^2\ 2p^6, 3s^2\ 3p^5$) accepts the electron from Na, it becomes Cl^- ($1s^2, 2s^2\ 2p^6, 3s^2\ 3p^6$). It becomes more stable by having completely filled 3p sub-shell. This results in a strong electronic attraction between the positive sodium ions and the negative chlorine ions. Ionic crystals are usually insulators, transparent, soluble in polar solvents like water and insoluble in non-polar solvents like benzene. It is difficult to deform ionic solids because of the strong electrostatic force between the ions. Thus, ceramic materials are very brittle and cannot be deformed easily in the solid state. The electrical conductivity of an ionic crystal, in general, is very low because there are no free electrons to conduct current.

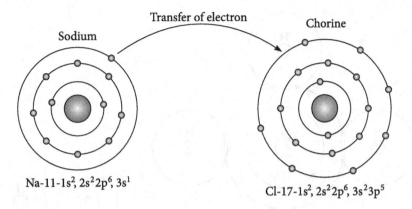

Figure 4.3 | Transfer of electron from sodium atom to chlorine atom

Covalent bonds

Covalent bonds are formed between neighbouring atoms by the sharing of electron(s) of the participating atoms. Two atoms that are covalently bonded with each other contribute at least one electron to the bond, and the shared electrons may be considered to belong to both atoms. In solids, covalent bonding is often found for elements with a roughly half-filled outer shell. A prominent example is carbon that forms solids as diamond or graphite as well as complex molecules such as Fullerene or carbon nanotubes. The boiling point, melting point, solubility in water, thermal conductivity and electrical conductivity of covalent solids are usually lower than the ionic solids. When the overlapping orbitals are directionally oriented and not spatially symmetric, good overlapping and substantial decrease in the potential energy can occur. This gives directionality to the covalent bonds as a result of which covalently bonded materials are less dense. When bonds are directional, the atoms cannot pack together in as dense a manner, yielding a lower mass density.

Many non-metallic elemental molecules such as H_2, Cl_2, F_2, as well as molecules containing dissimilar atoms, such as CH_4, H_2O, HNO_3, are covalently bonded. Furthermore, this type of bonding is found in elemental solids such as diamond (carbon), silicon, germanium and other solid compounds. Covalent bonding is schematically illustrated in Fig. 4.4 for a molecule of methane CH_4 and water H_2O. The carbon atom has four valence electrons, whereas each of the four hydrogen atoms has a single valence electron. Each hydrogen atom can acquire a helium electron configuration (two $1s$ valence electrons) when the carbon atom shares with it one electron. The carbon now has four additional shared electrons, one from each hydrogen, for a total of eight valence electrons, and the electron structure of neon.

The number of covalent bonds that is possible for a particular atom is determined by the number of valence electrons. For N valence electrons, an atom can covalently bond with at most $8 - N$ other atoms. For example, $N = 7$ for chlorine, and $8 - N = 1$ which means that one Cl atom can bond to only one other atom, as in Cl_2. Similarly, for carbon,

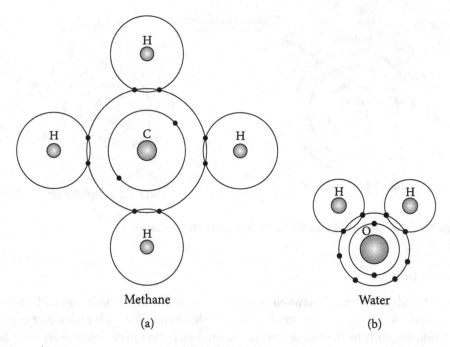

Methane
(a)

Water
(b)

Figure 4.4 | Schematic representation of covalent bonding in a molecule of methane and water

$N = 4$ and each carbon atom has $8 - 4 = 4$, electrons to share. Diamond is simply a three-dimensional interconnecting structure wherein each carbon atom covalently bonds with four other carbon atoms.

Metallic bonds

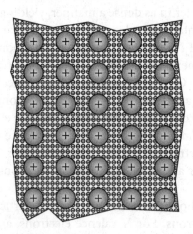

Figure 4.5 | Schematic illustration of metallic bonding

Metals are formed from elements on the left-hand side of the periodic table. A relatively simple model has been proposed that very nearly approximates the bonding scheme. Metallic bonding can only occur among a large aggregate of atoms, such as in a crystal. On the other hand, a covalent bond can occur between only two atoms, as in an isolated molecule. Metallic materials have one, two, or at most, three valence electrons. Having generally low electronegativity (the tendency of an atom or a functional group to attract electrons itself), they tend to lose their valence electrons easily. With this model, these valence electrons are not bound to any particular atom in the solid and are more or less free to drift randomly throughout the entire metal. In other words, these valence electrons are free from any particular atom and are only held collectively by the entire assemblage of atoms. They may be thought of as belonging to the metal as a whole, or forming a "sea of electrons" or an "electron cloud." These de-localized valence electrons are involved in the conduction of electricity and are therefore often called conduction electrons. The remaining non-valence electrons and atomic nuclei form what are called ion cores, which possess a net positive charge. In a metal, the ion cores are held more or less at fixed places in the crystal lattice. Figure 4.5 is a schematic illustration of metallic bonding. The free electrons shield the positively charged ion cores from mutually repulsive electrostatic forces, which they would otherwise exert upon one another; consequently, the metallic bond is non-directional in character. In addition, these free electrons act as a "glue" to hold the ion cores together. The random motion of valence electrons vanishes when stimulations are applied, such as electric fields, magnetic field or heat.

4.4.2 Secondary bonds

The two types of secondary bonds are discussed in detail in this section.

Hydrogen bonds

Hydrogen bonding, a special type of secondary bonding, is found to exist between some molecules that have hydrogen as one of the constituents. The hydrogen bond is stronger than a Van der Waals interaction, but weaker than covalent or ionic bonds. This type of bond occurs in both inorganic molecules such as water and organic molecules like proteins and DNA.

If the bond is to an electronegative atom like nitrogen, oxygen or fluorine that comes from another molecule or chemical group, the electron is mostly located close to that atom and the hydrogen nucleus represents an isolated positive (partial) charge. These bonds can occur between molecules (intermolecular), or within different parts of a single molecule (intramolecular). In 2011, an IUPAC Task Group gave the following modern definition of hydrogen bonding: "The hydrogen bond is an attractive interaction between a hydrogen atom from a molecule or a molecular fragment X–H in which X is more electronegative than H, and an atom or a group of atoms in the same or a different molecule, in which there is evidence of bond formation".

A hydrogen atom attached to a relatively electronegative atom is a hydrogen bond donor. An electronegative atom such as fluorine, oxygen, or nitrogen is a hydrogen bond acceptor, regardless of whether it is bonded to a hydrogen atom or not. An example of a hydrogen bond donor is ethanol C_2H_5OH, which has a hydrogen bonded to oxygen. An example of a hydrogen bond acceptor which does not have a hydrogen atom bonded to it is the oxygen atom on diethyl ether $(C_2H_5)_2O$. A hydrogen attached to carbon can also participate in hydrogen bonding when the carbon atom is bound to electronegative atoms, as is the case in chloroform, $CHCl_3$. The electronegative atom attracts the electron cloud from around the hydrogen nucleus and by decentralizing the cloud, leaves the atom with a partial positive charge. Because of the small size of hydrogen relative to other atoms and molecules, the resulting charge, though only partial, represents a large charge density. A hydrogen bond results when this strong positive charge density attracts a lone pair of electrons on another heteroatom (in organic chemistry, a heteroatom is any atom that is not carbon or hydrogen). Typical heteroatoms are nitrogen, oxygen, sulphur, phosphorus, chlorine, bromine and iodine, all of which become hydrogen bond acceptors.

The length of hydrogen bonds depends on bond strength, temperature and pressure. The bond strength itself is dependent on temperature, pressure, bond angle and dielectric constant. The typical length of a hydrogen bond in water is 197×10^{-12} m. The ideal bond angle depends on the nature of the hydrogen bond donor.

Intermolecular hydrogen bonding is responsible for the high boiling point of water (100°C) compared to the other group 16 hydrides that have no hydrogen bonds. Intramolecular hydrogen bonding is partly responsible for the secondary, tertiary and quaternary structures of proteins and nucleic acids. It also plays an important role in the structure of polymers, both synthetic and natural.

Hydrogen bonds in water

The simplest example of a hydrogen bond is found between water molecules. In a discrete water molecule, there are two hydrogen atoms and one oxygen atom. Two molecules of water can form a hydrogen bond between them; the simplest case, when only two molecules are present, is called the water dimer and is often used as a model system. When more molecules are present, as is the case of liquid water, more bonds are possible because the oxygen of one water molecule has two lone pairs of electrons, each of which can form a hydrogen bond with a hydrogen on another water molecule. This can repeat such that every water molecule is H-bonded with up to four other molecules. Hydrogen bonding strongly affects the crystal structure of ice, helping to create an open hexagonal lattice. The density of ice is less than water at the same temperature; thus, the solid phase of water floats on the liquid, unlike most other substances. Liquid water's high boiling point is due to the high number of hydrogen bonds each molecule can form relative to its low molecular mass. Owing to the difficulty of breaking these bonds, water has a very high boiling point, melting point and viscosity compared to otherwise similar liquids not conjoined by hydrogen bonds. Water is unique because its oxygen atom has two lone pairs and two hydrogen atoms, meaning that the total number of bonds of a water molecule is up to four. The exact number of hydrogen bonds formed by a molecule of liquid water fluctuates with time and depends on the temperature. It was found that at 25°C, there are, in an average, 3.59 hydrogen bonds in water. At 100°C, this number decreases to 3.24 due to the increased molecular motion and decreased density, while at 0°C, the average number of hydrogen bonds increases to 3.69.

Our discussion on hydrogen bonding will remain incomplete unless we state some interesting consequences of it.

i.   Increase in the melting point, boiling point, solubility and viscosity of many compounds can be explained by the concept of hydrogen bonding.

ii.   The presence of hydrogen bonds can cause an anomaly in the normal succession of states of matter for certain mixtures of chemical compounds as temperature increases or decreases. These compounds can be liquid until a certain temperature, then solid even as the temperature increases, and finally liquid again as the temperature rises over the "anomaly interval".

iii. Smart rubber is a polymer that is able to "heal" when torn. It can heal itself at room temperature and can be repeatedly used multiple times. The supramolecular self-healing rubber can be processed, re-used, and recycled the rubber can be torn apart and if lightly placed together again, it will start to self-heal. Smart rubber depends only on hydrogen bonds to make this all possible. It does not depend on covalent bonding or ionic bonding, which are present in normal rubber. Unlike covalent and ionic bonding, hydrogen bonding can occur simply by pressing the two faces of a substance together.

iv.   The strength of nylon and cellulose fibers is attributed to hydrogen bonding.

v.   Wool, being a protein fiber is held together by hydrogen bonds, causing wool to recoil when stretched. However, washing at high temperatures can permanently break the hydrogen bonds and a garment may permanently lose its shape.

Van der Waals bonds

The term van der Waals bonding refers to a weak, purely quantum mechanical effect. Van der Waals' forces include all intermolecular forces that act between electrically neutral molecules. They are effective only up to several hundred angstroms. Intermolecular forces are feeble; however, without them, life as we know it would be impossible. Water would not condense from vapour into solid or liquid forms if its molecules did not attract each other. Intermolecular forces are responsible for many properties of molecular compounds, including crystal structures (e.g., the shapes of snowflakes), melting points, boiling points, heats of fusion and vapourization, surface tension and densities. Intermolecular forces pin gigantic molecules like enzymes, proteins and DNA into the shapes required for biological activity. This type of interaction is present in any solid but it is much weaker than ionic, covalent, or metallic bonding and therefore, Van der Waals bonding is only observable for solids that do not show other bonding behaviour, for example, noble gases. Pure Van der Waals crystals can only exist at very low temperatures.

Broadly speaking, the Van der Waals force is the sum of the attractive or repulsive forces between molecules (or between parts of the same molecule) other than those due to covalent bonds, the hydrogen bonds, or the electrostatic interaction of ions with one another or with neutral molecules. The term includes (i) the force between two permanent dipoles

(Van der Waals–Keesom force), (ii) the force between a permanent dipole and a corresponding induced dipole (Van der Waals–Debye force), (iii) the force between two instantaneously induced dipoles (Van der Waals–London dispersion force). Van der Waals interactions have been illustrated in Fig. 4.6.

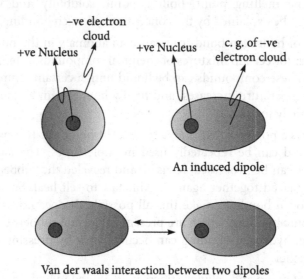

Figure 4.6 │ Van der Waals interactions

Van der Waals–Keesom forces are the attractive or repulsive electrostatic interactions between permanent dipoles, quadrupoles and in general, between multipoles. These interactions are very weak, hence no long-range alignment can be observed in liquids. When two polar molecules (permanent dipoles) are near each other, a dipole–dipole interaction is developed between them. The angle-averaged interaction energy $W(r)$ between two permanent dipoles is given by

$$W(r) = \frac{\mu_1^2 \mu_2^2}{3(4\pi\varepsilon\varepsilon_0)^2 k_B T r^6} \qquad (4.5)$$

μ_1, μ_2 = dipole moment of first and second dipoles.

r = distance between the centres of the two dipoles.

T = temperature in Kelvin

k_B = Boltzmann constant

The second largest contribution to the Van der Waals force is the Van der Waals–Debye force. It is the interaction between a permanent dipole and an induced dipole. The angle-averaged interaction energy $W(r)$ between them is given by

$$W(r) = \frac{\mu_1^2 \alpha_{02} + \mu_2^2 \alpha_{01}}{\left(4\pi\varepsilon_0\right)^2 r^6} \qquad (4.6)$$

α_{01}, α_{02} = polarizability of non-polar molecules. The interaction energy is independent of the temperature because the induced dipole follows immediately after the motion of the permanent dipole and is thus not affected by thermal motion.

The Van der Waals–London dispersion force is generally an attractive short range force and decays rapidly to zero away from a surface. It is omnipresent irrespective of the properties of the molecules. Van der Waals–London dispersion forces arise due to the fluctuation of the electron cloud surrounding the nucleus of electrically neutral atoms. These fluctuations create transient/instantaneous dipoles that induce transient attractive dipoles in adjacent atoms/molecules. The interaction of the two neighbouring dipoles reduces the total energy and lead to bonding. The energy of interaction resulting from Van der Waals–London type interactions between two dissimilar and non-polar molecules in free space (or air) is given by

$$W(r) = -\frac{3}{2} \frac{\alpha_{01} \alpha_{02}}{\left(4\pi\varepsilon_0\right)^2 r^6} \frac{I_1 I_2}{I_1 + I_2} \qquad (4.7)$$

I_1, I_2 = ionization potentials of molecules.

This expression for two identical atoms or molecules will reduce to

$$W(r) = -\frac{3\alpha_0^2 I}{64\pi^2 \varepsilon_0^2 r^6}. \qquad (4.8)$$

Each of the aforementioned three interactions varies inversely with the sixth power of distance. Thus, finally, the equation that captures all three contributions of the Van der Waals interactions may be obtained by combining all the three interactions.

The Van der Waal bonds occur to some extent in all materials but are particularly important in plastics and polymers. Polymers are often classified as being either a thermoplastic or a thermosetting material. Thermoplastic materials can be easily re-melted for forming or recycling while thermosetting material cannot be easily re-melted. Thermoplastic materials consist of long chainlike molecules. Heat can be used to break the Van der Waal forces between the molecules and change the form of the material from a solid to a liquid. By contrast, thermosetting materials have a three-dimensional network of covalent bonds. These bonds cannot be easily broken by heating and, therefore, cannot be re-melted and formed as easily as thermoplastics.

The Van der Waals forces between two macroscopic bodies can be calculated approximately by integrating over all molecules/atoms in one body with all molecules/atoms in the other body. We are citing here a few formulae in terms of the Hamaker constant A to calculate the Van der Waals interaction energy W in case of a few regular surfaces.

i. $$W = \frac{-AR_1 R_2}{6d(R_1 + R_2)}$$ for two spheres   (4.9)

ii. $$W = \frac{-AR}{6d}$$ for sphere and plane surface   (4.10)

iii. $$W = \frac{-A\sqrt{R_1 R_2}}{6d}$$ for two crossed cylinders   (4.11)

iv. $$W = \frac{-A}{12\pi d^2}$$ per unit area. For two surfaces   (4.12)

The values of the Hamaker constant A are in the range of 10^{-20} J to 10^{-19} J. The symbols used here are depicted in Fig. 4.7.

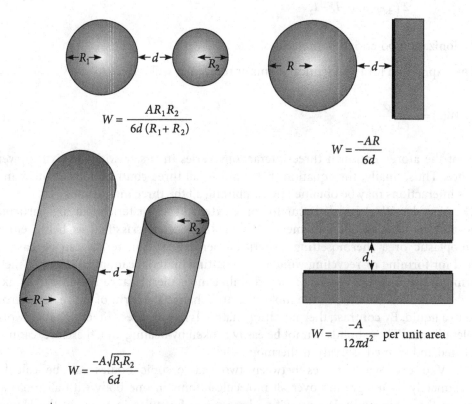

Figure 4.7 | Meaning of symbols used in Eqs (4.9) to (4.12)

4.5 Comparison of Different Types of Bonds

	Ionic	Covalent	Metallic	Van der Waal	Hydrogen
Principle cause of binding	Transfer of electrons between atoms and electrostatic attraction between them.	Mutual sharing of valence electrons between atoms	Attraction between the lattice of ion cores and the free electron gas.	Mutual polarization of atoms due to each other	Lowering of K.E of proton by the arrangement O- H-O.
Properties	Very strong binding	Strong binding	Moderate strong binding	Weak binding	Weak binding
	Poor electrical and thermal conductors	Conductivities over a wide range	High electrical and thermal conductivities	Poor electrical conductors	Low electrical and thermal conductivities.
	Transparent over a wide range of frequencies Closed packed structure.	Transparent to long wavelength radiation but opaque to shorter wavelength. Loose packed structure.	Opaque to all electromagnetic radiations from very low frequency to middle ultraviolet where they become transparent. Closed packed structure.	Transparent to electromagnet ic radiation. Closed packed structure	Transparent Loose structure
	Non directional bond	Strongly directional bond	Non-directional bond	Non-directional bond	Peculiar directional properties
Examples	NaCl, KCl, NaBr, KBr, MgO, MgCl	CH, Cl, H, Si, Ge.	Na, K, Mg, Al, Pb.	Na, N, He, Ar, Kr, Xe	Ice, KH, PO

4.6 Allotropy and Polymorphism

Under different conditions of temperature and pressure, a substance/element can exist in more than one physical form. Existence of an element in more than one physical form is known as allotropy and the different forms of an element are called allotropes. Existence of a substance in more than one form/crystal structure is known as polymorphism and the different forms of a substance are called polymorphs.

The term "allotropy" is used for elements only, not for compounds. The more general term, used for any crystalline material, is polymorphism. Allotropy refers only to different forms of an element within the same phase (i.e., different solid, liquid or gas forms); the changes of state between solid, liquid and gas in themselves are not considered allotropy. Allotropes are different structural modifications of an element; the atoms of the element are bonded together in a different manner. For example, the allotropes of carbon include diamond (where the carbon atoms are bonded together in a tetrahedral lattice arrangement), graphite (where the carbon atoms are bonded together in sheets of a hexagonal lattice), graphene (single sheets of graphite) and fullerenes (where the carbon atoms are bonded together in spherical, tubular, or ellipsoidal formations). Non-crystalline materials do not display the phenomenon of allotropy, because a non-crystalline material does not have a defined crystal structure; it cannot have more than one crystal structure

Allotropes are different structural forms of the same element and can exhibit quite different physical and chemical properties. The change between allotropic forms is triggered by stimuli like pressure, light and temperature. Therefore, the stability of the particular allotropes depends on particular conditions. For instance, depending upon temperatures, different allotropes of iron are shown in Fig. 4.8. When iron crystallizes at 2800°F, it is B.C.C. (δ-iron), at 2554°F, the structure changes to F.C.C. (γ-iron or austenite), and at 1670°F, it again becomes B.C.C. (α-iron or ferrite). The allotropic behaviour of elemental iron is shown in Fig. 4.8.

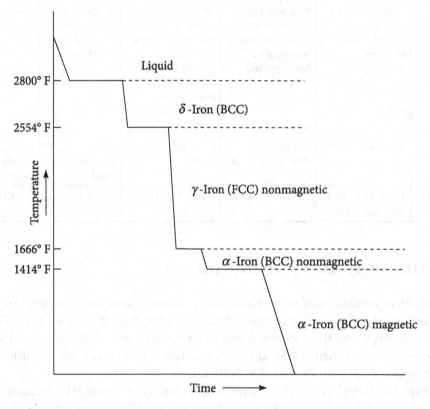

Figure 4.8 | Allotropic behaviour of pure iron

Polymorphism can potentially be found in any crystalline material including polymers, minerals and metals. An example of an organic polymorph is glycine, which is able to form monoclinic and hexagonal crystals. Mercuric iodide (HgI_2) forms two types of crystals:

(i) orthorhombic and (ii) trigonal. Calcium carbonate ($CaCO_3$) also exists in two types of crystalline forms: (i) orthorhombic (aragonite) and (ii) trigonal.

Polymorphous substances have similar chemical properties but different physical properties. Polymorphs have different stabilities and may spontaneously convert from a metastable form (unstable form) to the stable form at a particular temperature. They also exhibit different melting points, solubilities (which affect the dissolution rate of drugs and consequently their bioavailability in the body) and X-ray diffraction patterns.

Various conditions in the crystallization process are responsible for the development of different polymorphic forms. These conditions include:

i.   Solvent effects (the packing of crystal may be different in polar and non-polar solvents).

ii. Certain impurities that inhibit growth pattern and favour the growth of a metastable polymorphs.

iii. Level of super saturation from which the material is crystallized (in which generally, the higher the concentration above the solubility, the more likelihood of metastable formation).

iv. Temperature at which crystallization is carried out.

v.   Geometry of covalent bonds (differences leading to conformational polymorphism).

vi. Change in stirring conditions.

Wilhelm Ostwald suggested that the solid first formed on crystallization of a solution or a melt would be the least stable polymorph. Ostwald's rule or Ostwald's step rule states that in general it is not the most stable but the least stable polymorph that crystallizes first. Ostwald's rule is not a universal law; it is only a possible tendency in nature. Despite the potential implications, polymorphism is not always well understood.

From this discussion, it is clear that the allotropy of elements is just a special case of the phenomenon of polymorphism known for compounds. Keeping in view this fact, Ostwald proposed that the terms allotrope and allotropy be replaced by polymorph and polymorphism. However, IUPAC and most chemistry texts still favour the usage of allotrope and allotropy for elements.

Questions

4.1   What are the characteristics of bonding force?

4.2   Explain briefly how the bonding force is responsible for different phases of matter.

4.3   What are the different types of bonds in solids?

4.4   Describe the mechanism of ionic bonding in solids with examples.

4.5   Why do ionic solids have low conductivity?

4.6   Describe the mechanism of covalent bonds in solids with examples.

4.7   Why do ionic solids have low mass density?

4.8   Describe the formation of metallic bonds in solids.

4.9   Explain why solids having metallic bonds are good conductors of electricity.

4.10   Describe the mechanism of hydrogen bonding in liquids, taking water as an example.

4.11   Explain how hydrogen bonding affects properties of liquids.

4.12   Write down a few of the interesting consequences of hydrogen bonding.

4.13   What is a Van der Waals bond? Discuss different types of Van der Waals interactions.

4.14   Van der Waals interactions can hold inert gas atoms together to form solids at low temperatures, but they cannot hold such atoms together to form molecules in the gaseous state. Why not?

4.15   Discuss how Van der Waals interactions determine material properties.

4.16   Explain how geckos (a lizard-like animal) walk on ceilings or vertical walls.

4.17   Give a comparative account of different types of bonds.

4.18   Differentiate between allotropy and polymorphism, giving a few examples.

4.19   What are the different factors that are responsible for the development of different polymorphic forms?

4.20   What is Ostwald's rule?

Multiple Choice Questions

1.   Which of the following is not the definition of bonding energy?

   (i)   Energy required to bind together the atoms or molecules of a substance

   (ii)   Energy released in separating the atoms or molecules of a substance

   (iii) minimum potential energy of the atoms or molecules at inter-atomic equilibrium distance

   (iv)   maximum potential energy of the atoms or molecules at inter-atomic equilibrium distance

2.   The attractive forces, which keep the atoms together forcing them to form a solid may be given by

   (i) $F = -\dfrac{A}{r^i}$ (ii) $F = \dfrac{A}{r^i}$

   (iii) $F = Ar^i$ (iv) $F = -Ar^i$

3.   The repulsive forces, which keep the atoms apart forcing them not to coalesce into a point is given by

   (i) $F = -\dfrac{A}{r^i}$ (ii) $F = \dfrac{A}{r^i}$

   (iii) $F = Ar^i$ (iv) $F = -Ar^i$

4.   The reason for the strong repulsion among atoms at very short distances is

   (i)   Heisenberg's uncertainty principle

   (ii)   Newton's rule

(iii) Pauli's exclusion principle

(iv) Einstein's principle

5. How many types of primary bonds are there?

   (i) 1   (ii) 2

   (iii) 3   (iv) 4

6. Which of the following bonds is not a non-directional bond?

   (i) Hydrogen   (ii) Ionic

   (iii) Metallic   (iv) Van der Waals bonds

7. Which of the following bonds is not a primary bond?

   (i) Hydrogen   (ii) Ionic

   (iii) Metallic   (iv) Covalent

8. The bond that is formed between neighbouring atoms by the transfer of electrons from one atom to other atom is called

   (i) Hydrogen   (ii) Ionic

   (iii) Metallic   (iv) Covalent

9. The bond that is formed between neighbouring atoms by the sharing of electrons among the participating atoms is called

   (i) Hydrogen   (ii) Ionic

   (iii) Metallic   (iv) Covalent

10. Ionic bonding involves the transfer of electrons from

   (i) an electropositive atom to an electronegative atom

   (ii) an electropositive atom to an electropositive atom

   (iii) an electronegative atom to an electronegative atom

   (iv) an electronegative atom to an electropositive atom.

11. What types of bonding is found in CH_4 and HNO_3?

   (i) Hydrogen   (ii) Ionic

   (iii) Metallic   (iv) Covalent

12. Which of the following is correct?

   (i) Van der Waals bond strength>covalent bond strength>hydrogen bond strength

   (ii) hydrogen bond strength>covalent bond strength>Van der Waals bond strength

   (iii) covalent bond strength>Van der Waals bond strength>hydrogen bond strength

   (iv) covalent bond strength>hydrogen bond strength>Van der Waals bond strength

13. What type of bond exists in H_2O?

   (i) Hydrogen   (ii) Ionic

   (iii) Metallic   (iv) Covalent

14. What type of bonding exists in alcohol in water?
    (i) Hydrogen
    (ii) Ionic
    (iii) Metallic
    (iv) Covalent

15. Pure Van der Waals crystals can only exist
    (i) at very high temperatures
    (ii) at very low temperatures
    (iii) at medium temperatures
    (iv) at room temperatures

16. Van der Waals' bond are formed between
    (i) electrically charged molecules
    (ii) electrically charged atoms
    (iii) electrically neutral molecules
    (iv) positive and negative molecules

17. Which type of bondings bind proteins or DNA molecules into the shapes required for biological activity?
    (i) ionic bonding
    (ii) covalent bonding
    (iii) hydrogen bonding
    (iv) Van der Waals bonding

18. Which of the followings is not a Van der Waals' interaction?
    (i) Van der Waals–Coulomb force
    (ii) Van der Waals–Keesom force
    (iii) Van der Waals–Debye force
    (iv) Van der Waals–London dispersion force

19. The values of the Hamaker constant lie in the range of
    (i) 10^{-21} J to 10^{-19} J
    (ii) 10^{-21} J to 10^{-20} J
    (iii) 10^{-20} J to 10^{-19} J
    (iv) 10^{-19} J to 10^{-18} J.

20. What type of forces help a gecko in its movement on smooth surfaces?
    (i) Coulombic interaction
    (ii) biological interaction
    (iii) Van der Waals' interaction
    (iv) gravitational interaction

Answers

1 (iv)	2 (ii)	3 (i)	4 (iii)	5 (iii)	6 (i)	7 (i)	8 (ii)
9 (iv)	10 (i)	11 (iv)	12 (iv)	13 (iv)	14 (i)	15 (ii)	16 (iii)
17 (iv)	18 (i)	19 (iii)	20 (iii)				

5 Magnetic Properties of Materials

5.1 Introduction

In contradiction to our day-to-day spoken language, without exception, scientifically, all materials are magnetic; there are no materials that can actually be called non-magnetic. The material which has the ability to respond to an externally applied magnetic field or can be magnetized is called a magnetic material. All materials can be magnetized to a varying degree of magnetization.

5.2 Magnetic Parameters

The terms which can be used to describe the concepts of magnetism are called magnetic parameters. The important magnetic parameters that are used to characterize the magnetic behaviour of materials are enumerated here.

i. **Magnetic dipole moment** $\vec{\mu}_m$: Any two equal and opposite magnetic poles separated by a small distance constitute a magnetic dipole. If pole strength of the dipole is m [Ampere × meter] and the distance from the south pole to the north pole is $\vec{\ell}$, the magnetic dipole moment $\vec{\mu}_m$ of a magnetic dipole is defined as

$$\vec{\mu}_m = \vec{\ell}m \qquad (5.1)$$

The magnetic dipole moment $\vec{\mu}_m$ is a vector quantity, the direction being from the south pole to the north pole. The magnetic dipole moment of the current carrying loop $\vec{\mu}_m$ is also given as

$$\vec{\mu}_m = ni\vec{a} \qquad (5.2)$$

where i is the current flowing in a loop of n turns having area $\vec{a}$. Here direction of the magnetic dipole moment μ_m may be found out by applying the screw rule: (i) Place the screw at the centre of the loop perpendicular to the plane of the loop. (ii) Rotate the screw in the direction of the current flowing in the loop. (iii) The direction of linear motion of the screw is the direction of the magnetic moment of the current-carrying loop. The torque τ experienced by a magnetic dipole placed in a magnetic field of induction $\vec{B}$ is given as

$$\vec{\tau} = \vec{\mu}_m \times \vec{B} \tag{5.3}$$

ii. **Intensity of magnetization** $\overline{M}$: Intensity of magnetization or simply the magnetization, a vector quantity, is defined as the net magnetic dipole moment per unit volume.

iii. **Magnetic induction** $\vec{B}$: The magnetic induction or the magnetic flux density in magnitude is defined as the magnetic flux φ per unit area, i.e.,

$$B = \frac{d\varphi}{dA} \tag{5.4}$$

Its direction being in the direction of magnetic flux.

iv. **Magnetic field intensity** $\overline{H}$: The magnetic field intensity or magnetic field strength is defined as the ratio of magnetic induction in free space to the magnetic permeability of the space. It may be thought of as the magnetic field produced by the flow of current in wires. The magnetic field strength depends upon the geometry of the circuits and the current, but not on the medium. Its unit is A/m.

v. **Magnetic permeability** μ: The magnetic permeability of a linear isotropic medium μ is defined by the relation

$$\vec{B} = \mu\vec{H} \tag{5.5}$$

Magnetic permeability is the measure of the ability of a material to support the formation of a magnetic field within itself. In other words, it is the degree of magnetization that a material obtains in response to an applied magnetic field. The reciprocal of magnetic permeability is magnetic reluctivity.

vi. **Relative magnetic permeability** μ_r: The relative magnetic permeability μ_r is defined as the ratio

$$\mu_r = \frac{\mu}{\mu_0} \tag{5.6}$$

Here, $\mu_0 = 4p \times 10^{-7}$ Henry/m is the permeability of free space. Physical meanings of relative magnetic permeability is the same as that of magnetic permeability μ.

vii. **Magnetic susceptibility** χ_m: Magnetic susceptibility χ_m is the quantitative measure of the extent to which a material may be magnetized in response to an applied magnetic field. The magnetic susceptibility χ_m per unit volume is defined as the magnetization produced in the material per unit applied magnetic field intensity, i.e.,

$$\chi_m = \frac{M}{H} = \frac{\mu_0 M}{B} \qquad\qquad (5.7)$$

Magnetic susceptibility χ_m is a measure of the change in the magnetic moments of the atoms caused by an external applied magnetic field.

Example 5.1

The magnetic susceptibility of a material at room temperature is 0.82×10^{-8}. Calculate its magnetization under the action of a magnetic induction of 0.25 Tesla.

Solution

The data given are

$\chi_m = 0.82 \times 10^{-8}$ and $B = 0.25$ T

The magnetization of the material is given by

$$M = \frac{\chi_m B}{\mu_0} = \frac{0.82 \times 10^{-8} \times 0.25}{4\pi \times 10^{-7}} = 1.63 \times 10^{-3} \, \text{A/m}$$

5.3 Magnetic Parameter Relations

$$\vec{B} = \mu \vec{H} = \mu_0 \mu_r \vec{H}$$

In terms of magnetization,

$$\vec{B} = \mu_0 \left(\vec{H} + \vec{M} \right) \qquad\qquad (5.8)$$

or $\mu_r \vec{H} = \vec{H} + \vec{M}$

or $\mu_r = 1 + \chi$ $\qquad\qquad (5.9)$

Example 5.2

A material attains a magnetization of 3400 A/m under the action of a magnetic induction of 0.006 Wb/m². Calculate the magnetizing field and susceptibility and relative permeability of the material.

Solution

The data given are

$M = 3400$ A/m and $B = 0.006$ Wb/m².

We know $\vec{B} = \mu_0 \left(\vec{H} + \vec{M} \right)$ $\chi_m = \dfrac{M}{H}$ and $\mu_r = 1 + \chi_m$

$$H = \frac{B}{\mu_0} - M$$

Assuming $\vec{B}$, $\vec{H}$, and $\vec{M}$ are in the same direction.

$$H = \frac{0.006}{4\pi \times 10^{-7}} - 3400 = 1375 \text{ A/m}$$

$$\chi_m = \frac{M}{H} = \frac{3400}{1375} = 2.47$$

$$\mu_r = 1 + \chi_m = 3.47$$

Example 5.3

Derive a relation between the orbital magnetic dipole moment $\vec{\mu}_\ell$ and the orbital angular momentum $\vec{L}_\ell$ of Bohr's hydrogen atom.

Solution

The centripetal force mv^2/r on the electron is produced by Coulomb's electrostatic force of attraction $e^2/4\pi\varepsilon_0 r^2$ on the electron. Thus, we have

$$\frac{e^2}{4\pi\varepsilon_0 r^2} = \frac{mv^2}{r}$$

Here r is the Bohr radius of the electron path and m is the mass of the electron. From this equation, we have

$$v = \sqrt{\frac{e^2}{4\pi\varepsilon_0 rm}} \qquad\qquad (A)$$

$$\text{or} \quad \omega r = \sqrt{\frac{e^2}{4\pi\varepsilon_0 rm}}$$

$$\text{or} \quad v = \sqrt{\frac{e^2}{16\pi^2\varepsilon_0 r^3 m}} \qquad\qquad (B)$$

The orbital current i associated with the electron's orbit from the definition of current is given by

$$i = ev = \sqrt{\frac{e^4}{16\pi^2\varepsilon_0 r^3 m}}$$

The magnitude of the orbital magnetic moment of the electron μ_ℓ is given by

$$\mu_\ell = i\pi r^2 = \frac{e^2}{4}\frac{\sqrt{r}}{\sqrt{\pi\varepsilon_0 m}} \qquad\qquad (C)$$

The orbital angular momentum of the electron L_ℓ from its definition is given by

$$L_\ell = mvr \ (= n\hbar \text{ of course!}, n = \text{principal quantum number})$$

$$\text{or} \quad L_\ell = \sqrt{\frac{e^2 mr}{4\pi\varepsilon_0}}$$

$$\text{or} \quad \sqrt{r} = L_\ell \frac{\sqrt{4\pi\varepsilon_0}}{e\sqrt{m}}$$

Putting this value of $\sqrt{r}$ into Eq. (C), we have

$$\mu_\ell = L_\ell \frac{e}{2m}$$

5.4 Classification of Materials from the Magnetic Point of View

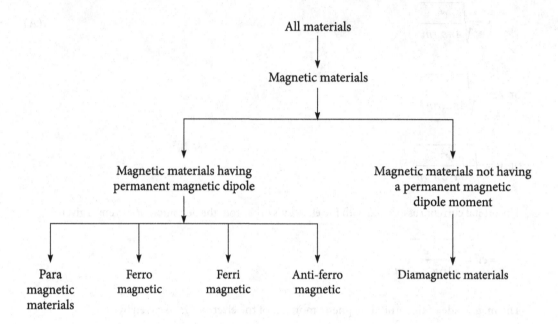

5.5 Origin of Magnetic Moments

The macroscopic magnetic properties of a substance are a consequence of the magnetic moments associated with its individual atoms or electrons. Unfortunately, a complete description of the origin of magnetism or magnetic moments is purely quantum mechanical which is beyond the scope of the book. However, here we try to give a classical picture of magnetic moments. Each electron in an atom has magnetic moments that originate from two sources. One source is the orbital motion of the electron around the positively charged nucleus producing a small current loop having a small magnetic moment along its axis of rotation as shown in Fig. 5.1(a).

The other source is the spin of the electron producing spin magnetic moment which is directed along the spin axis as shown in Fig. 5.1(b). The spin magnetic moments may be only in the "up" direction or in the anti-parallel "down" direction. Thus, each electron in an atom may be thought of as being a small magnet having permanent orbital and spin magnetic moments.

The most fundamental magnetic moment is the spin magnetic moment of an electron known as the Bohr magneton

$$\mu_B \left(= \frac{e\hbar}{2m} = 9.27 \times 10^{-24}\,\text{Am}^2 \right).$$

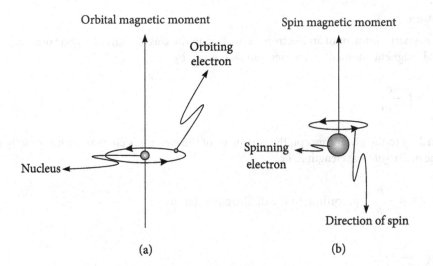

Figure 5.1 Gross picture of magnetic moment. (a) Orbiting electron producing orbital magnetic moment, (b) Spinning electron producing spin magnetic moment

For each electron in an atom, the spin magnetic moment is either $+\mu_B$ (spin up) or $-\mu_B$ (spin down). The orbital magnetic moment of an electron is equal to $m_\ell \mu_B$, m_ℓ, being the magnetic quantum number of the electron.

In each individual atom, the orbital magnetic moments of some electron pairs cancel each other; this also holds good for spin moments. For example, the spin moment of an electron with spin up will cancel that of one with spin down. The net magnetic moment for an atom is just then the sum of the magnetic moments of each of the constituent electrons, including both orbital and spin contributions and taking into account moment cancellation. For an atom having completely filled electron shells or sub-shells, when all the electrons are considered, there is total cancellation of both orbital and spin moments. Thus, materials composed of atoms having completely filled electron shells (inert gases) are not capable of being permanently magnetized.

Example 5.4

One Bohr magneton $1\mu_B$ may be defined as the magnetic moment of an electron circulating in the classical circular Bohr orbit of hydrogen atom of perimeter exactly equal to one de Broglie wavelength. Prove that one Bohr magneton

$$= \frac{e\hbar}{2m},$$

m being the mass of the electron.

Solution

The magnetic moment of an electron circulating in the classical circular Bohr orbit called the orbital magnetic moment of the electron μ_ℓ is given by

$$\mu_\ell = L_\ell \frac{e}{2m}$$

According to the given concept, the perimeter of the electron's circular orbit is exactly equal to one de Broglie wavelength λ, i.e.,

$$2\pi r = \lambda = \frac{h}{p} \quad \text{(according to the de Broglie relation)}$$

or $pr = \dfrac{h}{2\pi} = \hbar$

or $L_\ell = \hbar$

Putting this value of L_ℓ into the expression for μ_ℓ we get

$$\mu_\ell = \frac{e\hbar}{2m}$$

Using the given definition of one Bohr magneton (as given in the question), we write

$$\mu_B = \frac{e\hbar}{2m}$$

5.6 Diamagnetism

The kind of magnetism characteristic of materials repelled by a strong magnet is called diamagnetism. Gold, germanium, bismuth, silicon and so on exhibit diamagnetism. These materials are repelled by the magnetic field, i.e., they move from a stronger magnetic field to a weaker one and if the material is like a rod, line up at right angles to a non-uniform magnetic field. The external magnetic field cannot enter into the materials. Since there is no permanent magnetic dipole moment, the magnetic effects exhibited by these materials are very weak. The magnetic susceptibility is negative and independent of the external magnetic field and temperature. Diamagnetic materials are characterized by constant, small negative susceptibilities.

Diamagnetism is induced by a change in the orbital motion of the electrons due to an applied magnetic field. The magnitudes of the induced magnetic moments are extremely small and their resultant is in a direction opposite to that of the applied magnetic field as shown in the Fig. 5.2.

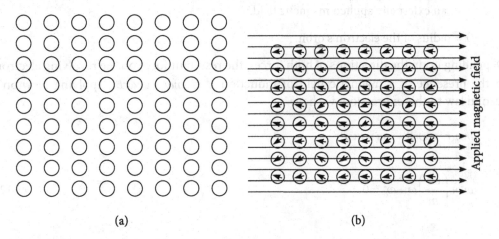

(a)   (b)

Figure 5.2 (a) In the absence of an external magnetic field, no atomic dipole moments exist in diamagnetic materials. (b) The atomic dipole configuration of diamagnetic materials under the action of the applied external field. Here atomic dipole moments are induced in the atoms of the diamagnetic materials and the magnetic dipole moment directions are nearly opposite to that of the applied magnetic field

5.6.1 Langevin's classical theory of diamagnetism

Diamagnetism is associated with the tendency of the electrical charges to shield partially the interior of a body from an applied magnetic field. Lenz's law states that when the flux through an electrical circuit is changed, an induced current is setup in such a direction as to oppose the flux change.

In the absence of an externally applied magnetic field, the centripetal force of the electrostatic origin, on an electron in Bohr's orbit is given by

$$F_E = m\omega_0^2 r \tag{5.10}$$

In the presence of an external magnetic field of induction B, magnetic force F_B on the revolving electron is given as

$$F_B = -evB = -e\omega rB \tag{5.11}$$

where

ω_0 = angular frequency of revolution of an electron in Bohr's orbit in the absence of an externally applied magnetic field.

ω = angular frequency of revolution of an electron in Bohr's orbit in the presence of an externally applied magnetic field.

r = radius of the electron's orbit.

Depending upon the directions of both forces, the resultant force on the revolving electron in the presence of a magnetic field of induction B is $m\omega_0^2 r \pm e\omega r B$. Applying Newton's second law to this expression, we have

$$m\omega_0^2 r - e\omega r B = ma = m\omega^2 r$$

or $$\omega^2 + \left(\frac{eB}{m}\right)\omega - \omega_0^2 = 0 \tag{5.12}$$

or $$\omega = -\frac{eB}{2m} \pm \sqrt{\left(\frac{eB}{2m}\right)^2 + \omega_0^2}$$

Even in strongest magnetic field, $\omega_0 \gg \dfrac{eB}{2m}$. Hence, we can have from Eq. (5.12),

$$\omega = -\frac{eB}{2m} \pm \omega_0 \tag{5.13}$$

Equation (5.13) shows that the external field B changes ω_0 into ω. The $\pm$ on ω_0 means that those electrons whose orbital magnetic moments are parallel to the field are slowed down and those whose moments are anti-parallel are speeded up by an amount $\dfrac{eBr}{2m}$. The result is called Larmor's theorem. The Larmor frequency ω_L is defined by

$$\omega_L = \frac{eB}{2m}$$

or $$v_L = \frac{1}{2\pi}\frac{eB}{2m} \tag{5.14}$$

Thus, due to the application of a magnetic field of induction B, the frequency of the revolving electron changes by $eB/2t$ and is called Larmor frequency ω_L. Since the current associated with an electron's orbit is equal to $e \times v$, the extra current due to Larmor frequency for Z electrons is given by

$$I = \left(-Ze\right)v_L = -\frac{Ze^2 B}{4\pi m} \qquad (5.15)$$

The magnetic dipole moment μ_m associated with the electron's orbit due this extra current $-Ze^2 B/4\pi m$ will be given by

$$\mu_m = -\frac{Ze^2 B}{4\pi m}\pi < \rho^2 > = -\frac{Ze^2 B}{4m} < \rho^2 > \qquad (5.16)$$

Here $<\rho^2> = <x^2> + <y^2>$ is the mean square of the perpendicular distance of the electron from the field axis (i.e., z-axis) through the nucleus. The negative sign in Eq. (5.16) shows that the direction of induced dipole moment is always opposite to the applied field. The mean square distance of the electrons from the nucleus is $<r^2> = <x^2> + <y^2> + <z^2>$. For a spherically symmetrical distribution of charge, $<x^2> = <y^2> = <z^2>$ so that

$$< r^2 > = \frac{3}{2} < \rho^2 >$$

Putting the value of $<\rho^2>$ from this equation into Eq. (5.16), we have

$$\mu_m = -\frac{Ze^2 B}{4m} \times \frac{2}{3} < r^2 >$$

or $\quad \mu_m = -\frac{Ze^2 B}{6m} < r^2 > = -4.68 \times 10^{-9} ZB < r^2 > \qquad (5.17)$

The average values of distances are used because the effective radii of all the electrons are not same.

If there are N numbers of atoms per unit volume, the magnetization $\overrightarrow{M}$ will be

$$\overrightarrow{M} = N\overrightarrow{\mu}_m$$

The diamagnetic susceptibility of the substance will be

$$\chi_{dia} = \frac{\mu_0 M}{B} = \frac{\mu_0 N \mu_m}{B} = -\frac{\mu_0 N}{B}\frac{Ze^2 B}{6m} < r^2 >$$

or $\chi_{dia} = -\dfrac{\mu_0 N Z e^2}{6m} <r^2> = -5.89 \times 10^{-15} NZ <r^2>$ (5.18)

[in SI system]

This is the classical Langevin result. This equation shows that diamagnetic susceptibility is independent of temperature and is negative. $<r^2>$ can be calculated by knowing the details of the wave functions for the electrons within the atom. For diamagnetic materials, the value of the susceptibility (a measure of the relative amount of induced magnetism) is always negative and very small; typically near -10^{-6}.

Example 5.5

Calculate the diamagnetic susceptibility of one mole of copper when the average atomic radius is 1.28 Å. The atomic number of copper is 29.

Solution

The data given are

average atomic radius = 1.28 Å and $Z = 29$.

According to Eq. (5.18), the diamagnetic susceptibility of the substance is

$$\chi_{dia} = \frac{\mu_0 M}{B} = \frac{\mu_0 N \mu_m}{B} = -\frac{\mu_0 N}{B} \frac{Z e^2 B}{6m} <r^2>$$

The diamagnetic susceptibility of one mole of the substance will be

$$\chi_{dia} = -\frac{\mu_0 N_A Z e^2}{6m} <r^2> = -3.55 \times 10^{12} Z <r^2> / (\text{mole.m}^2)$$

The diamagnetic susceptibility of one mole of copper will be

$$\chi_{dia} = -3.55 \times 10^{12} Z <r^2> / (\text{mole.m}^2)$$

$$= -3.55 \times 10^{12} \times 29 \times (1.28 \times 10^{-10}) m^2 / (\text{mole.m}^2)$$

$$= 1.69 \times 10^{-6} / \text{mole}$$

The susceptibility per one mole of substance is called molar susceptibility.

Example 5.6

A magnetic field of induction $B = 2.0$ Wb/m^2 is applied perpendicularly to the plane of the electron's path of radius 5.1×10^{-11} m. Calculate the change in magnetic moment of a circulating electron.

Solution

The data given are

$B = 2.0$ Wb/m^2 and $\langle \rho^2 \rangle = (5.1 \times 10^{-11} \text{m})^2$

The change in magnetic moment of a circulating electron according to Eq. (5.16) is given by

$$= -\frac{Ze^2 B}{4m} \langle \rho^2 \rangle = -\frac{1 \times 2.56 \times 10^{-38} \times 2}{4 \times 9.1 \times 10^{-31}} \times \left(5.1 \times 10^{-11}\right)^2 = \pm 3.7 \times 10^{-29} \text{ Am}^2$$

The change in magnetic moment of a circulating electron due to the application of a strong magnetic field is very small!

5.7 Paramagnetism

The kind of magnetism characteristic of materials weakly attracted by a strong magnet is called paramagnetism. Most elements and some compounds are paramagnetic. Strong paramagnetism is exhibited by compounds containing iron, palladium, platinum and the rare-earth elements. In such compounds, atoms of these elements have some inner electron shells that are incomplete, causing their unpaired electrons to spin like tops and orbit like satellites, thus making the atoms or molecules a permanent magnet tending to align with and hence strengthen an applied magnetic field. Strong paramagnetism decreases with rising temperature because of the de-alignment of magnetic moments produced by the greater thermally generated random motion of the atomic magnets. Weak paramagnetism, independent of temperature, is found in many metallic elements in the solid state, such as sodium and the other alkali metals, because an applied magnetic field affects the spin of some of the loosely bound conduction electrons. The value of susceptibility for paramagnetic materials is always positive and at room temperature is typically about 10^{-5} to 10^{-4} for weakly paramagnetic substances and about 10^{-4} to 10^{-2} for strongly paramagnetic substances.

In case of paramagnetic materials each atom possesses a permanent dipole moment by virtue of incomplete cancellation of spin and/or orbital magnetic moments. In the absence of an external magnetic field, these atomic magnetic moments are randomly oriented as a result of which the material possesses no net macroscopic magnetization. These atomic dipoles are free to rotate and paramagnetism results when they are preferentially aligned, by rotation with an external field as shown in Fig. 5.3.

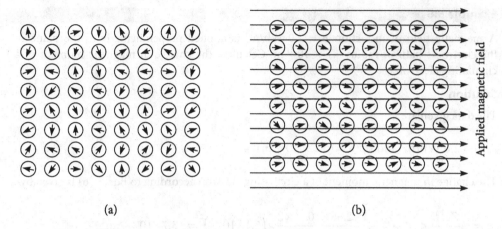

Figure 5.3 (a) The atomic dipole configuration of paramagnetic materials in the absence of an applied external field. Here magnetic dipole moments are randomly oriented. (b) The atomic dipole configuration of paramagnetic materials under the action of an applied external field. Here magnetic moment directions are nearly in the same direction as that of an applied magnetic field

5.7.1 Langevin's classical theory of paramagnetism

As discussed earlier, paramagnetic materials possess permanent atomic dipole moment that which is the vector sum of orbital and spin moments of various electrons in the atoms or molecules even in the absence of an external magnetic field. Due to thermal energy, these dipole moments are randomly oriented. When a magnetic field is applied to these materials, the permanent magnetic dipole moments align themselves in the direction of the field. Langevin explained the properties of diamagnetic substances on the basis of the Maxwell–Boltzmann distribution since at ordinary temperatures, the particles are subjected to ordinary thermal agitations and would thus be prevented from taking exact alignment – a kind of statistical equilibrium will be set up. According to the Maxwell–Boltzmann distribution, the number of dipoles n present at any temperature T is given by

$$n = n_0 e^{-\frac{E}{kT}} \tag{5.19}$$

The energy E possessed by a dipole under the influence of an external magnetic field of induction $\vec{B}$ is

$$E = -\vec{\mu}_m \cdot \vec{B} = -\mu_m B \cos\theta \tag{5.20}$$

Thus, upon substitution in Eq. (5.19), the number of dipoles n present at any temperature T is obtained as

$$n = n_0 e^{\frac{\mu_m B \cos\theta}{kT}} \tag{5.21}$$

Average value of $\overline{\mu}_m$

The average value of $\overline{\mu}_m$ denoted by $<\overline{\mu}_m>$ can be calculated in the following way. The component of $\overline{\mu}_m$ in the direction of $\vec{B}$ is $\mu_m \cos \theta$ as shown in Fig. 5.4.

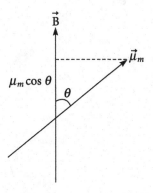

Figure 5.4 | The component of $\overline{\mu}_m$ in the direction of $\vec{B}$ is $\mu_m \cos \theta$

The average value of $\overline{\mu}_m$ denoted by $<\overline{\mu}_m>$ in the external field direction will be given by

$$<\overline{\mu}_m> = \frac{\int\limits_0^n \mu_m \cos\theta \, dn}{\int\limits_0^n dn} \qquad (5.22)$$

Differentiating Eq. (5.21), we get $dn = -\dfrac{n_0 \mu_m B}{kT} e^{\frac{\mu_m B \cos\theta}{kT}} \sin\theta \, d\theta$ and when substituted into Eq. (5.22), we get

$$<\overline{\mu}_m> = \frac{\int\limits_0^\pi \mu_m \cos\theta \, e^{\frac{\mu_m B \cos\theta}{kT}} \sin\theta \, d\theta}{\int\limits_0^\pi e^{\frac{\mu_m B \cos\theta}{kT}} \sin\theta \, d\theta} \qquad (5.23)$$

Now let

$$a = \frac{\mu_m B}{kT}$$

$$x = \cos\theta \, ; \; dx = -\sin\theta \, d\theta$$

Upon these substitutions into Eq. (5.23), we get

$$< \overline{\mu}_m >= \frac{\int_{-1}^{1} \mu_m x e^{ax} dx}{\int_{-1}^{1} e^{ax} dx} = \frac{\mu_m \left(x e^{ax} \right)\Big|_{-1}^{1} - \mu_m \int_{-1}^{1} e^{ax} dx}{e^{ax}\Big|_{-1}^{1}}$$

$$= \mu_m \left(\frac{e^a + e^{-a}}{e^a - e^{-a}} - \frac{1}{a} \right)$$

or $$< \overline{\mu}_m >= \mu_m \left(\coth a - \frac{1}{a} \right)$$

or $$< \overline{\mu}_m >= \mu_m L(a) \qquad\qquad (5.24)$$

where $L(a) = \coth a - \dfrac{1}{a}$ is called the Langevin function and is plotted in Fig. 5.5.

If there are N numbers of dipoles per unit volume, the magnetization M will be given by

$$M = N < \overline{\mu}_m >= N \mu_m L(a)$$

or $$M = M_s L(a) \text{ with } a = \frac{\mu_m B}{kT} \text{ and } M_s = N \mu_m \qquad\qquad (5.25)$$

where M_s is called the saturation magnetization and is characteristic of the material.

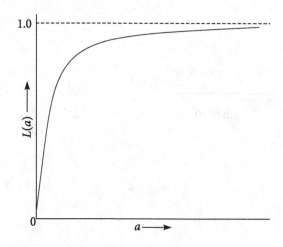

Figure 5.5 | Plot of the Langevin function L(a)

Case i: $a = \dfrac{\mu_m B}{kT}$ is large

For a high magnetic field and low temperature,

$$a = \frac{\mu_m B}{kT}$$

is large. $\lim_{a \to \infty} L(a) = 1$ or from Fig. 5.5 for large values of a, the Langevin function tends to unity. In other words,

$$\frac{M}{M_s} \to 1 \ \text{ or } \ M \to M_s$$

for a high magnetic field and low temperature [See Eq. (5.25)]. This shows that at very low temperature and high magnetic field, normalized magnetization approaches saturation magnetization. Thus, a complete alignment of dipoles in paramagnetic substances occurs at very low temperature and high magnetic field.

Case ii: $a = \dfrac{\mu_m B}{kT}$ is small

For a low magnetic field and high temperature, $a = \dfrac{\mu_m B}{kT}$ is small. Expanding the Langevin function into a power series, we have

$$L(a) = \coth a - \frac{1}{a} = \frac{1}{a} + \frac{a}{3} - \frac{a^2}{45} + \ldots - \frac{1}{a} = \frac{a}{3} - \frac{a^2}{45} + \ldots$$

or $L(a) \approx \dfrac{a}{3}$ for small values of $a = \dfrac{\mu_m B}{kT}$.

For low values of a, the Langevin function is linear as shown in Fig. 5.5. Thus, at a low magnetic field and high temperature, from Eq. (5.24), the average value of dipole moments is given as

$$< \mu_m > = \frac{1}{3} \mu_m a \tag{5.26}$$

If there are N number of dipoles per unit volume, the magnetization M will be given by

$$M = \frac{N}{3} \mu_m \frac{\mu_m B}{kT}$$

or $\quad \dfrac{\mu_0 M}{B} = \dfrac{N\mu_0 \mu_m^2}{3kT}$

or $\quad \chi_p = \dfrac{N\mu_0 \mu_m^2}{3kT}$ $\hfill$ (5.27)

Equation (5.27) is the expression for paramagnetic susceptibility of the substance. It can be written as

$$\chi_p = \frac{N\mu_0 \mu_m^2}{3k}\frac{1}{T} = C\frac{1}{T} \hfill (5.28)$$

where $C = \dfrac{N\mu_0 \mu_m^2}{3k}$ is known as the Curie constant.

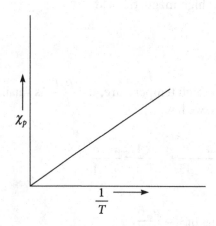

Figure 5.6 | Curie's experimental results on paramagnetism performed on oxygen gas at different temperatures

Equation (5.28) is in perfect agreement with the experimental result of Curie performed on oxygen gas at different temperatures.

Example 5.7

A paramagnetic material contains 8.7×10^{26} ions/m³ with magnetic moment 0.3 μ_B. Calculate the magnetization under the action of a magnetic induction of 1.5 Tesla at 300 K.

Solution

The data given are

$N = 8.7 \times 10^{26}$ ions/m³, $\mu_m = 0.3\ \mu_B = 0.3 \times 9.27 \times 10^{-24}$ A/m² and $B = 1.5\ T$

The magnetization paramagnetic material according to Langevin's theory is given as

$$M = \frac{N\mu_m^2 B}{3kT} = \frac{8.7\times10^{26}\times\left(0.3\times9.27\times10^{-24}\right)^2\times1.5}{3\times1.38\times10^{-23}\times300} \text{ A/m} = 0.81 \text{ A/m}.$$

Example 5.8

A paramagnetic material has a BCC structure with a unit cell edge of 2.8Å. If the saturation value of the magnetization is 1.5×10^6 A/m, calculate the average magnetization contributed per atom in Bohr magneton.

Solution

The data given are

$$M = 1.5 \times 10^6 \text{ A/m } \quad a = 2.8\text{Å} = 2.8 \times 10^{-10} \text{ m}$$

and the number of atoms per unit cell is 2 since the material has a BCC structure.
   The number of atoms per unit volume

$$= \frac{\text{Number per unit cell}}{\text{Volume of the unit cell}} = \frac{2}{a^3}$$

$$= \frac{2}{(2.8\times10^{-10})^3} = 9.11\times10^{28}$$

The dipole moments per unit volume, i.e., magnetization is 1.5×10^6 A/m. That means 9.11×10^{28} number of atoms have total magnetic moments 1.5×10^6 A/m. Therefore, one atom has on the average $\dfrac{1.5\times10^6}{9.11\times10^{28}}$ dipole moments, i.e.,

$$\mu_m = \frac{1.5\times10^6}{9.11\times10^{28}} = 1.65\times10^{-23} \text{ A/m}^2$$

Now $9.27\times10^{-24} \text{ A/m}^2 = 1\mu_B$

Hence, $1.65\times10^{-23} \text{ A/m}^2 = \dfrac{1.65\times10^{-23}}{9.27\times10^{-24}} \mu_B = 1.78\mu_B$

Thus, the average magnetization contributed per atom is 1.78 μ_B.

Objections raised by Leeuwan

i.   If the permanent magnetic moment is associated with and is proportional to the angular momentum of the moving electrons, then the magnetic moment cannot have a fixed magnitude but must take values from $-\infty$ to $+\infty$.

ii.   If the particle of magnetic moment μ_m is placed in a weak magnetic field of induction $\vec{B}$, it cannot place itself at any angle θ to the field as proposed by Langevin.

To overcome these, quantum theory of paramagnetism was proposed.

5.7.2 Quantum theory of paramagnetism

In quantum mechanics, the total angular momentum $\hbar\vec{J}$ is defined as the vector sum of the total orbital angular momentum $\hbar\vec{L}$ and the total spin angular momentum $\hbar\vec{S}$, i.e.,

$$\vec{J} = \vec{L} + \vec{S} \tag{5.29}$$

The magnitudes of $\vec{L}$ and $\vec{S}$ are given by respectively $\sqrt{\ell(\ell+1)}$ and $\sqrt{s(s+1)}$ with $\ell = 0,1,2,3, \ldots, (n-1)$, n = principal quantum number and $s = \pm\frac{1}{2}$; the general expression for total magnetic moment $\vec{\mu}_m$ of an atom or ion in free space including both orbital and spin angular momenta contribution is given by

$$\vec{\mu}_m = -g\frac{e\hbar}{2m}\vec{J} = -g\mu_B\vec{J} \tag{5.30}$$

The constant g is called the spectroscopic splitting factor or Lande's g factor and is given for a free atom as

$$g = 1 + \frac{J(J+1)+S(S+1)-L(L+1)}{2J(J+1)} \quad \text{for a free atom} \tag{5.31}$$

$$= 2.0023 \approx 2 \quad \text{for an electron.}$$

J = total angular momentum quantum number.

L = total orbital angular momentum quantum number.

S = total spin angular momentum quantum number.

Equation (5.31) is called the Lande equation. $\frac{e\hbar}{2m} = \mu_B$ is defined as the Bohr magneton. It is closely equal to the spin magnetic moment of a free electron.

Since the magnetic moment of an electron is observable only in terms of its response to an external field, the component of the magnetic moment in the direction of the applied field, i.e., in the z direction will be

$$\mu_{mz} = -g\mu_B m_j \tag{5.32}$$

where m_j, called the azimuthal quantum number, can take $2J + 1$ values starting from $-J$ to $+J$ through 0.

The number of dipoles with maximum energy (at $\theta = 0$, i.e., in the field direction) according to Eq. (5.21) is

$$n = n_0 e^{\frac{\mu_{mz} B}{kT}}$$

Since dipole moment of a single dipole is μ_{mz}, n number of dipoles will have

$$n\mu_{mz} = n_0 \mu_{mz} e^{\frac{\mu_{mz} B}{kT}} = -n_0 g \mu_B m_j e^{\frac{-g \mu_B m_j B}{kT}} \quad \left(\because \mu_{mz} = -g\mu_B m_j \right)$$

dipole moment. Therefore, the average dipole moment $<\vec{\mu}_{mz}>$ of a single dipole will be given by

$$<\vec{\mu}_{mz}> = \frac{\text{Total dipole moments of all the dipoles}}{\text{Total number of dipoles per unit volume}}$$

$$= \frac{\displaystyle\sum_{m_j=-J}^{J} -n_0 g\mu_B m_j e^{\frac{g\mu_B m_j B}{kT}}}{\displaystyle\sum_{m_j=-J}^{J} n_0 e^{\frac{g\mu_B m_j B}{kT}}} = -\frac{\displaystyle\sum_{m_j=-J}^{J} g\mu_B m_j e^{\frac{g\mu_B m_j B}{kT}}}{\displaystyle\sum_{m_j=-J}^{J} e^{\frac{g\mu_B m_j B}{kT}}}$$

If there are N numbers of dipoles per unit volume, then the average dipole moment per unit volume will be $N<\vec{\mu}_{mz}>$ which is nothing but the magnetization M of the substance. Thus, magnetization of the substance will be

$$M = N<\mu_{mz}> = -Ng\mu_B \frac{\displaystyle\sum_{m_j=-J}^{J} m_j e^{\frac{g\mu_B m_j B}{kT}}}{\displaystyle\sum_{m_j=-J}^{J} e^{\frac{g\mu_B m_j B}{kT}}} \tag{5.33}$$

Now letting $\dfrac{g\mu_B B}{kT} = x$ in this equation, we get

$$M = -Ng\mu_B \frac{\displaystyle\sum_{m_j=-J}^{J} m_j e^{-m_j x}}{\displaystyle\sum_{m_j=-J}^{J} e^{-m_j x}} = Ng\mu_B \frac{d}{dx}\left[\ln\left(\sum_{m_j=-J}^{J} e^{-m_j x} \right) \right]$$

$$= Ng\mu_B \frac{d}{dx}\left[\ell n\, e^{Jx}\left(1+e^{-x}+\ldots+e^{-2Jx}\right)\right]$$

$$= Ng\mu_B \frac{d}{dx}\left[\ell n\left(e^{Jx}\frac{1-e^{-(2J+1)x}}{1-e^{-x}}\right)\right]$$

$$= Ng\mu_B \frac{d}{dx}\left[\ell n\left(\frac{e^{\left(J+\frac{1}{2}\right)x}-e^{-\left(J+\frac{1}{2}\right)x}}{e^{\frac{x}{2}}-e^{-\frac{x}{2}}}\right)\right]$$

$$= Ng\mu_B \frac{d}{dx}\left[\ell n\left(\sinh\left(J+\frac{1}{2}\right)x\right)-\ell n\left(\sin\left(\frac{x}{2}\right)\right)\right]$$

or $$M = Ng\mu_B\left[\frac{2J+1}{2}\coth\frac{2J+1}{2}x-\frac{1}{2}\coth\frac{x}{2}\right]$$

Letting $x = a/J$ in this equation, we have

$$M = Ng\mu_B\left[\frac{2J+1}{2}\coth\frac{2J+1}{2J}a-\frac{1}{2}\coth\frac{a}{2J}\right]$$

or $$M = Ng\mu_B JB_J(a)$$ (5.34)

and $<\vec{\mu}_{mz}> = g\mu_B JB_J(a)$ (Since $M = M = N <\vec{\mu}_{mz}>$)   (5.35)

where

$$B_J(a) = \frac{2J+1}{2J}\coth\frac{2J+1}{2J}a-\frac{1}{2J}\coth\frac{a}{2J} \text{ with } a = \frac{g\mu_B JB}{kT} \text{ is called Brillouin's function}^*.$$

---

* Properties of Brillouin's function:

    (i) $\lim\limits_{J\to\infty}$ Brillouin's function = Langevin's function, i.e.,

    (ii) $\lim\limits_{J\to\infty}\dfrac{2J+1}{2J}\coth\dfrac{2J+1}{2J}a-\dfrac{1}{2J}\coth\dfrac{a}{2J}=\coth a-\dfrac{1}{a}$

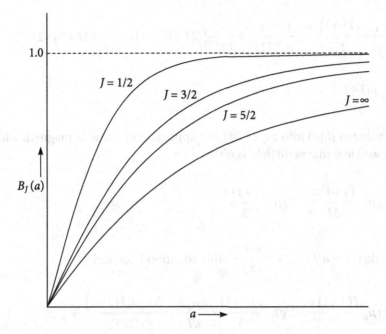

Figure 5.7 | Plotting of Brillouin's function* for different values of J. It is obvious from this figure that $\lim\limits_{a\to\infty}\dfrac{2J+1}{2J}\coth\dfrac{2J+1}{2J}a-\dfrac{1}{2J}\coth\dfrac{a}{2J}=1$, 1 is the maximum value of $B_J(a)$

Equation (5.34) is known as the Curie–Brillouin law. It expresses very satisfactorily the results obtained in case of hydrated gadolinium sulphate $Gd_2(SO_4)_3 8H_2O$ at liquid helium temperature down to 1.3 K and magnetic field up to 20000 Oestered.

For a low magnetic field and/or high temperatures, i.e., $a << 1$, Brillouin's function $B_J(a)$ will be reduced to

$$B_J(a)=\frac{2J+1}{2J}\coth\frac{2J+1}{2J}a-\frac{1}{2J}\coth\frac{a}{2J}$$

$$\approx\frac{2J+1}{2J}\left[\frac{2J}{(2J+1)a}-\frac{(2J+1)a}{6J}\right]-\frac{1}{2J}\left[\frac{2J}{a}+\frac{a}{6J}\right]\quad \text{for } a<<1^{**}$$

(iii) $\lim\limits_{J\to\frac{1}{2}}\dfrac{2J+1}{2J}\coth\dfrac{2J+1}{2J}a-\dfrac{1}{2J}\coth\dfrac{a}{2J}=\tanh a$

(iv) $\lim\limits_{a\to\infty}\dfrac{2J+1}{2J}\coth\dfrac{2J+1}{2J}a-\dfrac{1}{2J}\coth\dfrac{a}{2J}=1$, 1 is the maximum value of $B_J(a)$.

** $\coth y=\dfrac{1}{y}+\dfrac{y}{3}-\dfrac{y^3}{45}+...;\ \coth y\approx\dfrac{1}{y}+\dfrac{y}{3}$ for $y<<1$.

$$= \frac{1}{a} + \frac{(2J+1)^2\, a}{12J^2} - \frac{1}{a} - \frac{a}{12J^2} = \frac{a}{12J^2}\left((2J+1)^2 - 1\right) = \frac{a}{12J^2}(2J+2)\times 2J$$

or $B_J(a) \approx \dfrac{a(J+1)}{3J}$ (5.36)

Putting this value of $B_J(a)$ into Eq. (5.34), the approximate value of magnetization at a high temperature and low magnetic field is obtained as

$$M = Ng\mu_B J\frac{(J+1)a}{3J} = Ng\mu_B\frac{(J+1)a}{3}$$ (5.37)

As we have taken $x = a/J$ and $x = \dfrac{g\mu_B B}{kT}$, this equation becomes

$$M = Ng\mu_B\frac{J(J+1)x}{3} = Ng\mu_B\frac{J(J+1)}{3}\times\frac{g\mu_B B}{kT} = \frac{Ng^2\mu_B^2 J(J+1)}{3kT}B$$

or $\dfrac{\mu_0 M}{B} = \dfrac{N\mu_0 g^2\mu_B^2 J(J+1)}{3k}\dfrac{1}{T}$

or $\chi_m = \dfrac{N\mu_0 g^2\mu_B^2 J(J+1)}{3k}\dfrac{1}{T} = \dfrac{C}{T}$ (5.38)

where

$$C = \frac{N\mu_0 g^2\mu_B^2 J(J+1)}{3k} = \text{Curie constant}$$ (5.39)

Equation (5.38) is nothing but Curie's law in magnetism. According to this law, magnetic susceptibility is inversely proportional to absolute temperature. Equation (5.39) represents the Curie constant.

In all these expressions, $g\sqrt{J(J+1)}$ is called the effective number of Bohr magneton p and is given as

$$p = g\sqrt{J(J+1)}$$

The effective number of Bohr magneton p determines the magnetic moment of an atom or ion. p can be determined experimentally by using Eq. (5.38) and theoretically by knowing the electronic configuration of atoms or ions.

The magnitude of total angular momentum J is calculated by using the following rules:

i. $J = |L - S|$ when the shell is less than half filled.

ii. $J = |L + S|$ when the shell is more than half filled.

iii. $J = |S|$ when the shell is exactly half filled. Here $L = 0$

5.8 Ferromagnetism

Ferromagnetism is a kind of magnetism that is associated with iron, cobalt, nickel, and some alloys or compounds containing one or more of these elements. It also occurs in gadolinium and a few other rare earth elements. In contrast to other substances, ferromagnetic materials are magnetized easily, and in strong magnetic fields, the magnetization approaches a definite limit called saturation magnetization, i.e., maximum value of magnetization. When a field is applied and then removed, the magnetization does not return to its original value; this phenomenon is referred to as hysteresis. A ferromagnetic substance possesses a spontaneous magnetic moment, i.e., magnetic moments in one direction even in the absence of an external magnetic field as shown in Fig. 5.8. The magnetization in the absence of external magnetic field is called spontaneous magnetization.

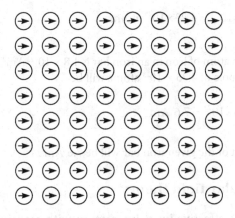

Figure 5.8 | The atomic dipole configuration of ferromagnetic materials in the absence of an applied external field. Here, atomic magnetic moments are aligned in one direction even in the absence of an externally applied magnetic field

Ferromagnetic substances are qualitatively different from other substances by the following properties.

a.   In ferromagnetic substances, the relation between magnetization $\overline{M}$ and external magnetic field $\overline{B}$ is not linear.

b. In ferromagnetic substances, magnetization $\vec{M}$ is not a unique function of the applied magnetic field. It also depends on the field to which they have been exposed previously.

c. Ferromagnetic substances can exist in a permanently magnetized state even when there is no external field.

d. Ferromagnetic substances become paramagnetic substances above a certain temperature called Curie temperature T_C, T_C being different for different substances.

Example 5.9

The dipole moment associated with an atom of iron in an iron bar is 1.8×10^{-23} Am². Assume that all the atoms in the bar, which is 5 cm long and has a 1 cm² cross-section, have their dipole moments aligned. What is the dipole moment of the bar? [For iron, density = 7.85 gm/cm³ and atomic molar mass = 55.85 gm.]

Solution

The volume of the given bar is 5×1 cm³ = 5 cm³
Mass of the bar is 5×7.85 gm = 39.25 gm = 39.25×10^{-3} kg

55.85 gm = 55.85×10^{-3} kg contains 6.023×10^{-23} atoms

39.25×10^{-3} kg of iron will contain

$$\frac{6.023 \times 10^{23}}{55.85 \times 10^{-3}} \times 39.25 \times 10^{-3} = 4.23 \times 10^{23} \text{ atoms.}$$

The dipole moment of an atom of iron in an iron bar is 1.8×10^{-23} Am². Therefore, since all the atoms are aligned, the dipole of 4.23×10^{23} atoms will be

$$1.8 \times 10^{-23} \text{ Am}^2 \times 4.23 \times 10^{23} = 7.6 \text{ Am}^2$$

Thus, the dipole moment of the given iron bar will be 7.6 Am².

5.8.1 Weiss's molecular field theory

The magnetic field that is effective in its interaction with atomic currents in an atom or molecule is called the molecular field or internal field or Weiss field. According to French physicist Pierre Ernest Weiss, the interaction between the atomic dipoles in a ferromagnetic substance produces an internal field $\vec{B}_i$. In the presence of the external magnetic field $\vec{B}_e$, the total field $\vec{B}_T$ in the substance will be given by

$$\vec{B}_T = \vec{B}_e + \vec{B}_i \tag{5.40}$$

The internal field $\vec{B}_i$ is proportional to magnetization, $\vec{M}$, i.e.,

$$\vec{B}_i = \lambda \vec{M} \tag{5.41}$$

where λ is a constant independent of temperature. Since $\vec{B}_e$ and $\vec{B}_i$ are parallel, putting Eq. (5.41) into Eq. (5.40), we have

$$B_T = B_e + \lambda M \tag{5.42}$$

Let us consider a ferromagnetic substance containing N atoms per unit volume each with a total angular momentum quantum number J. The magnitude of magnetization M of the substance according to Eq. (5.34) is given as

$$M = Ng\mu_B J B_J(a) \tag{5.43}$$

where $\quad a = \dfrac{g\mu_B J B_T}{kT} = \dfrac{g\mu_B J (B_e + \lambda M)}{kT} \tag{5.44}$

For spontaneous magnetization, $B_e = 0$ (by definition). Therefore, for spontaneous magnetization, Eq. (5.44) becomes

$$a = \frac{g\mu_B J \lambda M(T)}{kT}$$

or $\quad M(T) = \dfrac{akT}{g\mu_B J\lambda} \tag{5.45}$

In Eq. (5.45), we have written $M(T)$ in place of M to show its temperature dependence, i.e., M is a function of temperature. Again using Eq. (5.34), the saturation magnetization M_s, i.e., maximum value of magnetization is obtained when $a \to \infty$ (i.e., $T \to 0$) since

$$\lim_{a\to\infty} \frac{2J+1}{2J}\coth\frac{2J+1}{2J}a - \frac{1}{2J}\coth\frac{a}{2J} = 1,\ 1 \text{ being the maximum value of } B_J(a). \text{ Thus, we have}$$

$$M_s(0) = Ng\mu_B J \times B_J(a)\big|_{max} = Ng\mu_B J \times 1 = Ng\mu_B J \tag{5.46}$$

Upon substitution of $Ng\mu_B J = M_s(0)$ into Eq. (5.34), it becomes

$$M(T) = M_s(0) B_J(a)$$

or $\quad \dfrac{M(T)}{M_s(0)} = B_J(a) \tag{5.47}$

Again from Eqs (5.45) and (5.46), we have

$$\frac{M(T)}{M_s(0)} = \frac{k}{N\lambda g^2 \mu_B^2 J^2} Ta \tag{5.48}$$

One can obtain the simultaneous solution of Eqs (5.47) and (5.48) by plotting both the functions $M(T)/M_s(0)$ against a as shown in Fig. 5.9.

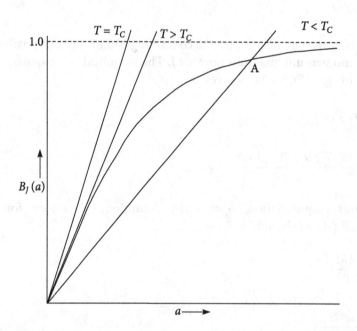

Figure 5.9 | Brillouin plot. The graphical solution of simultaneous Eqs (5.47) and (5.48). The point of intersection shows the existence of magnetization even when the external magnetic field is withdrawn. The magnetization vanishes at $T \geq T_c$ since intersection of the two curves is not possible then. It should be remembered that with increase of a, T decreases

The plot of Eq. (5.48) gives straight line and in Fig. 5.9, it has been plotted for three different temperatures. The following observations can be observed from Fig. 5.9.

i.   The straight line at a characteristic temperature called Curie point (also called the Curie temperature) T_c is tangential to the plot of Eq. (5.47) (Brillouin plot) at the origin $a = 0$. This proves that spontaneous magnetization vanishes at $T = T_c$. The Curie point is defined as the temperature at which spontaneous magnetization vanishes.

ii.   The straight line for any temperature above the Curie point T_c ($T > T_c$) intersects the Brillouin plot only at the origin $a = 0$. This shows that spontaneous magnetization does not exist for $T > T_c$.

iii. The straight line for any temperature below the Curie point T_C ($T < T_C$) intersects the Brillouin plot at the point A as well as at the origin $a = 0$. That is, a simultaneous solution of the equation exists only for temperatures below the Curie point T_C or spontaneous magnetization exists only at temperatures below Curie point T_C.

5.8.2 Temperature effects on spontaneous magnetization

For a low magnetic field and/or for high temperatures, i.e., for $a << 1$, Brillouin's function $B_J(a)$ reduces approximately to

$$B_J(a) \approx \frac{a(J+1)}{3J} \quad \text{(see Eq. (5.36))}$$

Therefore, the slope of the Brillouin plot at the origin

$$\left. \frac{dB_J(a)}{da} \right|_{a=0} = \frac{J+1}{3J}$$

From Eq. (5.47), we have

$$\left. \frac{dB_J(a)}{da} \right|_{a=0} = \left. \frac{d}{da} \left(\frac{M(T)}{M_s(0)} \right) \right|_{a=0} = \frac{kT}{N\lambda g^2 \mu_B^2 J^2}$$

The plots of $M(T)/M_s(0)$ and $B_J(a)$ (see Eq. (5.46)) are the same. Therefore, their slopes at the origin should be equal. Thus, we have

$$\frac{J+1}{3J} = \frac{kT}{N\lambda g^2 \mu_B^2 J^2} \tag{5.49}$$

At $T = T_C$, Eq. (5.49) becomes

$$\frac{J+1}{3J} = \frac{kT_C}{N\lambda g^2 \mu_B^2 J^2}$$

or $$T_C = \frac{Ng^2 \mu_B^2 J(J+1)}{3k} \lambda = \frac{N\mu}{3k} \lambda \tag{5.50}$$

In this equation, $\mu = g^2 \mu_B^2 J(J+1)$ is the total magnetic moment per atom. Eq. (5.50) reveals that the Curie point T_C is proportional to λ, the molecular field constant.

Now combining Eqs (5.48) and (5.50), we get

$$\frac{M(T)}{M_s(0)} = \frac{J+1}{3J}\left(\frac{T}{T_c}\right)a \tag{5.51}$$

Here,

$M_s(0)$ = saturation magnetization

$M(T)$ = spontaneous magnetization at any temperature T.

$M(T)/M_s(0)$ must satisfy both the values given by Eqs (5.47) and (5.51) and it can be obtained by the intersection method described in the previous section (See Fig. 5.9). Fig. 5.10 shows the plot of $M(T)/M_s(0)$ versus T/T_c for $J = 1/2$, 1 and ∞.

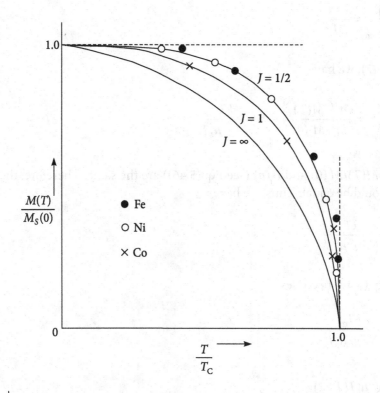

Figure 5.10 The plot of $M(T)/M_s(0)$ versus T/T_c for $J = 1/2$, 1, and ∞. Since the plot for $J = 1/2$ fits the experimental data best, it indicates that the magnetization comes from spin moments rather than orbital moments of the electrons

5.8.3 Paramagnetic region

Above the Curie temperature, the spontaneous magnetization is zero. For small values of a, the Brillouin function $B_J(a)$ is given by

$$B_J(a) \approx \frac{a(J+1)}{3J}$$

Putting this value of the Brillouin function $B_J(a)$ into Eq. (5.43), we have

$$M = Ng\mu_B J \frac{a(J+1)}{3J} \tag{5.52}$$

In Eq. (5.52),

$$a = \frac{g\mu_B (B+\lambda M)J}{kT} \tag{5.53}$$

Putting Eq. (5.53) into Eq. (5.52), we have

$$M = Ng\mu_B J \frac{(J+1)}{3J} \times \frac{g\mu_B (B+\lambda M)}{kT}$$

or

$$M\left(\frac{3kT - Ng^2\mu_B^2 (J+1)\lambda}{3kT}\right) = \frac{Ng^2\mu_B^2 (J+1)B}{3kT}$$

or

$$\frac{\mu_0 M}{B} = \frac{Ng^2\mu_B^2 (J+1)\mu_0}{3kT - Ng^2\mu_B^2 (J+1)\lambda}$$

or

$$\chi_m = \frac{\dfrac{Ng^2\mu_B^2 (J+1)\mu_0}{3k}}{T - \dfrac{Ng^2\mu_B^2 (J+1)\lambda}{3k}}$$

or

$$\chi_m = \frac{C}{T-T_C} \quad \text{(Curie's law)} \tag{5.54}$$

where

$$C = \frac{Ng^2\mu_B^2(J+1)\mu_0^2}{3k} \quad \text{(Curie constant)} \tag{5.55}$$

and $\quad T_C = \frac{Ng^2\mu_B^2(J+1)\lambda}{3k} \quad$ (Curie point/Curie temperature) $\tag{5.56}$

Equation (5.54) is the well-known Curie law or the Curie–Weiss law that accounts for the temperature behaviour of many paramagnetic substances. It also gives excellent agreement with data for ferromagnetic substances above the Curie point, with a small deviation. The Curie point defined from the spontaneous magnetization theory differs by a few degrees from the experimental value found for the paramagnetic region. A plot of $1/\chi_m$ versus T reveals this fact.

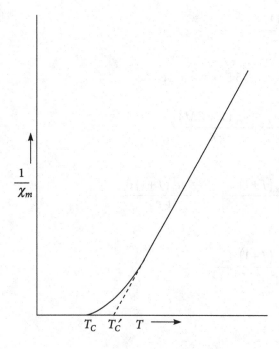

Figure 5.11 | The plot of $1/\chi_m$ versus T. From Eq. (5.54), we have $\frac{1}{\chi_m} = \frac{1}{C}T - \frac{T_C}{C}$, equation of a straight line. The dashed straight line is obtained by extrapolation. T_C' is the theoretical Curie point and T_C is the experimental Curie point

5.8.4 Criticisms of Weiss's molecular field theory

i.   There are discrepancies between theoretical and experimental curves below the Curie point.

ii. There is a break down of the linear relation expressed by the Curie–Weiss law close to the Curie point.

A first attempt to account for these discrepancies was based on the experimentally observed variability of the molecular magnetic moment which was assumed to be constant. Heisenberg's exchange interaction, a quantum concept according to which the variation of magnetic moment takes place discontinuously, accounts for this observation.

iii. Weiss's molecular field theory is silent about the origin of molecular field.

According to Heisenberg's exchange interaction, the molecular field is purely of electrical origin due to an interchange interaction of an electron.

5.8.5 Ferromagnetic domains

A ferromagnetic domain is a region of a ferromagnetic material in which all the spin magnetic moments of all the ions are aligned in the same direction. The domains are of microscopic size. They are separated from other domains by a wall known as a domain wall or Bloch wall. A domain wall is nothing but the thin boundary of a domain within which successive spins are gradually changing orientations in small steps up to the orientation of the adjacent domains as shown in Fig. 5.12.

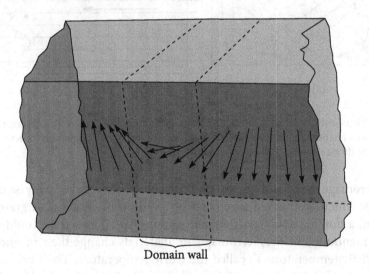

Domain wall

Figure 5.12 | The gradual change in magnetic dipole orientation across a domain wall

The thickness of a domain wall as shown in Fig. 5.12 is of the order of 10^{-2} μm. A domain thus acts as a single magnetic dipole. A ferromagnetic material may be composed of many domains, each having a different magnetic orientation, i.e., the domains have moments that point in different directions, so the material has no net magnetic moment in the absence of an external magnetic field. Figure 5.13(a) shows the presence of domains in a ferromagnetic material in the absence of a magnetic field and Fig. 5.13(b) shows the growth of a domains under the action of a magnetic field. If the ferromagnetic material is placed in a magnetic field, however, the material becomes magnetic. The external field causes the material to become a single domain with all moments aligned in the direction of the external field. The domains do not rotate their moments; instead, the walls between the domains, the Bloch walls, move. The domain with a magnetic moment along the field grows in size, while the others become smaller in size. If removed from the magnetic field, the material will remain magnetized for a considerable period of time.

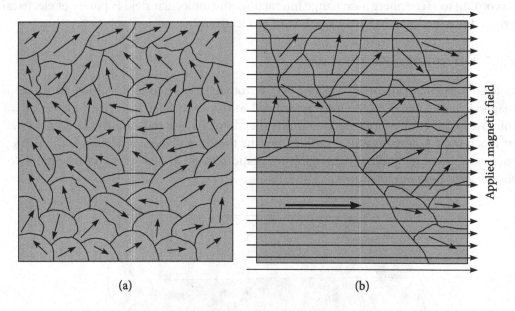

(a)   (b)

Figure 5.13 | Visualization of domains in a ferromagnetic material. (a) Shows the presence of domains in a ferromagnetic material in the absence of a magnetic field and (b) and the growth of domains under the action of a magnetic field

Nearly all ferromagnetic materials are polycrystalline; they have many small grains of single crystals, which are packed together with random orientation. A grain could be a single domain, a domain could include many grains, or a large grain could have several domains. As mentioned before, ferromagnetic materials change their magnetic ordering at a characteristic temperature T_C called the Curie temperature. The Curie temperatures for three common ferromagnetics – iron, cobalt, and nickel – are 1,043 K; 1,394 K and 631 K, respectively. For temperatures below T_C, the magnetic moments of the ions are aligned. For temperatures above T_C, the crystal is not ferromagnetic, since the individual atomic moments are no longer aligned. Above T_C, the moments have short-range order but no long-range order. Short-range order means that there is local ordering. If a dipole moment points in one direction, its neighbours have a tendency to point in the same direction. This tendency is maintained over several lattice sites but not for long distances. Long-range order is the tendency of dipole moments to align for large distances.

For temperatures a few degrees below T_C, the moments have strong short-range order but only a small amount of long-range order. The tendency for long-range order increases at lower temperatures. The Curie temperature is the point where long-range order begins as the temperature is lowered. If an iron bar is heated to a temperature above T_C, the bar is no longer ferromagnetic. If the bar is then cooled to a temperature below T_C, the grains become magnetic, but they orient their moments in random directions, so the bar as a whole is not ferromagnetic. A bar can be demagnetized by heating and then cooling it. It can be remagnetized by inserting it in a large magnetic field. Ferromagnetism is found in many insulators as well as metals. Chromium bromide ($CrBr_3$; $T_C = 37$ K), europium oxide (EuO; $T_C = 77$ K) and Gadolinium chloride ($Gdcl_3$; $T_C = 2.2$ K) are insulators as well as ferromagnetics.

One can observe magnetic domains by preparing a colloidal solution of the ferromagnetic material and then spraying it over the ferromagnetic material under study. The colloidal particles accumulate near the domain walls of the material. The same can be observed by using a microscope having a high resolving power. This method of observing domains is called the Bitter powder pattern method.

5.8.6 Hysteresis

Hysteresis is the irreversible variation of magnetization with applied field and illustrates the ability of a material to retain its magnetization, even after an applied field is removed. As discussed earlier, ferromagnetic materials normally do not possess net magnetization due to the random orientation of dipole moments of domains.

A ferromagnetic material below Curie temperature can be magnetized by two methods – one by rotating the domains in the direction of the applied magnetic field and the other is to permit the growth of the domain initially in the direction of the applied field. Actually, magnetization takes place by these two methods depending upon the strength of the applied field. The variation of magnetization with applied field, called a hysteresis loop, is shown in Fig. 5.14.

When a weak magnetic field is applied, it is found that domains which are oriented parallel to each other and are in "easy direction"* of magnetization, grow in size at the expense of less favourably oriented ones. This means that less energy is required for the domain growth initially than is required for domain rotation. Domain growth requires Bloch wall movement which involves expenditure of energy. Bloch wall movements have been observed by the Bitter pattern. The movements of Bloch walls are mostly reversible. This is indicated by Oa in Fig. 5.14. With increase of the applied field, the Bloch wall movement continues sharply and they are mostly irreversible. This is indicated by aa' in Fig. 5.14. The irreversibility explains the existence of hysteresis phenomena in ferromagnetic materials. At a', each domain is magnetized along the favourable direction*** .

*** Ferromagnetic materials are magnetized easily (i.e., saturation magnetization is achieved at low field) in certain preferred crystallographic directions. Such a direction in which a crystal can be easily magnetized is called favourable direction or easy direction. The direction in which a crystal cannot be easily magnetized is called unfavourable direction or hard direction.

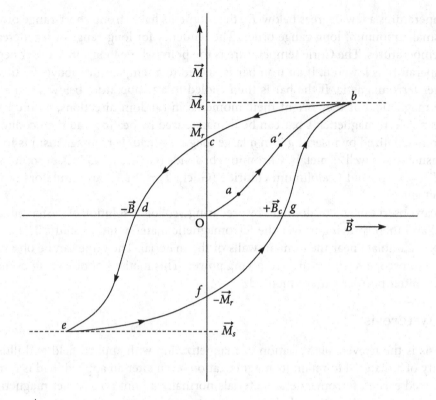

Figure 5.14 | Hysteresis loop for a ferromagnetic material

Application of a still higher field rotates the domains into field direction and this may be different from favourable directions. At *b*, all the domains are aligned in the field direction and the specimen is magnetized to its maximum extent. The magnetization attains its maximum value called saturation magnetization. We know that saturation magnetization is a function of the temperature of the specimen. Beyond *b*, the magnetization does not increase further with increase of the applied field strength.

After saturation magnetization of the specimen, when the applied field is gradually decreased, magnetization decreases along *bc* rather than along *ba'aO*. This means that magnetization lags behind the applied field, i.e., for each value of applied field, magnetization has higher values. When the applied field is completely removed, the specimen retains some magnetization called remnant magnetization $M_r = Oc$. The remnant magnetization measures the magnetization of the material in the absence of the applied field. In other words, this parameter measures the strength of the ferromagnetic material as a permanent material.

Again when the applied field strength is increased in the reverse direction, the orientation of domains changes in the opposite manner and magnetization decreases along *cd* as shown in Fig. 5.14 and becomes zero for a certain value of the applied field. The value of the reversed applied field for which magnetization of the specimen reduces to zero is called the coercive field B_C. Here in Fig. 5.14 it is Od. Now if the field is increased further in the negative direction, complete alignment of domains in the field direction occurs and the specimen attains saturation magnetization in the negative direction. When a reversed applied field is reduced to zero, the specimen retains some magnetization called remnant magnetization in the negative direction. As the applied field strength increases from zero, magnetization tends to zero and then increases in magnitude as shown in Fig. 5.14. The closed path of increase and decrease of magnetization with applied field is the hysteresis loop.

The alignment and de-alignment of domains and/or the to and fro movements of the domain walls or Bloch walls with change of direction of the applied field will result in energy loss called hysteresis loss. The hysteresis loss will be more if there are impurities or imperfections in the crystal. The area of the hysteresis loop is a measure of the energy loss. Each time a hysteresis loop is traversed, energy equal to the area of the loop is dissipated as heat. The power loss due to hysteresis in a transformer core depends on the number of times the full loop is traversed per second.

5.9 Hard and Soft Magnetic Materials

Ferromagnetic materials can be broadly classified into two categories, namely, hard ferromagnetic materials and soft ferromagnetic materials depending upon the amount of magnetization by the applied magnetic field.

5.9.1 Hard ferromagnetic materials

Materials which are very difficult to magnetize and demagnetize are said to be hard ferromagnetic materials. In hard ferromagnetic materials, the movement of domain walls is very difficult.

Properties of hard ferromagnetic materials

i.   In case of hard ferromagnetic materials, the hysteresis loop is very broad.

ii.   The hysteresis loop area is very large.

iii.   Since the area of the hysteresis loop is very large, hysteresis loss in these materials will be more.

iv.   These materials have small values of magnetic susceptibility.

v.   They have small values of permeability.

vi.   Hard ferromagnetic materials have high coercivity values.

vii.   They have high retentivity values.

viii. The loss due to eddy currents is more in these materials.

Examples of hard ferromagnetic materials are carbon steels, tungsten steels, alnico, and chromium steels.

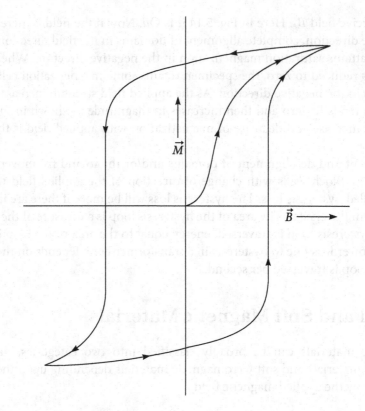

Figure 5.15 | Broad hysteresis curve for hard ferromagnetic materials. Area of the hysteresis loop is large implying large hysteresis loss

5.9.2 Soft ferromagnetic materials

Materials which are very easy to magnetize and demagnetize are said to be soft ferromagnetic materials. In soft ferromagnetic materials, movement of domain walls is very easy.

Properties of soft ferromagnetic materials

i.   In case of soft ferromagnetic materials, the hysteresis loop is very steep.

ii.   The hysteresis loop area is very small.

iii. Since the area of the hysteresis loop is very small, hysteresis loss in these materials will be less.

iv.   These materials have large values of magnetic susceptibility.

v.   They have large values of permeability.

vi.   Soft ferromagnetic materials have low coercivity values.

vii. They have low retentivity values.

viii. The loss due to eddy currents is less in these materials.

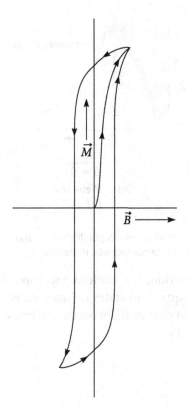

Figure 5.16 Steep hysteresis curve for soft ferromagnetic materials. Area of the hysteresis loop is small implying small hysteresis loss

Examples of soft ferromagnetic materials are silicon steels, nickel–iron alloys and iron–cobalt alloys.

5.10 Anti-Ferromagnetism

The complete description of the origin of magnetism or magnetic moments is purely quantum mechanical. Ferromagnetism arises due to only spin magnetic moments and not due to orbital magnetic moment. In ferromagnetic or anti-ferromagnetic materials, the magnitudes of spin magnetic moments are equal. The susceptibility of anti-ferromagnetic materials is small and positive. Initially, the susceptibility increases slightly as the temperature increases and beyond a particular temperature, known as the Neel temperature T_N, the susceptibility decreases as shown in Fig. 5.17. Below the Neel temperature T_N of an anti-ferromagnet, the spins have antiparallel orientations; the susceptibility attains its maximum value at Neel temperature T_N, where there is a well-defined kink in the curve of χ_m versus T.

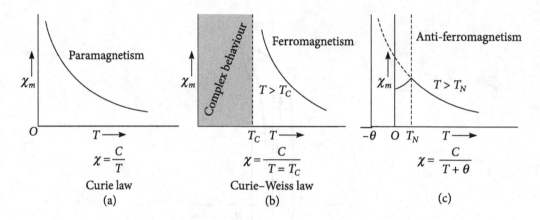

Figure 5.17 | The variation of magnetic susceptibility with the temperature of paramagnetic, ferromagnetic and anti-ferromagnetic materials

Heisenberg's exchange interaction, a quantum mechanical concept between any two electrons present in different quantum states reduces the energy when they have parallel spin. The exchange interaction energy E_{ex} between any two electrons is represented using Heisenberg's exchange integral as

$$E_{ex} = -2J_e \vec{s}_i \cdot \vec{s}_j \tag{5.57}$$

where

J_e = numerical value of Heisenberg's exchange integral. (see Fig. 5.18)

$\vec{s}_i$ = spin magnetic moment of the ith electron

$\vec{s}_j$ = spin magnetic moment of the jth electron

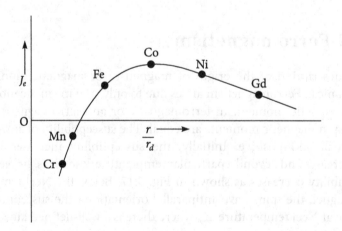

Figure 5.18 | Plot between J_e and r/r_d; r is the inter-atomic distance and r_d is the radius of the 3d orbital

Generally, the exchange interaction energy E_{ex} is negative and depends on the inter-atomic distance between the two interacting atoms. For ferromagnetic materials, $J_e > 0$ and E_{ex} is negative as a result of which $\vec{s_i}$ and $\vec{s_j}$ has to be parallel (see Eq. (5.57)). That means in ferromagnetic materials, spin magnetic moments of atoms or ions are parallel. For anti-ferromagnetic materials, $J_e < 0$ and E_{ex} is negative as a result of which $\vec{s_i}$ and $\vec{s_j}$ has to be antiparallel (see Eq. (5.57)). That means in anti-ferromagnetic materials, spin magnetic moments of atoms or ions are antiparallel.

5.10.1 Anti-ferromagnetic susceptibility

The susceptibility of an anti-ferromagnetic material at higher temperatures, $T > T_N$ is given by

$$\chi_m = \frac{2C}{T + \theta} \text{ for temperatures } T > T_N$$

where

$$\theta = \frac{N \mu_m^2 (\alpha + \beta)}{3k}$$

α = interaction parameter for two atoms having like spins.

β = interaction parameter for two unlike atoms.

$$C = \frac{2N \mu_m^2}{3k}$$

The susceptibility of an anti-ferromagnetic material at lower temperatures, $T < T_N$ and when the applied magnetic field is perpendicular to the spin axis is given by

$$\chi_{m\perp} \approx \frac{1}{\beta} \text{ for temperatures } T < T_N$$

$\chi_{m\perp}$ is independent of temperature. The susceptibility of an anti-ferromagnetic material at $T = 0$ K and when the applied magnetic field is parallel to the spin axis is

$$\chi_{m\|} \approx 0$$

A typical example of an anti-ferromagnetic material is manganese oxide (MnO). Manganese oxide is a ceramic material that is ionic in character, having both Mn^{+2} and O^{-2} ions. No net magnetic moment is associated with the O^{-2} ions since there is a total cancellation of both spin and orbital magnetic moments. However, Mn^{+2} ions possess a net magnetic moment that is predominantly of spin origin. These Mn^{+2} ions are arrayed in the crystal structure such that the moments of adjacent ions are antiparallel. This arrangement is represented schematically in Fig. 5.19. Obviously the opposite magnetic moments cancel one another and as a consequence, the solid as a whole possesses no net magnetic moment.

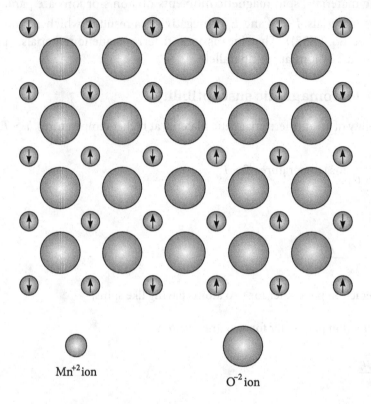

Mn^{+2} ion

O^{-2} ion

Figure 5.19 | Schematic representation of antiparallel alignment of spin magnetic moments for anti-ferromagnetic manganese oxide (MnO)

Other examples of anti-ferromagnetic materials are ferrous oxide and chromium oxide etc.

5.11 Ferrimagnetism

The term "ferrimagnetism" was coined by Neel to describe the properties of certain materials (later known as ferrimagnetic materials.) which below a certain temperature, exhibit spontaneous magnetization in sub-lattices arising from non-parallel alignments of atomic dipole moments as shown in Fig. 5.20.

Ferrimagnetic materials are a special class of ferromagnetic materials. The most important properties of ferrimagnetic materials include high magnetic permeability and high electrical resistance. The magnetic moments of two sub-lattices in anti-ferromagnetic materials are equal in magnitude and opposite in direction with the result that they completely compensate one another. However, there are materials in which the magnitudes

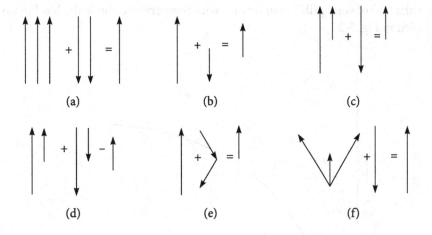

Figure 5.20 | Illustration of possible types of sub-lattices arrangements which can give rise to spontaneous magnetization. (a) Different number of similar magnetic moments in the sub-lattices; (b) Equal number of dissimilar magnetic moments in the sub-lattices; (c) One sub-lattice contains two different types of magnetic moments one of which also occurs in equal numbers on the second sub-lattice so that the net effect is that of just one type of moment; (d) Extrapolation of (c); (e) When there are strong interactions between the ions within a given sub-lattice, then a triangular arrangement is possible; (f) One sub-lattice consists of three different types of moments, each of which is bound by some anisotropic

of magnetic moments in the two sub-lattices are not the same due to the difference in their numbers or by the nature of the atoms that make up the sub-lattices. This leads to the appearance of a finite difference in the magnetic moments of the sub-lattices and to an appropriate spontaneous magnetization of the materials. Such an uncompensated anti-ferromagnetism is defined as ferrimagnetism. The macroscopic magnetic characteristics of ferromagnetics and ferrimagnetics are very similar. Ferrimagnetic materials are popularly known as ferrites. [Ferrites in a magnetic sense should not be confused with the ferrite α–iron.].

Ferrites are hard, brittle, iron-containing, generally grey or black and are polycrystalline. They are composed of iron oxide and one or more other metals in chemical combination. Ferrites are represented by the molecular formula $M^{2+}Fe^{3+}O_4^{2-}$ or equivalently $M^{2+}O^{2-}.Fe^{3+}O_3^{2-}$ where M stands for divalent metal such as Fe, Mn, Co, Ni, Cu, Mg, Zn, or Cd. In ferrites, the Fe ions exist both in +2 and +3 valence states in the ratio 1:2. Examples of ferrites are ferrous ferrites, manganese ferrites and cadmium ferrites, zinc ferrites and cobalt ferrites.

5.11.1 Properties of ferrites

i.　Ferrites have high electrical resistance.

ii.　They have high magnetic permeability.

iii. The variation of susceptibility of ferrites with temperature above the Neel temperature is shown in Fig. 5.21.

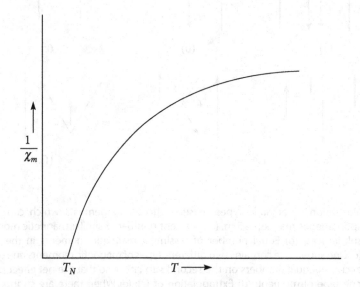

Figure 5.21 Susceptibility–temperature diagram for a typical ferrite above its Neel temperature

5.11.2 Applications of ferrites

Ferrites are very versatile magnetic materials having many modern-day potential applications due to their characteristic properties.

i. Soft ferrites are used in the manufacture of pot core inductors.

ii. Small size antennas used in small size radio receivers are made by winding a suitable coil on a ferrite rod.

iii. Computer hard discs, floppy drives, credit cards, ATM cards, audio and video cassettes, and so on are made of ferrites.

iv. Ferrites are used in the manufacture of microwave devices like circulators, isolators and phase shifters.

v. Hard ferrites are used in the manufacture of permanent magnets.

vi. Ferrites are used in the manufacture of television line out put transformers and television tube scanning yokes.

vii. They are used in the production of low frequency ultrasonic waves by magnetostriction effect.

5.12 Comparisons

Dia magnetism	Para magnetism	Ferro magnetism	Anti-ferro magnetism	Ferri magnetism
There are no permanent dipoles.	A paramagnetic substance has a non-vanishing angular momentum and hence it possesses permanent dipoles.	The permanent dipoles are strongly aligned in the same direction and consequently a large spontaneous magnetization results even in the absence of applied field. This alignment is due to strong internal field called molecular Weiss field.	The magnetic moments in one sub-lattice will be aligned antiparallel to those in other sub-lattice because molecular field is negative.	The magnetic moments in one sublattice will be aligned antiparallel to those in other sub-lattice because molecular field is negative.
The diamagnetic susceptibility is small and negative.	The paramagnetic susceptibility is small and positive.	The ferromagnetic susceptibility is very large and positive.	The Anti-ferromagnetic susceptibility is small and positive.	The ferrimagnetic susceptibility is very large and positive
It is almost independent of temperature.	At a finite temperature each atom has a thermal energy on the average equal to kT and the resulting thermal agitation tends to make the dipole orientation random. To align the dipoles in a given direction a field strong enough is needed to compete with the force due to thermal agitation. The field commonly available in the laboratory is far too small to turn the dipoles slightly in the direction of field so that on the average there are slightly more dipoles pointing in the direction of the applied field	As long as molecular field is strong enough to compete with thermal agitation, nearly all the dipoles have parallel alignment and the substance is ferromagnetic. As the temperature is raised thermal agitation becomes stronger and fewer dipoles are aligned. Consequently spontaneous magnetization decreases with increasing temperature. At critical temperature called Curie temperature the internal field is no longer sufficient to overcome the thermal agitation. Above the Curie temperature	At critical temperature called Neel temperature below which magnetic moments on both sub-lattices have an orderly arrangement, while above the Neel temperature the magnetic moments becomes randomly oriented as in paramagnetic substance. The Neel temperature is a measure of the strength of internal field, higher the Neel temperature stronger is the internal field.	At critical temperature called Neel temperature below which magnetic moments on both sub-lattices have an orderly arrangement, while above the Neel temperature the magnetic moments becomes randomly oriented as in paramagnetic substance. The Neel temperature is a measure of the strength of internal field, higher the Neel temperature stronger is the internal field.

	than in the opposite direction. Since the applied field has to compete with the thermal agitation the paramagnetic susceptibility decreases with increasing temperature	the substance loses its spontaneous magnetization and becomes paramagnetic.		
Organic materials, light elements, Alkali earths, Bismuth	Superconducting metals	Some transitions and rare earth metals	Salts of transition elements	Ferrites

Questions

5.1 What is a magnetic dipole?

5.2 What do you mean by magnetic dipole moment of a current-carrying loop?

5.3 What is Bohr magneton?

5.4 How can you find out the direction of magnetic moment of a current-carrying loop?

5.5 What do you mean by intensity of magnetization?

5.6 What do you mean by magnetic flux?

5.7 What is magnetic induction?

5.8 What do you mean by magnetic field intensity?

5.9 What do you mean by magnetic permeability?

5.10 What is relative magnetic permeability?

5.11 Define magnetic susceptibility.

5.12 What is the physical meaning of magnetic susceptibility?

5.13 What are different types of materials from the magnetic point of view?

5.14 Assume that an electron is a small sphere of radius R, its charge and mass being distributed uniformly through out its volume. Derive an expression for its spin magnetic moment. [This model of the electron is too mechanistic to be in the spirit of the quantum physics view of this particle]

5.15 What is ferromagnetic domain?

5.16 Give a classical picture of the origin of magnetic moments.

5.17 What are the properties of diamagnetic materials?

5.18 Give Langevin's classical theory of diamagnetism.

5.19 What is the origin of diamagnetism?

5.20 Prove that diamagnetic susceptibility is negative.

5.21 Prove that the direction of induced dipole moment in a diamagnetic material is always opposite to the applied field.

5.22 Derive Curie's law of paramagnetism.

5.23 Give Langevin's classical theory of paramagnetism.

5.24 What is the origin of paramagnetism?

5.25 What are the objections raised by Leeuwan against Langevin's classical theory of paramagnetism.

5.26 Derive an expression for the average magnetic moment of an electron in the direction of the applied field in terms of the Brillouin function.

5.27 Show that magnetic susceptibility of a paramagnetic substance is inversely proportional to the absolute temperature.

5.28 What do you mean by the effective number of Bohr magneton?

5.29 What is the origin of ferromagnetism?

5.30 Show that spontaneous magnetization in a ferromagnetic material exists for temperatures below the Curie point T_c.

5.31 What is the temperature effect on spontaneous magnetization?

5.32 What is Curie temperature? Derive an expression for it.

5.33 What are the criticisms against Weiss's molecular field theory?

5.34 Explain how an external magnetic field affects ferromagnetic domains.

5.35 How does temperature affect ferromagnetic domains?

5.36 How is ferromagnetism lost in ferromagnetic materials?

5.37 How can a ferromagnetic material below Curie temperature be magnetized?

5.38 Materials composed of atoms having a completely filled electron shell are not capable of being permanently magnetized. Explain why?

5.39 Explain the origin of the hysteresis curve of ferromagnetic materials.

5.40 Explain the hysteresis curve of ferromagnetic materials.

5.41 What is coercive field?

5.42 What is remnant magnetization?

5.43 What is a hard magnetic material? What are its properties?

5.44 What is hysteresis loss in ferromagnetism?

5.45 What is a soft magnetic material? What are its properties?

5.46 What are the properties of anti-ferromagnetic materials?

5.47 How does Heisenberg's exchange interaction explain the origin of anti-ferromagnetism?

5.48 Distinguish between ferromagnetic and ferrimagnetic materials.

5.49 What are the practical applications of ferrites?

5.50 Give a comparative chart of diamagnetic, paramagnetic, ferromagnetic anti-ferromagnetic and ferrimagnetic materials.

Problems

5.1   The magnetic susceptibility of a material at room temperature is 0.87×10^{-8}. Calculate its magnetization under the action of a magnetic induction of 0.35 Tesla.

[Ans 2.42×10^{-3} A/M]

5.2   A magnetic field of induction $B = 4.0$ Wb/m^2 is applied perpendicularly to the plane of the electron's path of radius 5.1×10^{-11} m. Calculate the change in magnetic moment of a circulating electron.

[Ans $\pm 7.4 \times 10^{-29}$ Am^2]

5.3   The dipole moment associated with an atom of iron in an iron bar is 1.8×10^{-23} Am^2. Assume that all the atoms in the bar, which is 10 cm long and has a 1 cm^2 cross-section, have their dipole moments aligned. What is the dipole moment of the bar? [For iron, density = 7.85 gm/cm^3 and atomic mass = 55.85 gm].

[Ans   15.2 Am^2]

5.4   Calculate the value of 1 Bohr magneton.

[Ans 9.27×10^{-24} Am^2]

5.5   A material attains a magnetization of 3200 Amp/met under the action of a magnetic induction of 0.005 Wb/m^2. Calculate the magnetizing field and susceptibility and relative permeability of the material.

[Ans   780 A/m, 4.1, 5.1]

5.6   Calculate the diamagnetic susceptibility of one mole of He atoms when the average atomic radius is 0.53 Å. The atomic number of a Helium atom is 2.

[Ans 1.994×10^{-8}/mole]

5.7   Calculate the diamagnetic susceptibility of one mole of gold when the average atomic radius is 1.44 Å. The atomic number of copper is 79.

[Ans 5.82×10^{-6}/mole]

5.8   A paramagnetic material contains 10^{-26} ions/m^3 with magnetic moment $0.1\mu_B$. Calculate the magnetization under the action of a magnetic induction 1.0 Tesla at 300 K.

[Ans   0.02 A/m.]

5.9   A paramagnetic material has a magnetic field intensity of 10^4 A/m. If the susceptibility of the material at room temperature is 3.7×10^{-3}, calculate the magnetization and magnetic induction of the material.

[Ans   37 A/m, 1.26×10^{-2} Wb/m^2]

5.10   A paramagnetic material has a bcc structure with a unit cell edge of 2.5 Å. If the saturation value of the magnetization is 1.8×10^6 A/m, calculate the average magnetization contributed per atom in Bohr magneton.

[Ans $1.517\mu_B$]

Multiple Choice Questions

1.   What is the unit of magnetic dipole moment?

(i)   Am

(ii)   A/m

(iii)   A/m^2

(iv)   None of the above

2.   What is the unit of magnetic field intensity?

(i)   Am

(ii)   A/m

(iii)   A/m^2

(iv)   None of the above

3. What is the relation between magnetic relative permeability and magnetic susceptibility?

    (i) $\mu_r = 1 + \chi_m$

    (ii) $\mu_r = 1 - \chi_m$

    (iii) $\dfrac{1}{\mu_r} = 1 + \chi_m$

    (iv) $\dfrac{1}{\mu_r} = 1 - \chi_m$

4. What is the direction of induced dipole moments in a diamagnetic material?

    (i)   In the direction of the external field

    (ii) In the direction opposite to that of the external field

    (iii) In the direction perpendicular to the external field

    (iv) Cannot be said

5. What is the unit of the relative magnetic permeability?

    (i)   Wb/Am   (ii) Am/Wb

    (iii) Henry/m   (iv) No units

6. Diamagnetic materials are characterized by

    (i)   Variable, small negative susceptibilities

    (ii) Constant, large negative susceptibilities

    (iii) Constant, small positive susceptibilities

    (iv) Constant, small negative susceptibilities

7. The value of susceptibility for paramagnetic materials is always

    (i)   Positive   (ii) Negative

    (iii) Zero   (iv) May be positive or negative

8. Langevin's function is given by

    (i) $L(a) = \coth a - a$ (ii) $L(a) = \tanh a - \dfrac{1}{a}$

    (iii) $L(a) = \tanh a - a$ (iv) $L(a) = \coth a - \dfrac{1}{a}$

9. In case of paramagnetic matter, susceptibility is

    (i)   Directly proportional to absolute temperature

    (ii) Directly proportional to square of absolute temperature

    (iii) Inversely proportional to absolute temperature

    (iv) Inversely proportional to square of absolute temperature

10. Out of the following materials, which has no permanent atomic dipole moments
    - (i) Diamagnetic
    - (ii) Paramagnetic
    - (iii) Ferromagnetic
    - (iv) All the three

11. The Curie temperature is defined as the temperature
    - (i) Above which a ferromagnetic material becomes diamagnetic
    - (ii) Above which a ferromagnetic material becomes paramagnetic
    - (iii) Above which a paramagnetic material becomes diamagnetic
    - (iv) Above which a paramagnetic material becomes ferromagnetic

12. According to Weiss's molecular field theory
    - (i) Theoretical Curie point and experimental Curie point are equal
    - (ii) Theoretical Curie point is less than experimental Curie point
    - (iii) Theoretical Curie point is more than experimental Curie point
    - (iv) There is no specific relation between the two

13. Ferromagnetic materials normally do not possess net magnetization.
    - (i) True
    - (ii) False

14. Hysteresis loss in a soft magnetic material is more than that in a hard magnetic material.
    - (i) True
    - (ii) False

15. Ferromagnetism arises due to
    - (i) Only spin magnetic moments
    - (ii) Only orbital magnetic moment
    - (iii) Both spin magnetic moments and orbital magnetic moments
    - (iv) Inherent property of the material

16. The susceptibility of an anti-ferromagnetic material at a temperature more than the Neel temperature is given by
    - (i) $\chi_m = \dfrac{2C}{T + \theta}$
    - (ii) $\chi_m = \dfrac{2C}{T - \theta}$
    - (iii) $\chi_m = \dfrac{C}{T + \theta}$
    - (iv) $\chi_m = \dfrac{C}{T - \theta}$

17. Ferrimagnetic materials are a subset of
    - (i) Anti-ferromagnetic materials
    - (ii) Paramagnetic materials
    - (iii) Diamagnetic materials
    - (iv) Ferromagnetic materials

18. The two most important properties of ferrimagnetic materials are
    - (i) High magnetic permeability and low electrical resistivity
    - (ii) Low magnetic permeability and high electrical resistivity
    - (iii) High magnetic permeability and high electrical resistivity
    - (iv) Low magnetic permeability and low electrical resistivity

19. In anti-ferromagnetic materials the spin moments associated with atoms are aligned
    (i)   Anti-parallel to each other   (ii) Parallel to each other
    (iii) Anti-parallel but of unequal magnitudes   (iv) Random to each other

20. Silicon steel is a soft ferromagnetic material.
    (i)   Yes   (ii) No

Answers

1 (i)	2 (ii)	3 (i)	4 (ii)	5 (iv)	6 (iv)	7 (i)	8 (iv)
9 (iii)	10 (i)	11 (ii)	12 (iii)	13 (i)	14 (ii)	15 (i)	16 (i)
17 (iv)	18 (iii)	19 (i)	20 (i)				

6 Superconductivity

6.1 Introduction

Superconductivity, a quantum phenomena on macroscopic scale, is a fascinating and challenging field of physics. Scientists and engineers throughout the world have been striving to develop an understanding of this remarkable phenomenon for many years. Yet, for nearly 75 years, superconductivity was a relatively obscure subject. Today however, superconductivity is being applied to many diverse areas such as medicine, theoretical and experimental science, the military, transportation, power production, electronics, and many more.

In 1911, Dutch physicist Heike Kamerlingh Onnes discovered that at 4.2 K, the electrical resistance of mercury became zero and thus laid the foundation stone for superconductivity. He immediately predicted many uses for superconductors. The phenomenon of complete disappearance of electrical resistance along with the exhibition of perfect diamagnetism by various solids when they are cooled below a characteristic temperature, the transition or critical temperature is called superconductivity. The perfect diamagnetic property of superconductors below the critical temperature implies a very strong repulsion in magnetic field. The disappearance of electrical resistivity in superconductors means electric currents can flow for infinite time without the expenditure of energy! According to one calculation, the decay time of supercurrent will not be less than 100,000 years! The problem with the commercial applications of superconductors is that they are superconductors at low temperatures, much below the room temperature. High temperature (much below the room temperature) superconductors (mainly ceramics) suffer from the problems of brittleness, instabilities in some chemical environments (such as moist air) and a tendency for impurities to segregate at the surfaces of the crystals (where they interfere with the flow of high currents in the superconducting state). These problems have yet to be overcome.

In 1995 at ambient pressure, the highest critical temperature of 139 K was achieved in $Hg_{0.2}Tl_{0.8}Ca_2Ba_2Cu_3O$. Very recently in 2015, at pressure of 90×10^9 Pa, a group of researchers in Germany achieved critical temperature of 203 K in the sulfur hydride system. Thus, the world record of highest accepted superconducting critical temperature as of 2015 is 203 K in sulfur hydride system.

6.2 Zero Resistivity

The resistivity of a material should remain constant as the temperature of the material tends to absolute zero, because at low temperatures, the lattice contributions to resistivity tend to zero and impurity contributions remain constant. Many metals behave in this manner and are called normal metals. The behaviour of normal metals in this regard is shown in Fig. 6.1.

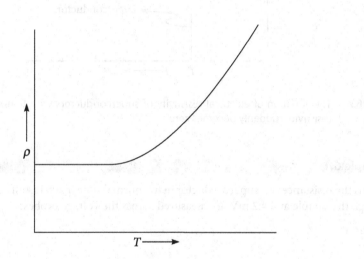

Figure 6.1 | The variation of electrical resistivity of normal conductors with temperature. As $T \rightarrow 0$ K, the resistivity decreases and then remains constant

The behaviour of another class of materials is quite different. As the temperature of the material is lowered, its resistivity decreases and at some critical temperature its resistivity suddenly vanishes as shown in Fig. 6.2 and on verification, it shows perfect diamagnetism. These types of materials are called superconductors. The materials in which superconductivity phenomenon is observable are called superconductors. In superconductors, electric current can flow even in the absence of applied voltage! Superconductors have no resistance at all! The critical temperatures T_C, varies from one superconductor to another. Mercury becomes a superconductor at or below 4.15 K, i.e. the critical temperature of mercury is 4.15 K. Similar behaviour has been found in approximately 28 other elements, including lead and tin, and in thousands of alloys and chemical compounds.

It is interesting to note that good conductors like copper, silver and gold do not become superconductors at low temperatures. In fact, superconductivity results due to the strong interactions between lattice points and electrons, whereas weak interactions exist between the lattice points and electrons in good conductors.

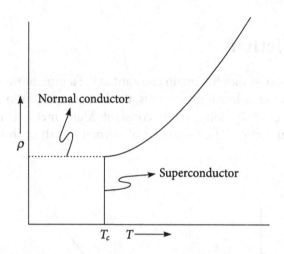

Figure 6.2 | The variation of electrical resistivity of superconductors with temperature. At $T = T_c$, resistivity suddenly becomes zero

Example 6.1

What is the resistance of a superconductor in the normal state if 300 mA of current are passing through the sample and 4.2 mV are measured across the voltage probes?

Solution

$$R = \frac{V}{I} = \frac{4.2 \times 10^{-3}\,V}{300 \times 10^{-3}\,A} = 0.014\,\Omega$$

6.3 Critical Temperature T_C

The critical temperature of a superconducting material is defined as the temperature at which the material becomes a superconductor. It is also called transition temperature or characteristic temperature. The critical temperature depends on the material and is a function of the strength of the surrounding magnetic field. It is also observed that the critical temperature varies with mass of the isotope according to the relation

$$M^{\alpha}T_C = \text{constant} \tag{6.1}$$

where M is the mass of the isotope and $0.15 \leq \alpha \leq 0.50$. For most materials, $\alpha = 0.5$ and in some cases, $\alpha = 0$, i.e., there is no isotope effect.

The present research on superconductors mainly focuses on developing engineering materials having high critical temperatures so that large-scale practical applications of the superconductivity phenomenon can be possible. In 1986, Karl Alex Müller and J. George Bednorz discovered that certain type-II superconductors (type-I and type-II superconductors will be explained in Section 6.9) could retain their superconductivity at critical temperatures as high as 35 K. Compounds retaining their superconductivity at critical temperatures as high as 134 K have been studied. Till date, the world record of highest accepted superconducting critical temperature is 203 K in sulfur hydride system.

Example 6.2

The critical temperature for mercury with isotopic mass 199.5 is 4.18 K. Calculate its critical temperature when its isotopic mass changes to 203.4.

Solution

The data given are $M_1 = 199.5$, $M_2 = 203.4$ and $T_{C1} = 4.18$ K.

The critical temperature in terms of its isotopic mass is given by

$$T_C = AM^{-\frac{1}{2}}$$

Therefore, we have

$$\frac{T_{C1}}{T_{C2}} = \left(\frac{M_1}{M_2}\right)^{-\frac{1}{2}} = \sqrt{\frac{M_2}{M_1}}$$

or $T_{C2} = T_{C1}\sqrt{\frac{M_1}{M_2}} = 4.18 \times \sqrt{\frac{199.5}{203.4}} = 4.14\,\text{K}$

6.4 Critical Magnetic Field B_C

The critical magnetic field at a certain temperature is defined as the magnetic field above which a superconductor becomes a normal conductor. The critical magnetic field depends on the material and the temperature of the material. Such materials are superconductors only for values of T and B below their respective curves and are normal conductors for values of T and B above these curves. The variation of the critical magnetic field with temperature for type-I superconductors is shown in Fig. 6.3. The critical magnetic field is maximum at 0 K as shown in the figure. The critical magnetic fields are quite low, less than 0.1 Tesla. Therefore, type-I superconductors cannot be used as coils of strong electromagnets.

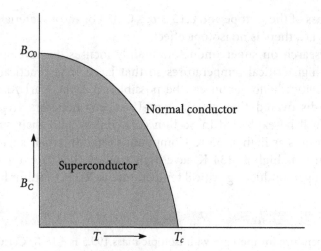

Figure 6.3 | The variation of critical magnetic field with temperature of a superconducting material. Inside the curve, the material is a superconductor and outside the curve, it is a normal conductor

The critical magnetic field for type-I superconductors appears to follow the relation

$$B_C = B_{C0}\left(1-\frac{T^2}{T_C^2}\right)$$ (6.2)

Example 6.3

Calculate the critical magnetic field for tin at 1.5 K and 2.5 K. The following data for tin are given: $T_C = 3.72$ K and $B_C = 30.5 \times 10^{-3}$ T at 0 K.

Solution

The data given are $T = 1.5$ K and 2.5 K, $T_C = 3.72$ K and $B_{C0} = 30.5 \times 10^{-3}$ T at 0 K.
The critical magnetic field at 1.5 K will be

$$B_C = B_{C0}\left(1-\frac{T^2}{T_C^2}\right) = 30.5\times10^{-3}\left(1-\frac{1.5^2}{3.72^2}\right) = 25.54\times10^{-3}T$$

The critical magnetic field at 2.5 K will be

$$B_C = 30.5\times10^{-3}\left(1-\frac{2.5^2}{3.72^2}\right) = 16.72\times10^{-3}T$$

6.5 Critical Current Density J_C

Since there is no loss in electrical energy when superconductors carry electric current, relatively narrow wires made of superconducting materials can be used to carry huge currents. However, there is a certain maximum current that these materials can be made to carry, above which they stop being superconductors. If too much current is passed through a superconductor, it will revert to the normal state even though it may be below its transition temperature. The critical current I_C is defined as the current above which a superconductor becomes a normal conductor. The critical current I_C for a cylindrical superconducting wire of radius r is given by

$$I_C = \frac{2\pi R B_C}{\mu_0} \quad \text{since } B = \frac{\mu_0 I}{2\pi r} \quad \text{(Biot-Savart law)} \tag{6.3}$$

The critical current produces a magnetic field more than the critical magnetic field on the surface of the superconductor, destroying the superconducting properties of the material. The value of critical current density is a function of temperature, i.e., the colder you keep the superconductor, the more current it can carry. The critical current density J_C from Eq. (6.3) for a cylindrical superconducting wire of radius r is given by

$$J_C = \frac{2B_C}{\mu_0 R} \tag{6.4}$$

Figure 6.4 is a plot of voltage versus current for a superconducting wire.

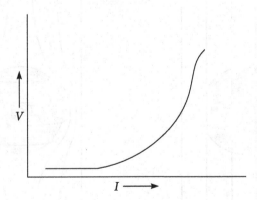

Figure 6.4 | Plot of voltage versus current for a superconducting wire

Example 6.4

Find the critical current density of an Indium wire of radius 6 mm at 2 K. The data given for Indium are $T_C = 3.4$ K and $B_C = 29.3 \times 10^{-3}$ T at 0 K.

Solution

The data given are $T = 2$ K, $T_C = 3.4$ K and $B_{C0} = 29.3 \times 10^{-3}\,T$ at 0 K.
The critical magnetic field at 2 K will be

$$B_C = B_{C0}\left(1 - \frac{T^2}{T_C^2}\right) = 29.3 \times 10^{-3}\left(1 - \frac{2^2}{3.4^2}\right) = 19.16 \times 10^{-3}\,T$$

The critical current density is given by

$$J_C = \frac{2B_C}{\mu_0 R} = \frac{2 \times 19.16 \times 10^{-3}}{4\pi \times 10^{-7} \times 6 \times 10^{-3}}\,\frac{A}{m^2} = 5.082 \times 10^{6}\,\frac{A}{m^2}$$

6.6 Meissner Effect

Figure 6.5 demonstrates what occurs when a superconductor is placed inside a magnetic field. When the temperature is lowered to below the critical temperature, T_C, the superconductor will "push" the magnetic field out of itself. It does this by creating surface currents in itself which in turn produces a magnetic field exactly countering the external field, producing a "magnetic mirror". The superconductor becomes perfectly diamagnetic, cancelling all

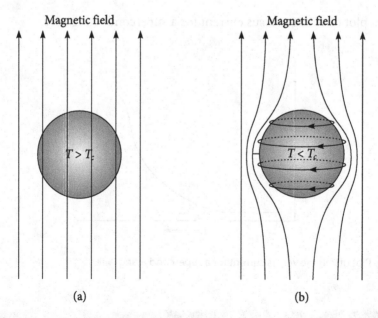

Figure 6.5 | Illustration of Meissner effect. A superconductor will push out the applied magnetic field from the interior of the superconducting material at or below the critical temperature

magnetic flux in its interior. This perfect diamagnetic property of superconductors is perhaps the most fundamental macroscopic property of a superconductor. Flux exclusion due to what is referred to as the Meissner effect, can be easily demonstrated in the classroom by lowering the temperature of the superconductor to below its critical temperature T_C and placing a small magnet over it. The magnet will begin to float above the superconductor. In most cases, the initial magnetic field from the magnet resting on the superconductor will be strong enough that some of the field will penetrate the material, resulting in a non-superconducting region. The magnet, therefore, will not levitate as high as one introduced after the superconductive state has been obtained.

6.7 Josephson Effect

Up to this point, those properties of superconductors which are commonly referred to as macroscopic properties, such as the Meissner effect and zero resistance have been discussed. We will now focus on those properties which can be explained by quantum mechanical concepts. An example of microscopic properties is the phenomenon of electron tunnelling in superconductors. Tunnelling is a process arising from the wave nature of the electron. It occurs because of the transport of electrons through spaces that are forbidden by classical physics because of a potential barrier. The tunnelling of a pair of electrons between superconductors separated by an insulating barrier was first discovered by Brian Josephson in 1962. Josephson discovered that if two superconducting materials were separated by a thin insulating barrier such as an oxide layer of 10 to 20 angstroms thickness, it is possible for electron pairs to pass through the barrier without resistance. The flow of electric current between two pieces of superconducting material separated by a thin layer of insulating material is known as the dc Josephson effect. Ordinary materials, where a potential difference must exist for a current to flow do not show the dc Josephson effect. A Josephson junction consists of two superconductors separated by a thin insulating barrier. The current that flows in through a d.c. Josephson junction has a critical current density which is characteristic of the junction material and geometry. Pairs of superconducting electrons will tunnel through the barrier. As long as the current is below the critical current for the junction, there will be zero resistance and no voltage drop across the junction. Figure 6.6 demonstrates the Josephson effect. Figure 6.7 is a graph of the current–voltage relation for a Josephson junction.

The Josephson junction can behave as a superfast switching device. Josephson junctions can perform switching functions such as switching voltages approximately ten times faster than ordinary semiconducting circuits. A Superconducting Quantum Interfernce Device (SQUID) is nothing but two Josephson junctions in parallel. It is the most sensitive magnetic flux sensor currently known. The extreme sensitiveness of the SQUID is useful in the fields of biomagnetism, materials science, metrology, astronomy, geophysics, and so on.

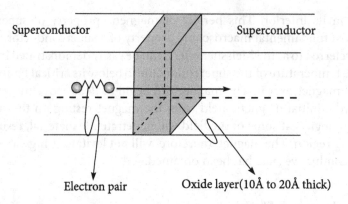

Superconductor   Superconductor

Electron pair   Oxide layer(10Å to 20Å thick)

Figure 6.6 | Illustration of the Josephson effect

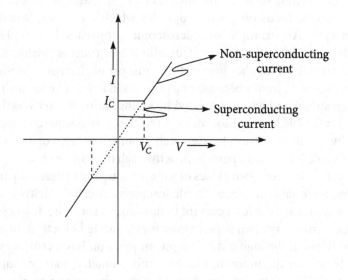

Figure 6.7 | Plotting of the current and voltage for a Josephson junction

6.8 Theory of Superconductivity: BCS Theory

The accepted theory of superconductors is the BCS (Bardeen, Cooper, Schrieffer) theory which proposes strong interaction between phonons (carriers of lattice vibrational energy) and electrons.

The superconductivity phenomenon was explained in 1957 by three American physicists – John Bardeen, Leon Cooper and John Schrieffer. Their theory of superconductivity is popularly known as the BCS theory. The BCS theory explains superconductivity at temperatures close to absolute zero. Cooper realized that atomic lattice vibrations are directly responsible for unifying the entire current. They forced the electrons to pair up into teams that could pass all obstacles which caused resistance in the conductor. These teams of electrons are known as Cooper pairs. They constitute a system that functions as a single entity. Cooper and his colleagues knew that electrons which normally repel one another must feel an overwhelming attraction in superconductors. The answer to this problem was found to be in phonons, packets of sound waves energy present in the lattice as it vibrates. Although this lattice vibration cannot be heard, its role as a moderator is indispensable.

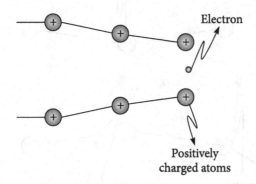

Figure 6.8 | A wave of lattice distortion due to attraction to a moving electron

According to BCS theory, as one negatively charged electron passes by positively charged ions in the lattice of the superconductor, the lattice distorts. This in turn causes phonons to be emitted which form a trough of positive charges around the electron. Figure 6.8 illustrates a wave of lattice distortion due to attraction to a moving electron. Before the electron passes by and before the lattice springs back to its normal position, a second electron is drawn into the trough. It is through this process that two electrons, which should repel one another, link up. The forces exerted by the phonons overcome the electrons' natural repulsion. The electron pairs are coherent with one another as they pass through the conductor in unison. They are screened by the phonons and are separated by some distance. When one of the electrons of the Cooper pair passes close to an ion in the crystal lattice, the attraction between the negative electron and the positive ion cause a vibration to pass from ion to ion until the other electron of the pair absorbs the vibration. The net effect is that one electron emits a phonon and the other electron absorbs the phonon. It is this exchange that keeps the Cooper pairs together. It is important to understand that the pairs are constantly breaking and reforming. Because electrons are indistinguishable particles, it is easier to think of them as permanently paired. Figure 6.9 illustrates how two electrons, the Cooper pair, become locked together.

By pairing off two by two, the electrons pass through the superconductor more smoothly. The electron may be thought of as a car racing down a highway. As it speeds along, the car cleaves the air in front of it. Trailing behind the car is a vacuum, a vacancy in the atmosphere quickly filled by inrushing air. A tailgating car would be drawn along with the returning air into this vacuum. The rear car is effectively attracted to the one in the front. As the negatively charged electrons pass through the crystal lattice of a material, they draw the surrounding positive ion cores toward them. As the distorted lattice returns to its normal state, another electron passing nearby will be attracted to the positive lattice in much the same way that a tailgater is drawn forward by the leading car.

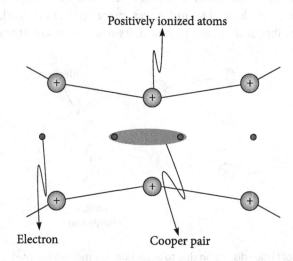

Positively ionized atoms

Electron   Cooper pair

Figure 6.9 | Electrons can be attracted to one another through interactions with the crystalline lattice. This figure illustrates how two electrons, called Cooper pair, become locked together

The electrons in the superconducting state are like an array of rapidly moving vehicles. Vacuum regions between cars lock them all into an ordered array as does the condensation of electrons into a macroscopic, quantum ground state. Random gusts of wind across the road can be envisioned to induce collisions, as thermally excited phonons break pairs. With each collision, one or two lanes are closed to traffic flow, as a number of single-particle quantum states are eliminated from the macroscopic, many-particle ground state.

The BCS theory successfully shows that electrons can be attracted to one another through interactions with the crystalline lattice. This occurs despite the fact that electrons have the same charge. When the atoms of the lattice oscillate as positive and negative regions, the electron pair is alternatively pulled together and pushed apart without a collision. The electron pairing is favourable because it has the effect of putting the material into a lower energy state. When electrons are linked together in pairs, they move through the superconductor in an orderly fashion.

As long as the superconductor is cooled to very low temperatures, the Cooper pairs stay intact, due to the reduced molecular motion. As the superconductor gains heat energy, the vibrations in the lattice become more violent and break the pairs. As they break, superconductivity diminishes. Superconducting materials and alloys have characteristic transition temperatures from normal conductors to superconductors called critical temperature T_C. Below the superconducting transition temperature, the resistivity of a material is exactly zero. Superconductors made from different materials have different T_C values. Figure 6.10 is a graph of the resistance of superconducting material $YBa_2Cu_3O_7$ versus its temperature. The critical temperature of $YBa_2Cu_3O_7$ is 93 K as shown in the figure.

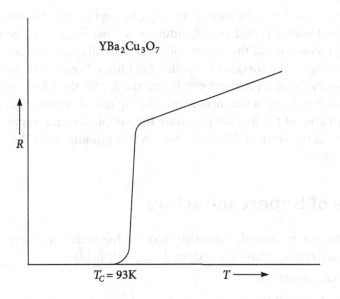

Figure 6.10 | The variation of resistance of superconducting material $YBa_2Cu_3O_7$ versus its temperature. (See also Fig. 6.2)

Since there is no loss in electrical energy when superconductors carry electrical current, relatively narrow wires made of superconducting materials can be used to carry huge currents. However, as mentioned earlier, there is a certain maximum current that these materials can be made to carry, above which they stop being superconductors. If too much current is pushed through a superconductor, it will revert to the normal state even though it may be below its transition temperature. The value of critical current density J_C is a function of temperature; i.e., the colder you keep the superconductor the more current it can carry. Figure 6.4 is a graph of voltage versus current for a superconductive wire. For practical applications, J_C values in excess of 1000 A/mm are preferred.

An electrical current in a wire creates a magnetic field around a wire. The strength of the magnetic field increases as the current in the wire increases. Because superconductors are able to carry large currents without loss of energy, they are well suited for making strong electromagnets. When a superconductor is cooled below its transition temperature T_C and a magnetic field is increased around it, the magnetic field remains around the superconductor. If the magnetic field is increased to a given limit, the superconductor will go to the normal resistive state.

The maximum value for the magnetic field at a given temperature is known as the critical magnetic field B_C. For all superconductors, there exist a region of temperatures and magnetic fields within which the material is superconducting. Outside this region, the material is normal. Figure 6.3 demonstrates the relationship between temperature and magnetic fields.

Type-II superconductor behaviour also helps to explain the Meissner effect. When levitating a magnet with a Type-I superconductor, a bowl shape must be used to prevent the magnet from shooting off the superconductor. The magnet is in a state of balanced forces while floating on the surface of expelled field lines. Because the field at the surface of a samarium–cobalt magnet is about 600 G and the B_{C1} for the YBCO superconductor is less that 200 G, the pellet is in the mixed state during the Meissner effect demonstration. Some of the field lines of the magnet penetrate the sample and are trapped in defects and grain boundaries in the crystals. This is known as flux pinning. It "locks" the magnet to a region above the pellet.

6.9 Types of Superconductors

Superconductors may be broadly classified into the following two categories depending upon their magnetization behavior in external magnetic fields.

i.   Type-I superconductors
ii.   Type-II superconductors

6.9.1 Type-I superconductors

Superconductors in which the magnetic field is totally excluded from the interior below a certain critical magnetic field B_C but which, at $B = B_C$, lose their superconductivity abruptly leading to the magnetic field penetrating fully, are termed as type-I or soft superconductors. Type-I superconductors exhibit the Meissner effect of magnetic flux expulsion. The magnetization curve for type-I superconductors is shown in Fig. 6.11. The magnetization curve shows that the transition at $B = B_C$ is reversible and means that if the applied magnetic field is reduced below the critical magnetic field B_C, the material again acquires superconducting properties and the field is expelled out. The highest critical magnetic field for these materials is of the order of $10^{-1}T$, making these materials unsuitable for use in high field superconducting materials. Type-I superconductors are called soft superconductors because of their tendency to expel out low magnetic fields. Type-1 category of superconductors mainly comprises metals and metalloids that show some conductivity at room temperature. Few examples of type-I superconductors are lead (Pb), mercury (Hg), chromium (Cr), aluminum (Al) and tin (Sn).

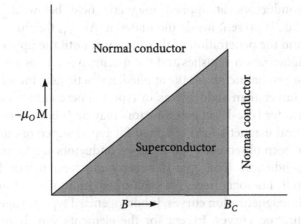

Figure 6.11 Magnetization (induced magnetic field) versus applied magnetic field for a type-I superconductor exhibiting a complete Meissner effect (perfect diamagnetism). Above the critical magnetic field B_c, the type-I superconductor behaves like a normal conductor. $-\mu_0 M$ is plotted along the vertical axis; the negative value of M corresponds to diamagnetism

6.9.2 Type-II superconductors

Materials which exhibit a magnetization curve similar to that shown in Fig. 6.12 are called type-II superconductors. Alloys and transition metals having high values of electrical resistivity fall under this category of superconductors. These superconductors have two critical fields–lower critical field B_{C1} and upper critical field B_{C2}.

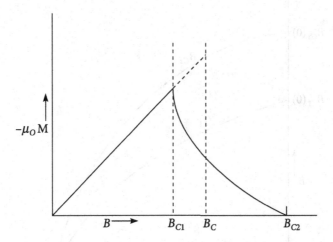

Figure 6.12 Superconducting magnetization (induced magnetic field) curve of type-II superconductors. The magnetic flux start to penetrate the specimen at a field B_{C1} less than the critical field B_c. The specimen is in vortex state between B_{C1} and B_{C2} and it has superconducting electrical properties up to B_{C2}. Above B_{C2}, the specimen is a normal conductor. For a given B_c, the area under the magnetization curve is the same for a type-II superconductor as for a type-I superconductor

For type-II superconductors, at applied magnetic field below B_{C1}, the specimen is diamagnetic as no flux is present inside the material. At B_{C1}, the flux begins to penetrate into the specimen and the penetration of flux increases until the upper critical field B_{C2} is reached. At B_{C2}, magnetization vanishes and the specimen becomes a normal conductor.

In this group of superconductors, as the applied magnetic field increases, magnetization vanishes gradually rather than suddenly as in type-I superconductors. The value of the critical magnetic field for type-II superconductors may be 100 times or more higher than the value of the critical magnetic field obtained for type-I superconductors. Critical field B_{C2} up to 30 T have been observed. Type-II superconductors are technically more useful than type-I superconductors. For type-II superconductors, metals like niobium and vanadium and carefully homogenized solid solutions of indium with lead and indium with tin exhibit reversible magnetization curves. Inhomogenized type-II superconductors show irreversible magnetization curves. Except for the elements vanadium, technetium, and niobium, type-II category of superconductors comprises metallic compounds and alloys like $Hg_{0.8}Tl_{0.2}Ba_2Ca_2Cu_3O_{8.33}$, $HgBa_2Ca_2Cu_3O_8$ $HgBa_2Ca_3Cu_4O_{10}{}^+$ $HgBa_2Ca_{1-x}Sr_xCu_2O_6{}^+$ and $HgBa_2CuO_4{}^+$.

A distinguishing characteristic of type-I and type-II superconductors is provided by a modification of the Meissner effect in type-II superconductors as shown in Fig. 6.12. This figure illustrates that superconductivity is only partially destroyed in type-II superconductors for $B_{C1} \leq B_C \leq B_{C2}$. The state of the specimen in this region is called the vortex state. The vortex state is really a mixture of the normal state and the superconducting state. In general, the vortex state is unstable for type-I superconductors where as it is stable for

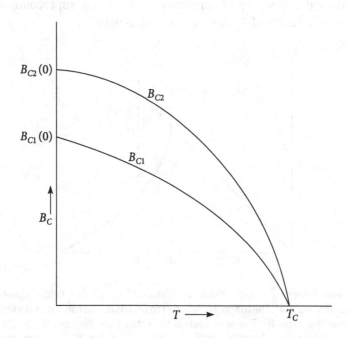

Figure 6.13 The variation of critical magnetic field with temperature for a type-II superconducting material

type-II superconductors. The variation of critical magnetic field with temperature for a type-II superconducting material is shown in Fig. 6.13.

6.10 Phase Diagram

The superconducting state is defined by three very important factors, namely, critical temperature T_C, critical field B_C and critical current density J_C. Each of these parameters is very much dependant on the other two parameters. Maintaining the superconducting state requires that both the magnetic field and the current density, as well as the temperature, remain below the critical values, all of which depend on the material. The phase diagram in Fig. 6.14 demonstrates the relationship between T_C, B_C and J_C. The highest values for B_C and J_C occur at 0 K, while the highest value for T_C occurs when B_C and J_C are zero. When considering all three parameters, the plot represents a critical surface. From this surface, and moving toward the origin, the material is superconducting. The material is normal in regions outside this surface. When electrons form Cooper pairs, they can share the same quantum wave function or energy state. This results in a lower energy state for the superconductor. T_C and B_C are values where it becomes favourable for the electron pairs to break apart. Current density larger than the critical value is forced to flow through a normal material. This flow through the normal material of the mixed state is connected with the motion of the magnetic field lines past pinning sites. For most practical applications, superconductors must be able to carry high currents and withstand high magnetic field without reverting to its normal state.

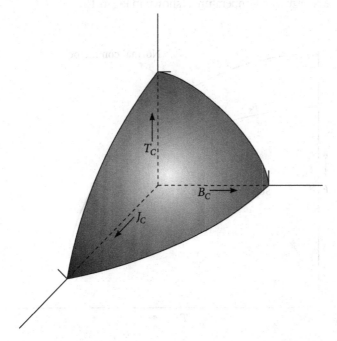

Figure 6.14 | Phase diagram of a superconductor. This is a 3D figure obtained by plotting critical temperature T_C, critical magnetic field B_C and critical current density J_C

Higher B_C and J_C values depend upon two important parameters which influence energy minimization, penetration depth and coherence length (distance upto which Cooper pairs moves without breaking). Penetration depth is the characteristic length of the decrease of a magnetic field due to surface currents.

6.11 Thermodynamic Properties of Superconductors

The thermal properties of superconductors such as entropy, specific heat and the like, change sharply as the temperature is lowered through the critical temperature.

The difference of the free energy between the normal state and the superconducting state in the absence of the magnetic field is given by

$$F_N(0) - F_s(0) = \frac{B_C^2}{2\mu_0} \tag{6.5}$$

Equation (6.5) gives the decrease in free energy when a material undergoes transition from the normal state to the superconducting state. Using this equation, we can calculate the change in entropy between normal state and superconducting state by using the fact

$$\frac{\partial}{\partial T}\left(F_N(0) - F_s(0)\right) = S_N - S_S.$$

The variation of free energy with temperature is shown in Fig. 6.15.

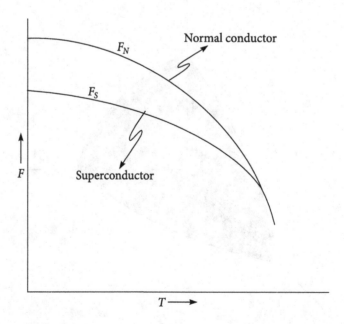

Figure 6.15 | Free energy as a function of temperature of a typical superconducting material in superconducting and normal state

6.11.1 Change in entropy

Entropy is a measure of disorderliness of a system. Differentiating Eq. (6.5) with respect to temperature, the change in entropy between normal state and superconducting state is obtained as

$$\frac{\partial}{\partial T}\left(F_N(0) - F_S(0)\right) = \frac{\partial}{\partial T}\left(\frac{B_C^2}{2\mu_0}\right)$$

or $$S_N - S_S = -\frac{B_C}{\mu_0}\left(\frac{\partial B_C}{\partial T}\right) = -\frac{B_C}{\mu_0} \times B_C(0)\left(-\frac{2T}{T_C^2}\right)$$

or $$S_N - S_S = \frac{2B_C(0)}{\mu_0 T_C^2}TB_C \tag{6.6}$$

Again putting the value of $B_C = B_{C0}\left(1 - \frac{T^2}{T_C^2}\right)$ from Eq. (6.2) into this equation, we get

$$S_N - S_S = \frac{2B_C(0)}{\mu_0 T_C^2}TB_C(0)\left(1 - \frac{T^2}{T_C^2}\right) = \frac{2B_C^2(0)}{\mu_0 T_C^2}\left(1 - \frac{T^2}{T_C^2}\right)T \tag{6.7}$$

Equation (6.7) shows that $S_N - S_S > 0$, i.e., entropy in the superconducting state is less than that of the normal state implying that the superconducting state is more ordered than the normal state. Using Eq. (6.7), we can calculate the change in specific heat between the normal state and the superconducting state by using the fact

$$T\frac{\partial}{\partial T}\left(S_N - S_S\right) = C_N - C_S.$$

The variation of entropy with temperature for both the normal and the superconductor state is shown in Fig. 6.16.

6.11.2 Change in specific heat

The change in specific heat between the normal state and the superconducting state is given by

$$C_N - C_S = T\frac{\partial}{\partial T}\left(S_N - S_S\right) \tag{6.8}$$

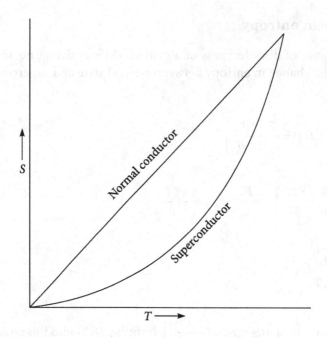

Figure 6.16 The variation of entropy with temperature for a typical superconducting material in the superconducting and normal state

or $\quad C_N - C_S = T\dfrac{\partial}{\partial T}\left(-\dfrac{B_C}{\mu_0}\dfrac{\partial B_C}{\partial T}\right) = -\dfrac{T}{\mu_0}\left[\left(\dfrac{\partial B_C}{\partial T}\right)^2 + B_C\left(\dfrac{\partial^2 B_C}{\partial T^2}\right)\right]$

According to Eq. (6.2), at $T = T_C$, $B_C = 0$. Therefore, the change in specific heat at transition temperature will be

$$C_N - C_S\big|_{T=T_C} = -\dfrac{T_C}{\mu_0}\left[\left(\dfrac{\partial B_C}{\partial T}\right)^2 + 0\times\left(\dfrac{\partial^2 B_C}{\partial T^2}\right)\right] = -\dfrac{T_C}{\mu_0}\left(\dfrac{\partial B_C}{\partial T}\right)^2$$

Putting the value of B_C from Eq. (6.2) into this equation, we get

$$C_S - C_N\big|_{T=T_C} = \dfrac{4B_C^2(0)}{\mu_0 T_C} > 0 \tag{6.9}$$

The positive value of $C_S - C_N\big|_{T=T_C}$ shows that the specific heat during the superconducting transition increases. The variation of the electronic specific heat of the superconductor with temperature at very low temperature is found to obey the following equation.

$$C_{Ve} = Ae^{-\frac{\alpha T_C}{T}} \tag{6.10}$$

Here A and α are constants. The variation of electronic specific heat with temperature is shown in Fig. 6.17.

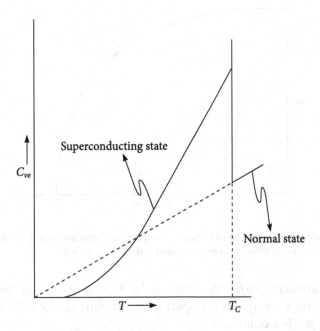

Figure 6.17 | Temperature dependence of the electronic specific heat of a typical superconducting material in the superconducting and the normal state

6.11.3 Thermal conductivity

For type-I superconductors, there is a marked drop in thermal conductivity during superconducting transition. This suggests that electronic contribution drops in and the superconducting electron plays no role in heat transfer.

6.11.4 Energy gap

The energy gap is a characteristic feature of all superconductors and determines their thermal properties as well as their response to high frequency electromagnetic fields. The variation of electronic specific heat of a superconductor with temperature at very low temperature is found to obey

$$C_{ve} = Ae^{-\frac{\alpha T_C}{T}}.$$

This suggests that an energy gap may exist in the superconducting electronic levels, separating the lowest excited state from the ground state. The existence of an energy gap in the superconducting electron levels has been confirmed by a number of experiments. Its variation with temperature is shown in Fig. 6.18.

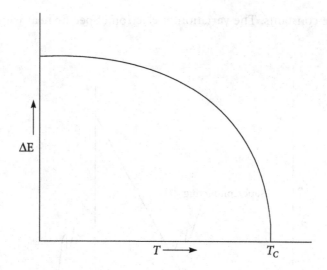

Figure 6.18 | Variation of energy gap of a superconductor with temperature. It increases from zero at critical temperature to a certain maximum value as $T \rightarrow 0$ K

The energy gap parameter plays a vital role in the BCS theory of superconductors. It is a measure of the energy needed to break apart a Cooper pair. The BCS theory estimates that at 0 K, the energy gap is around $3.53 \, kT_C$.

6.12 London Equations

The London theory is based on the rather old two-fluid model of superconductors. According to this theory, the superconductor can be thought of being composed of both normal electrons and superfluid electrons or superconducting electrons. The concentrations and speed of normal electrons and superfluid electrons are different in a superconductor. Let N_n = concentration of normal electrons in the superconductor, N_s = concentration of superfluid electrons or superconducting electrons in the superconductor, $\vec{v_n}$ = drift velocity of normal electrons in the superconductor, $\vec{v_s}$ = drift velocity of superfluid electrons in the superconductor.

If N_0 is the concentration of all types of electrons in the superconductor then we have

$$N_0 = N_n + N_s \tag{6.11}$$

The equation of motion of a superfluid electron under the action of an externally applied electric field $\vec{E}$ according to Newton's law is given by

$$m^* \frac{d\vec{v_s}}{dt} = -e\vec{E}, \quad (m^* = \text{effective mass of the electron}) \tag{6.12}$$

The current density vector due to superfluid electrons $\vec{J}_s$ as we know is given by

$$\vec{J}_s = -eN_s\vec{v}_s$$

From this equation, we have

$$\vec{v}_s = -\frac{\vec{J}_s}{eN_s}.$$

When this is put into Eq. (6.12), it yields

$$\frac{d\vec{J}_s}{dt} = \frac{N_s e^2}{m^*}\vec{E} \qquad (6.13)$$

This equation is called the first London equation.

Taking the curl of the first London equation, we have

$$\vec{\nabla}\times\frac{d\vec{J}_s}{dt} = \frac{N_s e^2}{m^*}\vec{\nabla}\times\vec{E} \qquad (6.14)$$

or $\qquad \dfrac{d}{dt}\left(\vec{\nabla}\times\vec{J}_s\right) = -\dfrac{N_s e^2}{m^*}\dfrac{\partial\vec{B}}{\partial t} \qquad (6.15)$

Integrating Eq. (6.15) with respect to time, and choosing the constant of integration to be zero so as to be in conformity with the Meissner effect, we have

$$\vec{\nabla}\times\vec{J}_s = -\frac{N_s e^2\vec{B}}{m^*} \qquad (6.16)$$

This equation is called the second London equation.

The second London equation in combination with Maxwell's equations can be used to explain the Meissner effect in the following manner.

One of Maxwell's equations is given by

$$\vec{\nabla}\times\vec{B} = \mu_0\vec{J}_s + \mu_0\varepsilon_0\frac{\partial\vec{E}}{\partial t} \qquad (6.17)$$

If the electric field does not vary with time, then $\dfrac{\partial\vec{E}}{\partial t}=0$ and under this condition, Eq. (6.17) boils down to

$$\vec{\nabla}\times\vec{B} = \mu_0\vec{J}_s \qquad (6.18)$$

Taking the curl of both sides of this equation, we get

$$\vec{\nabla} \times (\vec{\nabla} \times \vec{B}) = \mu_0 \vec{\nabla} \times \vec{J}_s$$

or $$\vec{\nabla}(\vec{\nabla} \cdot \vec{B}) - \nabla^2 \vec{B} = \mu_0 \vec{\nabla} \times \vec{J}_s$$

or $$-\nabla^2 \vec{B} = \mu_0 \vec{\nabla} \times \vec{J}_s,$$ (6.19)

Putting the second London equation into this equation, we have

$$\nabla^2 \vec{B} = \frac{\mu_0 N_s e^2}{m^*} \vec{B} = \frac{1}{\lambda^2} \vec{B}$$ (6.20)

where

$$\lambda = \left(\frac{m^*}{\mu_0 e^2 N_s} \right)^{\frac{1}{2}}$$ (6.21)

has the dimension of length and is called London's penetration depth. The value of the penetration depth λ is of the order of 1000 Å for a typical superconductor.

The magnitude form of Eq. (6.20) for a uniform magnetic field becomes

$$\nabla^2 B = \frac{1}{\lambda^2} B$$ (6.22)

If the uniform magnetic field is applied to the superconductor, only along one direction, say x direction, Eq. (6.22) becomes

$$\frac{d^2 B}{dx^2} = \frac{1}{\lambda^2} B$$

The solution of this second order differential equation gives

$$B(x) = C e^{-\frac{x}{\lambda}}, C = \text{integration constant.}$$ (6.23)

In Eq. (6.23), $B(x)$ is the value of magnetic induction at a depth of x beneath the surface. If B_0 is the magnetic induction at $x = 0$, i.e., on the surface of the superconductor, then this equation becomes

$$B(0) = C e^{-\frac{0}{\lambda}}$$

or $B(0) = B_0 = C \times 1$

Thus, $C = B_0$ and Eq. (6.23) becomes

$$B(x) = B_0 e^{-\frac{x}{\lambda}} \tag{6.24}$$

Now the penetration depth λ can be defined, using this equation, as the distance at which magnetic field B is reduced to $1/e$ of its initial value at the surface. Equation (6.24) shows that magnetic induction can penetrate only into a very thin layer of the superconductor and vanishes in the interior.

The penetration depth λ is also found to depend strongly on temperature of the superconductor and to become much larger as $T \to T_C$. This observation can be fitted extremely well by a simple expression of the form

$$\lambda(T) = \frac{\lambda(0)}{\sqrt{1 - \left(\dfrac{T}{T_C}\right)^4}} \tag{6.25}$$

In this equation

$\lambda(T)$ = penetration depth at temperature T K.

$\lambda(0)$ = penetration depth at temperature 0 K $= \left(\dfrac{m^*}{\mu_0 e^2 N_s}\right)^{\frac{1}{2}}$

Equation (6.25) implies that

$$N_s = N_0 \left[1 - \left(\frac{T}{T_C}\right)^4\right] \tag{6.26}$$

This equation shows that the concentration of superconducting electrons N_s increases from zero at T_C to N_0 at 0 K as shown in Fig. 6.19, which also depicts the temperature variation of penetration depth λ. N_s is called order parameter because it characterizes the order in the superconducting state.

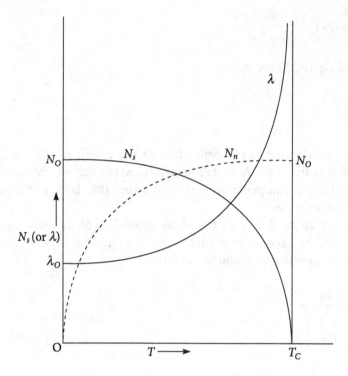

Figure 6.19 | Temperature variation of N_s, N_n, and λ

6.13 Applications of Superconductivity

Superconducting materials find applications in the area of medical magnetic-imaging devices, magnetic energy-storage systems, motors, generators, transformers, computer parts, and very sensitive devices for measuring magnetic fields (SQUID), voltages, or currents. The main advantages of devices made from superconductors are low power dissipation, high-speed operation and high sensitivity.

6.13.1 Transportation

i.   Magnetic-levitation is an application where superconductors perform extremely well. Transport vehicles such as trains can be made to "float" on strong superconducting magnets, virtually eliminating friction between the train and its tracks. Not only would conventional electromagnets waste much of the electrical energy as heat, they would have to be physically much larger than superconducting magnets. Recently, in 2015 April, Japan's maglev train broke the world speed record with 603 km/h in a test run.

6.13.2 Medical

i.   An area where superconductors can perform a life-saving function is in the field of biomagnetism. Doctors need a non-invasive means of determining what is going on inside the human body. By impinging a strong superconductor-derived magnetic field into the body, hydrogen atoms that exist in the body's water and fat molecules are forced to accept energy from the magnetic field. They then release this energy at a frequency that can be detected and displayed graphically by a computer.

ii.   The double-relaxation oscillation SQUID (superconducting quantum interference device) has been developed for use in magnetoencephalography. SQUIDs are capable of sensing a change in a magnetic field over a billion times weaker than the force that moves the needle on a compass. With this technology, the body can be probed to certain depths without the need for the strong magnetic fields associated with MRI's.

6.13.3 Fundamental research

i.   The Josephson effect in superconductivity resulted in an upward revision of Planck's constant from 6.62559×10^{-34} to 6.626196×10^{-34}

ii.   Superconductivity has become an essential tool in research work relating to elementary particles which will ultimately lead to the door of creation of the universe. High-energy particle research hinges on being able to accelerate sub-atomic particles to nearly the speed of light. Superconductor magnets make this possible.

6.13.4 Power systems

i.   Electric generators made with superconducting wire are far more efficient than conventional generators wound with copper wire. In fact, their efficiency is above 99% and their size about half that of conventional generators. These facts make them very lucrative ventures for power utilities.

ii.   Other commercial power projects in the works that employ superconductor technology include energy storage to enhance power stability which can provide instantaneous reactive power support.

iii. Recently, power utilities have also begun to use superconductor-based transformers and "fault limiters". The superconducting transformer can be connected to utility power network. In future, the MVA fault current limiter can be constructed with superconducting materials and used most effectively due to their ability to respond in just thousandths of a second to limit tens of thousands of amperes of current. Similarly, underground copper power cables can be replaced with superconducting cable, cooled with liquid nitrogen. By doing this, more current can be routed through existing cable tunnels. For example, 114 kg of superconducting wire can replace 8200 kg of vintage copper wire, making it over 7000% more space-efficient.

6.13.5 Computers

i.   The petaflop computer, a new age technology, has been envisioned and here superconductors play the vital role. A petaflop computer can do one quadrillion (10^{15}) floating point operations per second. It has been conjectured that devices on the order of 50 nanometers in size along with unconventional switching mechanisms, such as the Josephson junctions associated with superconductors, will be necessary to achieve such blistering speeds. These Josephson junctions would be incorporated into field effect transistors which then become part of the logic circuits within the processors. The tiny magnetic fields that penetrate type-II superconductors can be used for storing and retrieving digital information. The computers of the future will be built around superconducting devices. Competing technologies, such as quantum transistors, high-density molecular processor and DNA-based processing also have the potential to achieve petaflop benchmarks.

6.13.6 Electronics

ii.   In the electronics industry, ultra-high-performance filters are now being built by superconductors. Since superconducting wires have near zero resistance, even at high frequencies, many more filter stages can be employed to achieve a desired frequency response. This translates into an ability to pass desired frequencies and block undesirable frequencies in high-congestion radio frequency applications such as mobile telephone systems.

6.13.7 Military

i.   Superconductors have also found widespread applications in the military. SQUIDS are being used to detect mines and submarines. Significantly, smaller motors are being built for ships using superconducting wire and tape. 5000 horse power motor has been made with superconducting wire.

ii.   The military is also looking at using superconductive tape as a means of reducing the length of very low frequency antennas employed on submarines. Normally, the lower the frequency, the longer an antenna must be. However, inserting a coil of wire ahead of the antenna will make it function as if it were much longer. However, this loading coil will increase the resistance in the coil. Using superconductive materials can significantly reduce losses in this coil. The construction of superconducting microwave antenna has already been successful. The superconducting carbon nanotubes might be an ideal nano-antenna for high-gigahertz and terahertz frequencies applications.

iii. The most ignominious military use of superconductors may come with the deployment of "E-bombs". These are devices that make use of strong, superconductor-derived magnetic fields to create a fast, high-intensity electro-magnetic pulse to disable an enemy's electronic equipment. Such a device saw its first use in wartime, in March 2003, when US Forces attacked an Iraqi broadcast centre.

6.13.8 Space research

i.   Earth-orbiting satellites employ the "flux-pinning" properties of imperfect superconductors in gyroscopes to reduce friction to near zero. Superconducting X-ray detectors and ultra-fast, superconducting light detectors are being developed due to their inherent ability to detect extremely weak amounts of energy. Optical cameras of phenomenal sensitivity can be constructed with superconducting materials.

6.13.9 Internet

i.   Superconductors may even play a role in Internet communications soon. Superconductivity can be used to develop a superconducting digital router for high-speed data communications. Since Internet traffic is increasing exponentially, superconductor technology is being called upon to meet this super need.

6.13.10 Pollution control

i.   Another impetus to the wider use of superconductors is in pollution control. The reduction of green-house gas can be accomplished by the use of superconducting devices in different manufacturing units. The heavy reduction of carbon dioxide emissions in power plants can be accomplished by the use of high-temperature superconductors.

6.13.11 Refrigeration

i.   The future melding of superconductors into our daily lives will also depend to a great degree on advancements in the field of cryogenic cooling. New, high-efficiency magnetocaloric-effect compounds such as gadolinium–silicon–germanium are expected to enter the marketplace soon. Such materials should make possible compact, refrigeration units that facilitate additional high temperature superconductor applications.

The market demand for superconductor products is increasing day by day. Low-temperature superconductors are expected to play a dominant role in well-established fields such as MRI and scientific research, with high-temperature superconductors enabling the newer industries. All of this is, of course, contingent upon a linear growth rate. Should new superconductors with higher transition temperatures be discovered, growth and development in this exciting field could explode virtually overnight.

Questions

6.1   What do you mean by normal conductors?

6.2   What do you mean by superconductors?

6.3   What do you mean by critical temperature in superconductivity?

6.4   What do you mean by critical magnetic field in superconductivity?

6.5   What are type-I superconductors?

6.6   Explain why type-I superconductors cannot be used as coils of strong electromagnets.

6.7   What do you mean by critical current?

6.8   What is the reason behind the existence of critical current?

6.9   What do you mean by critical current density?

6.10   Prove the expression for critical current density from the expression for critical current.

6.11   What is the Meissner effect?

6.12   How does the Meissner effect occur?

6.13   Show that the electric field inside a super conductor is zero

6.14   Show that superconductors exhibit perfect diamagnetism.

6.15   What is the Josephson effect?

6.16   What is a Josephson junction?

6.17   Give few applications of the Josephson effect.

6.18   What is the role of phonons in superconductivity?

6.19   What are Cooper pairs and how are they formed?

6.20   How are Cooper pairs kept together?

6.21   Explain how phonons act as moderators in superconductivity.

6.22   Explain on the basis of the BCS theory how superconductivity is affected by temperatures.

6.23   Can very high currents be made to flow in superconducting wires? (There is no resistance at all!)

6.24   Why are superconducting wires well suited for making strong electromagnets?

6.25   What is flux pinning?

6.26   Distinguish between superconductors and normal conductors.

6.27   Distinguish between flow of electrons in superconductors and normal conductors.

6.28   Plot the voltage versus current graph in superconductors and normal conductors.

6.29   Show the variation of resistance versus the temperature of a superconductor and a normal conductor.

6.30   What are the properties of type-I superconductors?

6.31   Why are type-I superconductors called soft superconductors?

6.32   Why are type-I superconductors unsuitable for use as high field superconducting materials?

6.33   What are type-II superconductors?

6.34   What are the properties of type-II superconductors?

6.35   Why are type-II superconductors called hard superconductors?

6.36   Explain the penetration of magnetic flux in type-II superconductors.

6.37 Why are type-I superconductors unsuitable for use as high field superconducting materials?

6.38 Distinguish between type-I and type-II superconductors.

6.39 What is the basis of the BCS theory?

6.40 Explain the BCS theory of superconductivity.

6.41 What is the vortex state of a superconductor?

6.42 What is phase diagram in superconductivity Accordingly, change the serial number of rest of questions.?

6.43 Define critical surface in superconductivity.

6.44 Explain the phase diagram in superconductivity.

6.45 Outline the use of superconductors in the area of medicine.

6.46 Explain magnetic resonance imaging (MRI).

6.47 Explain how SQUIDs are used in medical imaging.

6.48 Explain the benefits and limitations of using superconductors in medicine.

6.49 Outline the use of superconductors in the area of transportation.

6.50 Explain Maglev trains.

6.51 Explain the benefits and limitations of Maglev trains using superconductors.

6.52 Outline the proposed uses of superconductors in the area of transportation.

6.53 Outline the use of superconductors in the area of computer development.

6.54 Outline the proposed uses of superconductors in the area of computer development.

6.55 Outline the use of superconductors in the area of power generation.

6.56 Outline the use of superconductors in the area of power distribution.

6.57 Outline the use of superconductors in the area of power storage.

6.58 Outline the proposed uses of superconductors in the area of power generation and storage.

6.59 Outline the use of superconductors in mining.

6.60 Outline the use of superconductors in space research.

6.61 Outline the use of superconductors in the area of communication.

6.62 Discuss the possible applications of superconductors in areas including computers, generators and motors, transmission of electricity through power grids.

6.63 Summarize the possible applications of superconductors in areas such as medicine, mining, geoscience and space research.

6.64 Explain how when a material undergoes transition from the normal state to the superconducting state, entropy of the material decreases.

6.65 Explain how when a material undergoes transition from the normal state to the superconducting state, specific heat of the material increases.

6.66 Explain the Meissner effect by using the London equation.

6.67 Prove that applied magnetic field decreases exponentially in a superconductor.

6.68   Define the London penetration depth? Derive an expression for it.

6.69   How does London's penetration depth vary with temperature?

6.70   How does concentration of superconducting electrons varies with temperature?

Problems

6.1   The critical temperature for a superconducting specimen with isotopic mass 196.5 is 4.18 K. Calculate its critical temperature when its isotopic mass changes to 203.4.
        [Ans   4.07 K]

6.2   The critical temperature for mercury with mass number 202 is 4.153 K. Calculate its critical temperature when its mass number changes to 200.   [Ans   4.174 K]

6.3   What is the resistance of a superconductor in the normal state if 400 milliamps of current are passing through the sample and 4.0 millivolts are measured across the voltage probes?   [Ans   0.010 W]

6.4   Calculate the critical magnetic field for aluminum at 1.0 K. The given data for aluminum are $T_c = 1.18$ K and $B_c = 10.5 \times 10^{-3}$ T at 0 K.   [Ans 2.96×10^{-3} T]

6.5   Find the critical current and critical current density for a lead wire of radius 5 mm at 4 K. The data given for lead are $T_c = 7.193$ K and $B_c = 80.3 \times 10^{-3}$ T at 0 K.
        [Ans   1387.5 A, 1.77×10^7 A/m²]

6.6   For a certain sample, critical fields are 4.8×10^5 A/m at 16 K and 7.2×10^5 A/m at 14 K. Calculate the transition temperature and critical field at 0 K and 8 K.
        [Ans 15×10^5 A/m, 12.48×10^5 A/m]

6.7   To what temperature must lead be cooled to be a superconductor in a magnetic field of 0.025 Tesla.   [Ans   5.96 K]

6.8   The critical temperature of lead is 7.193 K. Calculate the energy gap for this element assuming that the BCS theory holds good.   [Ans 2.1×10^{-3} eV]

6.9   The London penetration depth for lead at 3 K is 396 Å and at 7.1 K 1730 Å. Calculate its critical temperature and penetration depth at 0 K.   [Ans   7.193 K, 396 Å]

6.10   The superconducting critical temperature, density and standard atomic weight of tin are respectively 5.7 K, 7.3 g/cm³ and 118.7. Calculate the London penetration depth for it.   [Ans   291Å]

Multiple Choice Questions

1.   In general, elements which are good conductors of electricity are also superconductors.

        (i)   True   (ii)   False

2.   In superconductors, electric current can flow even in the absence of applied voltage.

        (i)   True   (ii)   False

3. Superconductivity was first discovered in
    (i) Mercury   (ii) Gold
    (iii) Platinum   (iv) Palladium

4. If the temperature of a superconductor is lowered below its critical temperature, the superconductor becomes
    (i) Diamagnetic   (ii) Paramagnetic
    (iii) Ferromagnetic   (iv) Ferrimagnetic

5. Which of the following superconducting elements has the highest critical temperature?
    (i) Indium   (ii) Lead
    (iii) Lanthanum   (iv) Niobium

6. The critical magnetic field of a superconductor
    (i) Increases with increasing temperature
    (ii) Decreases with increasing temperature
    (iii) Varies linearly with temperature
    (iv) Is independent of temperature

7. The Meissner effect implies magnetic susceptibility of a superconductor
    (i) Lies between –1 and 0   (ii) Lies between 1 and 0
    (iii) Is equal to –1   (iv) Is equal to 1

8. The most suitable materials for making coils of strong electromagnets are
    (i) Elemental superconductors
    (ii) Type-II superconductors
    (iii) Type-I superconductors
    (iv) Both type-I and type-II superconductors

9. The highest critical magnetic field for type-I materials is of the order of
    (i) 10^1 T   (ii) 10^0 T
    (iii) 10^{-1} T   (iv) 10^{-2} T

10. The two electrons of a Cooper pair have
    (i) Same energy but different momentum values
    (ii) Same energy but equal and opposite momentum values
    (iii) Different energy but same momentum values
    (iv) Same energy and same momentum values

11. Which of the following is correct?
    (i) $B_c < B_{c1} < B_{c2}$ (ii) $B_{c2} < B_{c1} < B_c$
    (iii) $B_{c1} < B_{c2} < B_c$ (iv) $B_{c1} < B_c < B_{c2}$

12. A Josephson junction emits microwaves when
    (i) The two superconductors are shorted
    (ii) An alternating current is passed through the junction

(iii) A direct current is passed through the junction

(iv) A photon is incident on the junction

13. Penetration depth is the distance at which the externally applied magnetic field reduces to

(i) 7% of the initial value (ii) 3% of the initial value

(iii) 37% of the initial value (iv) None of the above

14. When a material undergoes transition from the normal state to the superconducting state, free energy of the material

(i) Increases (ii) Decreases

(iii) Remains constant (iv) First increases then decreases

15. When a material undergoes transition from the normal state to the superconducting state, entropy of the material

(i) Increases (ii) Decreases

(iii) Remains constant (iv) First increases then decreases

16. When a material undergoes transition from the normal state to the superconducting state, specific heat of the material

(i) Increases (ii) Decreases

(iii) Remains constant (iv) First increases then decreases

17. When a material undergoes transition from the normal state to the superconducting state, thermal conductivity of the material

(i) Increases (ii) Decreases

(iii) Remains constant (iv) First increases then decreases

18. The energy gap in a superconductor is maximum at

(i) 0 K (ii) $3T_c$

(iii) $2T_c$ (iv) T_c

19. The energy gap in a superconductor is zero at

(i) 0 K (ii) $3T_c$

(iii) $2T_c$ (iv) T_c

20. Does London's penetration depth vary with temperature?

(i) yes (ii) no

Answers

1 (ii)	2 (i)	3 (i)	4 (i)	5 (iv)	6 (ii)	7 (iii)	8 (ii)
9 (iii)	10 (ii)	11 (iv)	12 (iii)	13 (iii)	14 (ii)	15 (ii)	16 (i)
17 (ii)	18 (iv)	19 (i)	20 (i)				

7.1 Introduction

The response of a material to electromagnetic radiation particularly in the visible range of 7×10^{-5} cm to 4×10^{-5} cm of wavelengths is called the optical property of the material. The optical characteristics of a material are determined solely by the interaction of the incident radiation with the electrons of the atoms constituting the material. Transparency and opaqueness of materials depend upon the frequency of the incident radiation. Reflection or absorption of radiation also depends upon the frequencies of the incident radiation. Most materials are transparent to X-ray radiations while they are opaque to visible radiation.

When light is incident on materials, any one or more of the following phenomena may occur:

i.   Scattering
ii.   Reflection
iii.   Refraction
iv.   Transmission
v.   Absorption
vi.   Luminescence

The knowledge of the mechanism of these phenomena is essential to understand the optical properties of materials. This knowledge is exploited in making window glasses, lenses, mirrors, anti-reflection coatings, lasers, optical fibers, photo diodes, optical memories, electro-optic modulators and many other optical devices.

7.2 Scattering

Light is a transverse electromagnetic wave in which electric and magnetic vectors, mutually perpendicular to each other, are also perpendicular to the propagation vector, i.e., to the direction of propagation. The magnitudes and directions of the electric and magnetic vectors change continuously in the light wave as it advances. The vibrating electric vector in the light wave is responsible for almost all optical phenomena. It is therefore called the light vector.

Suppose light is incident on a material. The molecules or atoms in the material are neutral containing equal amount of opposite electrical charges. The vibrating light vector of the light wave will exert a force on the positive charge of the molecules in the direction of the field and a force on the negative charge of the molecules in the opposite direction. The direction of the forces on the positive and negative charges changes rapidly since the electric vector in the light wave is vibrating. Since the charges in the molecules are not rigidly bound to them, vibrating electric vectors of the light wave can make the charges oscillate. The frequency of the oscillation of charges and the vibration frequency of light vectors of the light wave is the same. Thus, the vibrating electric vectors of the light wave produce accelerations of the charges. According to the electromagnetic theory, accelerated charges emit radiations in all direction which form scattered light. The scattered light is strongest in directions perpendicular to the axis of the dipole, which is the same as the direction of the electric vector in the incident wave. It is polarized with its electric vector in the plane containing the dipole. Almost all objects scatter radiation this is what makes them visible.

7.2.1 Applications

The colours of the sky (blue), the colours of the horizon during sun rise and sun set (red), colours of fog or clouds (white) is the result of scattering of sun rays by air molecules.

When light strikes fine particles or an irregular surface, it is reflected in all directions and is said to be scattered. When the scattering particles are very small compared to the wavelength of light, the intensity of the scattered light is related to that of the incident light by the inverse fourth power of the wavelength ($1/\lambda^4$) (Rayleigh scattering). As a result, light at the blue end of the spectrum is scattered much more intensely than that at the red end. The light from the sun is scattered by dust particles and clusters of gas molecules, and the scattered blue rays seen against the dark background of outer space cause the sky to appear blue. At sunrise and sunset, when sunlight travels the farthest, almost all of the blue rays are scattered and the light that reaches the earth directly is seen as predominantly red or orange. Scattering also causes that epitome of rare occurrences, the blue moon (seen when forest fires produce clouds composed of small droplets of organic compounds). If the size of the scattering particles approaches the wavelength of light or exceeds it, the complex Mie scattering theory applies and explains colours other than blue; white is scattered at the largest sizes, as in fog and clouds.

Raman and Rayleigh scattering occur when the dimensions of the scattering particles are less than 5 per cent of the wavelength of the incident radiation. Tyndall scattering occurs when the dimensions of the particles that are causing the scattering are larger than the wavelength of the scattered radiation. It is caused by the reflection of the incident radiation from the surfaces of the particles, reflection from the interior walls of the particles, and refraction and diffraction of the radiation as it passes through the particles.

7.3 Reflection

When light is incident on the surface of a material, a part of it is reflected. The reflectivity or reflection coefficient R of a material is defined as the fraction of the light intensity reflected. Mathematically, reflectivity R of a material is defined as the ratio of the intensity of the light reflected I_R to the intensity of the incident light I_0, i.e.,

$$R = \frac{I_R}{I_0} \tag{7.1}$$

7.3.1 Reflection by a dielectric surface

The reflection coefficient of a dielectric surface is given by

$$R = \left(\frac{\cos\theta_I - \dfrac{n_2}{n_1}\cos\theta_T}{\cos\theta_I + \dfrac{n_2}{n_1}\cos\theta_T} \right)^2$$

when the electric vector of the incident light wave is normal to the plane of incidence and

$$R = \left(\frac{\cos\theta_I - \dfrac{n_1}{n_2}\cos\theta_T}{\cos\theta_I + \dfrac{n_1}{n_2}\cos\theta_T} \right)^2$$

when the electric vector of the incident light wave is parallel to the plane of incidence. Here, θ_I and θ_T are the angle of incidence and the angle of refraction respectively. For normal incidence of light wave $\theta_I = 0 = \theta_T$, the reflection coefficient R of the dielectric material is given by

$$R = \left| \frac{\dfrac{n_2}{n_1} - 1}{\dfrac{n_2}{n_1} + 1} \right|^2 = \left| \frac{n-1}{n+1} \right|^2 \tag{7.2}$$

[Details of these equations have been dealt in Chapter 6, Part-I] where $n = (n_2/n_1)$ is the refractive index of the second medium with respect to first medium.] The refractive index may be complex. In terms of a complex refractive index $n_C = n + ib$, we can write Eq. (7.2) as

$$R = \left| \frac{n_C - 1}{n_C + 1} \right|^2 = \left(\frac{n_C - 1}{n_C + 1} \right) \times \left(\frac{n_C - 1}{n_C + 1} \right)^* \tag{7.3}$$

$$= \frac{(n_C - 1)(n_C - 1)^*}{(n_C + 1)(n_C + 1)^*}$$

Putting $n_C = n + ib$ into this equation, we get

$$R = \frac{(n + ib - 1)(n + ib - 1)^*}{(n + ib + 1)(n + ib + 1)^*} = \frac{(n - 1 + ib)(n - 1 - ib)}{(n + 1 + ib)(n + 1 - ib)}$$

$$= \frac{(n - 1)^2 + b^2}{(n + 1)^2 + b^2} = \frac{(n - 1)^2 + 4n + b^2 - 4n}{(n + 1)^2 + b^2} = \frac{(n + 1)^2 + b^2 - 4n}{(n + 1)^2 + b^2}$$

or $\quad R = 1 - \dfrac{4n}{(n + 1)^2 + b^2}$ $\hfill (7.4)$

In Eq. (7.4), n is the real part and b is the imaginary part of the complex refractive index.

Reflectivity in terms of the dielectric constant

The relation between refractive index and dielectric constant for a dielectric material is given by

$$n_C^2 = \varepsilon_{rC} \tag{7.5}$$

Putting $n_C = n + ib$ into Eq. (7.5), we get

$$\varepsilon_{rC} = (n + ib)^2 = n^2 - b^2 + 2inb$$

Since the dielectric constant is complex, it can be expressed in the form $\varepsilon_{rC} = \varepsilon_r + i\varepsilon_{ri}$ with ε_r and ε_{ri} being the real part and the imaginary part of the complex dielectric constant. Upon substitution, the aforementioned equation becomes

$$\varepsilon_r + i\varepsilon_{ri} = n^2 - b^2 + 2inb \tag{7.6}$$

Thus, we have

$$\varepsilon_r = n^2 - b^2; \text{ real part of } \varepsilon_{rC} \tag{7.7}$$

$$\varepsilon_{ri} = 2nb; \text{ imaginary part of } \varepsilon_{rC} \tag{7.8}$$

Thus, $\varepsilon_r^2 + \varepsilon_{ri}^2 = \left(n^2 - b^2\right)^2 + 4n^2 b^2 = \left(n^2 + b^2\right)^2$

or $\quad n^2 + b^2 = \sqrt{\varepsilon_r^2 + \varepsilon_{ri}^2} \tag{7.9}$

$$2n = \sqrt{4n^2} = \sqrt{2\left(n^2 + b^2 + n^2 - b^2\right)} = \sqrt{2\left(\sqrt{\varepsilon_r^2 + \varepsilon_{ri}^2} + \varepsilon_r\right)} \tag{7.10}$$

From Eq. (7.4), we can have

$$R = \frac{n^2 + b^2 + 1 - 2n}{n^2 + b^2 + 1 + 2n}$$

Putting Eqs (7.9) and (7.10) into this equation, we have

$$R = \frac{1 + \sqrt{\varepsilon_r^2 + \varepsilon_{ri}^2} - \sqrt{2\left(\sqrt{\varepsilon_r^2 + \varepsilon_{ri}^2} + \varepsilon_r\right)}}{1 + \sqrt{\varepsilon_r^2 + \varepsilon_{ri}^2} + \sqrt{2\left(\sqrt{\varepsilon_r^2 + \varepsilon_{ri}^2} + \varepsilon_r\right)}} \tag{7.11}$$

7.3.2 Reflection by a metallic surface

The reflection coefficient R at a metal surface when the electric vector of the light wave is normal to the plane of incidence is given by

$$R = \left(\frac{\cos\theta_I - (1+i)\sqrt{\dfrac{\mu_1 \sigma_2}{2\omega\mu_2\varepsilon_1}}}{\cos\theta_I + (1+i)\sqrt{\dfrac{\mu_1 \sigma_2}{2\omega\mu_2\varepsilon_1}}}\right) \times \left(\frac{\cos\theta_I - (1+i)\sqrt{\dfrac{\mu_1 \sigma_2}{2\omega\mu_2\varepsilon_1}}}{\cos\theta_I + (1+i)\sqrt{\dfrac{\mu_1 \sigma_2}{2\omega\mu_2\varepsilon_1}}}\right)^*$$

where* stands for the complex conjugate of the quantity.

Here

ω = angular frequency of the incident light wave.

μ_1 and μ_2 = the magnetic permeability of the first and the second medium.

ε_1 = permittivity of the first medium.

σ_2 = conductivity of the second medium.

On simplification of this equation, we get

$$R \approx 1 - 4\cos\theta_I \sqrt{\frac{\omega\mu_2\varepsilon_1}{2\mu_1\sigma_2}} \tag{7.12}$$

For a good conductor, since conductivity σ is very high, the second term of the LHS of Eq. (7.12) becomes negligibly small. Hence, for a good conductor

$$R \approx 1 \tag{7.13}$$

For normal incidence of the electromagnetic wave, for which $\theta_I = 0 = \theta_R$, Eq. (7.12) boils down to

$$R \approx 1 - 4\sqrt{\frac{\omega\mu_2\varepsilon_1}{2\mu_1\sigma_2}} \tag{7.14}$$

The reflection coefficient R at the metal surface when the electric vector of the incident light wave is parallel to the plane of incidence is given by

$$R = \frac{\left(\cos\theta_I - \cos\theta_T \dfrac{1-i}{2}\sqrt{\dfrac{2\omega\mu_2\varepsilon_1}{\mu_1\sigma_2}}\right)}{\cos\theta_I + \cos\theta_T \dfrac{1-i}{2}\sqrt{\dfrac{2\omega\mu_2\varepsilon_1}{\mu_1\sigma_2}}} \times \frac{\left(\cos\theta_I - \cos\theta_T \dfrac{1-i}{2}\sqrt{\dfrac{2\omega\mu_2\varepsilon_1}{\mu_1\sigma_2}}\right)^{*}}{\cos\theta_I + \cos\theta_T \dfrac{1-i}{2}\sqrt{\dfrac{2\omega\mu_2\varepsilon_1}{\mu_1\sigma_2}}} \tag{7.15}$$

where* stands for the complex conjugate of the quantity.
  On simplification of this equation, we get

$$R \approx 1 - \frac{4}{\cos\theta_I}\sqrt{\frac{\omega\mu_2\varepsilon_1}{2\mu_1\sigma_2}} \tag{7.16}$$

For normal incidence of the electromagnetic wave for which $\theta_I = 0 = \theta_R$, Eq. (7.16) boils down to

$$R \approx 1 - 4\sqrt{\frac{\omega\mu_2\varepsilon_1}{2\mu_1\sigma_2}} \qquad (7.17)$$

For a good conductor, since conductivity σ is very high, the second term of the LHS of Eq. (7.17) becomes negligibly small. Hence, for a good conductor

$$R \approx 1 \qquad (7.18)$$

For any angle of reflection, in case of good conductors, the reflection coefficient is very close to unity $[R \approx 1]$. That means almost the entire light energy incident on a metallic surface is reflected back. Therefore, metals are opaque to light. The extremely small amount of electromagnetic energy that flows into the metal is dissipated rapidly by heat loss associated with eddy currents. The skin depths (explained later in Section 7.5.1) of good conducting materials are very small implying that they absorb electromagnetic energy very strongly. Hence, we conclude that good conductors are good reflectors as well as good absorbers.

7.4 Refraction

When light is incident on the surface of a material, a part of it is refracted. The refractivity or refraction coefficient T of a material is defined as the fraction of the light intensity refracted. Mathematically, refractivity T of a material is defined as the ratio of the intensity of the light refracted I_T to the intensity of the incident light I_0, i.e.,

$$T = \frac{I_T}{I_0} \qquad (7.19)$$

7.4.1 Refraction by a dielectric surface

The refraction coefficient T at a dielectric surface when the electric vector of the light wave is normal to the plane of incidence is given by

$$T = \left(\frac{2\cos\theta_I}{\cos\theta_I + \dfrac{n_2}{n_1}\cos\theta_T}\right)^2 \times \frac{n_2}{n_1} \times \frac{\cos\theta_T}{\cos\theta_R} = \frac{4\dfrac{n_1}{n_2}\cos\theta_I\cos\theta_T}{\left[\dfrac{n_1}{n_2}\cos\theta_I + \cos\theta_T\right]^2} \qquad (7.20)$$

For normal incidence of the light wave for which $\theta_I = 0 = \theta_R$, this equation becomes

$$T = \frac{4\dfrac{n_1}{n_2}}{\left[\dfrac{n_1}{n_2}+1\right]^2} \tag{7.21}$$

The refraction coefficient T at the dielectric surface media when the electric vector of the light wave is parallel to the plane of incidence will be given by

$$T = \left(\frac{2\cos\theta_I}{\cos\theta_T + \dfrac{n_2}{n_1}\cos\theta_I}\right)^2 \times \frac{n_2}{n_1} \times \frac{\cos\theta_T}{\cos\theta_R} = \frac{4\dfrac{n_1}{n_2}\cos\theta_I\cos\theta_T}{\left[\dfrac{n_1}{n_2}\cos\theta_T + \cos\theta_I\right]^2} \tag{7.22}$$

For normal incidence of the light wave for which $\theta_I = 0 = \theta_R$, this equation becomes

$$T = \frac{4\dfrac{n_1}{n_2}}{\left[\dfrac{n_1}{n_2}+1\right]^2} \tag{7.23}$$

7.4.2 Refraction by a metallic surface

The refraction coefficient T at a metallic surface when the electric vector of the light wave is normal to the plane of incidence is given approximately by

$$T \approx 4\cos\theta_I \sqrt{\frac{\omega\mu_2\varepsilon_1}{2\mu_1\sigma_2}} \tag{7.24}$$

For a good conductor, since conductivity σ is very high, the second term of the LHS of this equation becomes negligibly small. Hence, for a good conductor

$$T \approx 0 \tag{7.25}$$

For normal incidence of the light wave for which $\theta_I = 0 = \theta_R$, Eq. (7.24) boils down to

$$T \approx 4\sqrt{\frac{\omega\mu_2\varepsilon_1}{2\mu_1\sigma_2}} \tag{7.26}$$

The refraction coefficient T at the metallic surface when the electric vector of the light wave is parallel to the plane of incidence will be given by

$$T \approx \frac{4}{\cos\theta_I}\sqrt{\frac{\omega\mu_2\varepsilon_1}{2\mu_1\sigma_2}} \qquad (7.27)$$

For normal incidence of the light wave for which $\theta_I = 0 = \theta_R$, Eq. (7.27) boils down to

$$T \approx 4\sqrt{\frac{\omega\mu_2\varepsilon_1}{2\mu_1\sigma_2}} \qquad (7.28)$$

For a good conductor, since conductivity σ is very high, the RHS of this equation becomes negligibly small. Hence, for a good conductor

$$R \approx 0 \qquad (7.29)$$

7.5 Absorption

In transmission through matter, the intensity of light decreases exponentially with distance; in effect, the fractional loss is the same for equal distances of penetration. The energy loss from the light appears as energy added to the medium, or what is known as absorption. A medium can be weakly absorbing at one region of the electromagnetic spectrum and strongly absorbing at another. If a medium is weakly absorbing, then it is dispersion [the variation of refractive index with frequency is called dispersion] and the absorption can be measured directly from the intensity of refracted, i.e., transmitted light. If it is strongly absorbing, on the other hand, the light does not survive even a few wavelengths of penetration. The refracted light is then so weak that measurements are at best difficult. Absorption and dispersion in such cases, nevertheless, may still be determined by studying only the reflected light. In the far ultraviolet region, it is the only practical means of studying absorption – a study that has revealed valuable information about electronic energy levels and collective energy losses in condensed material.

7.5.1 Macroscopic theory of absorption

The electromagnetic wave equation for the electric field vector in a conducting medium is given by

$$\nabla^2 E - \mu\varepsilon\frac{\partial^2 E}{\partial t^2} - \mu\sigma\frac{\partial E}{\partial t} = 0 \qquad (7.30)$$

The origin of the term

$$\mu\varepsilon\frac{\partial^2 E}{\partial t^2}$$

lies in the displacement current while the origin of the term

$$\mu\sigma\frac{\partial E}{\partial t}$$

lies in the conduction current. In almost all conducting media, the conduction current dominates the displacement current. Therefore, for a good conducting medium, Eq. (7.30) can be written as

$$\nabla^2 E - \mu\sigma\frac{\partial E}{\partial t} = 0 \tag{7.31}$$

The solution of this equation may be given as

$$E(r,t) = E_0 e^{-\beta r} e^{i(\alpha r - \omega r)} \tag{7.32}$$

The magnitude of the propagation vector k in a conducting medium is of the form

$$k = \alpha + i\beta \tag{7.33}$$

where $\alpha = \sqrt{\dfrac{\mu\sigma\omega}{2}}$ and $\beta = \sqrt{\dfrac{\mu\sigma\omega}{2}}$

Let $\delta = \dfrac{1}{\alpha} = \dfrac{1}{\beta} = \sqrt{\dfrac{2}{\mu\sigma\omega}}$ $\tag{7.34}$

Equation (7.32) becomes

$$E(r,t) = E_0 e^{-\frac{r}{\delta}} e^{i\left(\frac{r}{\delta} - \omega \cdot r\right)} \tag{7.35}$$

This equation shows that the amplitude of the light wave is $E_0 e^{\frac{-r}{\delta}}$ at a depth of r. Therefore, the intensity of the refracted light will be given as

$$I = I_0 e^{-\frac{2r}{\delta}} \tag{7.36}$$

When $r = \delta$, the intensity of the light wave decreases in magnitude to $(1/e^2)$ times its value at the surface (at the surface $r = 0$). The quantity δ is a measure of the distance of penetration of electromagnetic waves into the good conducting medium and is called skin depth. It is given by the expression

$$\delta = \sqrt{\frac{2}{\mu\sigma\omega}} \tag{7.37}$$

Equation (7.37) shows that the skin depth is a property of the medium and also depends upon the frequency of the incident wave. If the frequency of the medium increases, the skin depth δ decreases. The more is the skin depth, the less will be the attenuation of the electromagnetic waves in the medium and vice versa. The skin depth of seawater is relatively high for radio waves. This is the reason why radio communication with submarines inside several meters of seawater is difficult.

Equation (7.36) can be written as

$$I = I_0 e^{-\frac{2r}{\delta}} = I_0 e^{-\gamma r} \tag{7.38}$$

where

$$\gamma = \frac{2}{\delta} = \text{absorption coefficient} \tag{7.39}$$

I = intensity of the light wave at a depth of r from the surface.

I_0 = intensity of the incident light wave – Intensity of the reflected light wave = intensity of non-reflected incident radiation.

Thus, combining Eqs (7.37) and (7.39), we have

$$\gamma = \frac{2}{\delta} = 2\sqrt{\frac{\mu\sigma\omega}{2}} = \sqrt{2\mu\sigma\omega} \tag{7.40}$$

Equation (7.40) shows that the absorption coefficient γ is a property of the medium depending upon the conductivity and magnetic permeability of the medium. It also depends upon the frequency of the incident wave. If the frequency of the light wave increases, the absorption coefficient γ of the medium increases. The more is the absorption coefficient, the more will be the attenuation of the electromagnetic wave or light waves in the medium and vice versa. Equation (7.38) shows that the absorption coefficient γ is inversely proportional to the skin depth δ. The absorption coefficient γ of a perfect insulator according to Eq. (7.40) is zero since $\sigma = 0$.

Example 7.1

The fraction of the non-reflected radiation that is transmitted through a 5 mm thick material is 0.95. If thickness is increased to 12 mm, what fraction of light will be transmitted?

Solution

The data given are

$$\frac{I}{I_0} = 0.95 \text{ for } x = 5 \text{ mm}$$

However, we know that $I = I_0 e^{-\gamma r}$ or $\frac{I}{I_0} = e^{-\gamma r}$

Thus, we have $e^{5\gamma} = \frac{1}{0.95}$ or $\gamma = \frac{1}{5} \ell n \frac{1}{0.95} = 0.0103 \text{ mm}^{-1}$

Thus, for $x = 12$ mm, we have $\frac{I}{I_0} = e^{-0.103 \times 12} = 0.884$ Ans.

7.5.2 Absorption by electronic polarization

The electric vector, otherwise called light vector of the electromagnetic waves in the visible range interacts with the electron clouds surrounding each atom in the materials. This interaction shifts the negatively charged electron cloud relative to the positively charged nucleus of the atom momentarily. We call this phenomenon the electronic polarization (see Section 9.5.1). Thus, some energy of the electromagnetic wave is absorbed i.e. used up in shifting the electron cloud relative to the nucleus of the atom.

7.5.3 Quantum theory of absorption

The absorption of electromagnetic waves by materials has already been discussed on the basis of Maxwell's electromagnetic wave theory. This classical theory of absorption has limitations in explaining absorption in the entire range of the electromagnetic spectrum. Absorption of electromagnetic radiation by materials in the entire range of the electromagnetic spectrum can be understood on the basis of the quantum theory of light and the concepts of energy bands of solids.

Basic concepts

According to the quantum theory of electromagnetic radiation, radiation comprising particles called photons, have energy $h\nu$, h and ν being Planck's constant and frequency of the incident radiation respectively. The energy of a photon depends upon the frequency of the radiation. According to the band theory of solids, there are the allowed energy bands and the forbidden energy band. The outermost energy band is called the conduction band and below it is the valence band separated by the forbidden energy band E_g. When electromagnetic radiation or photons are incident on a material, the electrons in the material are, by absorbing the energy of the incident photon, excited to the higher energy bands from the lower energy bands. The incident radiation is thus absorbed by the material. An electron may be excited from an occupied state at energy E_1 to a higher energy unoccupied state at energy E_2 by absorbing the photon of energy $h\nu$ provided the energy difference between two states $E_2 - E_1$ $(= \Delta E)$ is equal to or less than the energy of the photon, i.e.,

$$E_2 - E_1 \le h\nu$$

or $\quad h\nu \ge \Delta E$ $\hfill$ (7.41)

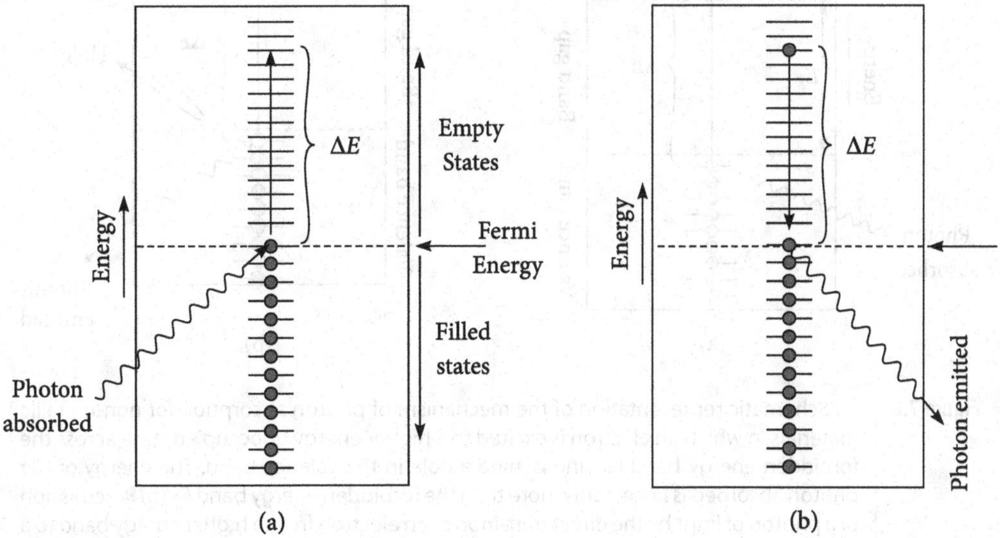

Figure 7.1 (a) Schematic representation of the mechanism of photon absorption for metallic materials in which an electron is excited to a higher energy unoccupied state. The change in energy of the electron ΔE is equal to the energy of the photon. (b) Re-emission of a photon of light by the direct transition of an electron from a high to a low energy state

The phenomena of absorption and emission of photons by metallic materials and non-metallic materials have been depicted schematically in Figs 7.1 and 7.2 respectively. It is important to know that in each excitation event, all the photon energy is absorbed. Photons simply vanish! When an electron in the valence band is excited to a state in the conduction band by absorbing photon energy more than or equal to the forbidden energy band E_g, a free electron in the conduction band and a hole in the valence band is created. These excitations with the accompanying absorption can take place only if the photon energy is equal to or more than the forbidden energy band E_g, i.e.,

$$h\nu \geq E_g$$

or $$\frac{hc}{\lambda} \geq E_g \qquad\qquad\qquad\qquad (7.42)$$

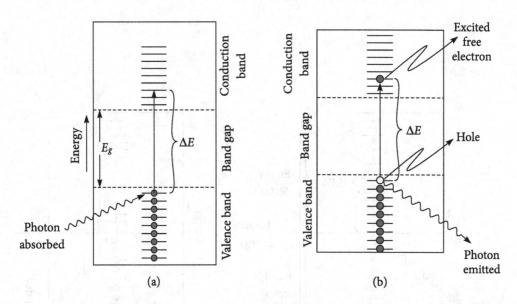

(a)   (b)

Figure 7.2 (a) Schematic representation of the mechanism of photon absorption for non-metallic materials in which an electron is excited to a higher energy unoccupied state across the forbidden energy band leaving behind a hole in the valence band. The energy of the photon absorbed is necessarily more than the forbidden energy band E_g. (b) Re-emission of a photon of light by the direct transition of an electron from a higher energy band to a low energy band

Example 7.2

Carbon has a band gap of 6 eV. Find the maximum wavelength for which carbon is opaque.

Solution

$$E_{g\,min} = \frac{hc}{\lambda_{max}}$$

or $\quad \lambda_{max} = \dfrac{hc}{E_{g\,min}} = \dfrac{6.63 \times 10^{-34} \times 3 \times 10^{8}}{6 \times 1.6 \times 10^{-19}}\,m = 2072\ nm$ $\hfill$ **Ans.**

Inter-band transition

Inter-band transition is defined as the transition of electrons from a lower energy valence band to a higher energy conduction band across the forbidden energy band. Inter-band transitions are of two types: (i) Direct inter-band transition and (ii) indirect inter-band transition.

i. **Direct inter-band transition**

In direct inter-band transition, an electron in the valence band, absorbing the energy of the incident photon, is excited to the conduction band leaving behind a hole in the valence band and making itself a free electron in the conduction band. In this process, the momentum of the electron does not change and no phonon (As a quantum of electromagnetic energy is a photon, so a quantum of lattice vibrational energy is a phonon. In analogy to a photon, the phonon can be viewed as a wave packet with particle-like properties.) is involved. The minimum energy of the incident photon is equal to the forbidden energy band for the process to occur.

ii. **Indirect inter-band transition**

In indirect inter-band transition, the energy of the incident photon is partly absorbed by an electron in the valence band that gets excited to the conduction band leaving behind a hole in the valence band and making itself a free electron in the conduction band. The rest of the incident photon energy is used in creating a phonon. Thus, in this process, the momentum of the electron changes and a phonon is created. The minimum energy of the incident photon is more than the forbidden energy band E_g for the process to occur.

Intra-band transition

Intra-band transition is defined as the transition of electrons from lower energy states to higher energy states within a single band. Such transitions are possible if there are partially filled energy bands or the valence band and the conduction band overlaps as in metals. In such cases, vacant higher energy states should be available for excitation within the band. This type of absorption can occur only in metals. Low energy infrared photons are absorbed by intra-band transition because to initiate this type of transition, only a small amount of energy is required. The maximum energy required for this transition to occur E_{max} is equal to the energy difference between the bottom and top of the valence band. Absorption of electromagnetic energy less than E_{max} by the metal takes place by intra-band transition. Intra-band transition occurs with the creation of a phonon.

7.5.4 Absorption by impurity

Up to now we have discussed briefly the absorption of electromagnetic radiation by materials through two processes: (i) Electronic polarization and (ii) valence band-conduction band transitions. Besides these two, there is a third process by which electromagnetic radiation can be absorbed by materials. If impurities or other electrically active defects are present, a new electron level within the band gap may be introduced, such as the donor and acceptor levels except that they lie closer to the centre of the band gap. Electromagnetic radiation of specific frequency may be absorbed as a result of electron transitions from or to these energy levels within the band gap as depicted in Fig. 7.3.

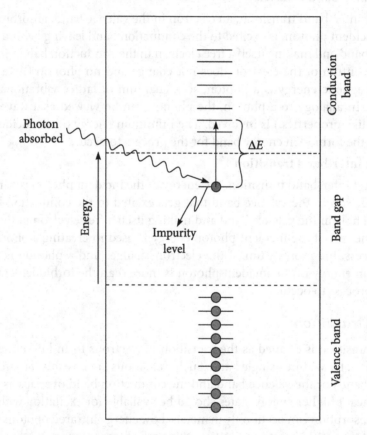

Figure 7.3 Mechanism of electron excitation, from an impurity level that lies within the band gap, by absorption of a photon

7.6 Transmission

The phenomenon of transmission is depicted in the Fig. 7.4.

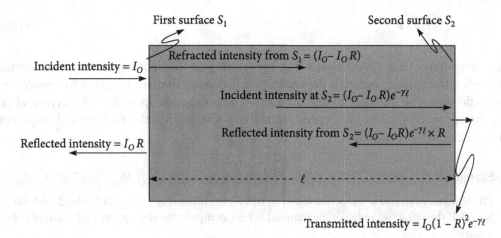

First surface S_1

Second surface S_2

Refracted intensity from $S_1 = (I_O - I_O R)$

Incident intensity = I_O

Incident intensity at $S_2 = (I_O - I_O R)e^{-\gamma \ell}$

Reflected intensity from $S_2 = (I_O - I_O R)e^{-\gamma \ell} \times R$

Reflected intensity = $I_O R$

ℓ

Transmitted intensity = $I_O (1 - R)^2 e^{-\gamma \ell}$

Figure 7.4 Transmission of light through a medium. ℓ = length of the absorbing medium, I_0 = incident intensity, R = reflectivity = reflectance = $\dfrac{\text{Reflected intensity}}{\text{Incident intensity}} = \dfrac{I_R}{I_0}$

Let I_0 be the intensity of the incident light. If R is the reflectivity of the medium, then the intensity of the reflected light I_R reflected from the first surface will be

$$I_R = I_0 R \tag{7.43}$$

The intensity of the refracted light at the first surface will be

$$I_0 - I_0 R \tag{7.44}$$

The intensity of the light incident on the second surface will be

$$(I_0 - I_0 R)e^{-\gamma \ell} \tag{7.45}$$

The intensity of the light reflected from the second surface will be

$$(I_0 - I_0 R)e^{-\gamma \ell} \times R$$

The amount of light refracted or transmitted from the second surface I_T will be

$$I_T = (I_0 - I_0 R)e^{-\gamma \ell} - (I_0 - I_0 R)e^{-\gamma \ell} \times R$$

or $\quad I_T = I_0(1 - R)e^{-\gamma \ell} - I_0(1 - R)e^{-\gamma \ell} \times R$

or $I_T = I_0(1-R)^2 e^{-\gamma \ell}$ (7.46)

The derivation of Eq. (7.46) shows that the fraction of incident light that is transmitted through a transparent medium depends upon the losses that are incurred by absorption and reflection. The intensity of the transmitted light depends upon the reflectivity and on the absorption coefficient, i.e., depends on the characteristics of the medium and frequency of the incident light wave.

Example 7.3

The transmissivity of a dielectric material of 10 mm thickness to normally incident light is 0.80. If the refractive index of the material is 1.55, compute the absorption coefficient of the material.

Solution

The data given are

$$T = \frac{I_T}{I_0} = 0.80 \text{ for } x = 10 \text{ mm,}$$

$$n = 1.5$$

The reflectivity R for a dielectric material for normal incidence is

$$R = \left|\frac{n-1}{n+1}\right|^2 = \left|\frac{1-1.55}{1+1.55}\right|^2 = 0.0465$$

We know that $I_T = I_0(1-R)^2 e^{-\gamma x}$

or $T = \dfrac{I_T}{I_0} = (1-R)^2 e^{-\gamma x}$

or $\dfrac{T}{(1-R)^2} = e^{-\gamma x}$

or $\gamma = \dfrac{1}{x}\ln\left(\dfrac{(1-R)^2}{T}\right) = 0.013 \text{ mm}^{-1}$

Example 7.4

The transmissivity of a dielectric material of 15 mm thickness to normally incident light is 0.80. If the refractive index of the material is 1.5, compute the thickness of the material that will yield a transmissivity of 0.70. All the reflection losses may be included.

Solution

The data given are

$$T = \frac{I_T}{I_0} = 0.80 \quad \text{for } x_1 = 15 \text{ mm,}$$

$$T = \frac{I_T}{I_0} = 0.70 \quad \text{for } x_2 = ?$$

$$n = 1.5$$

Reflectivity for a dielectric material for normal incidence is

$$R = \left| \frac{n-1}{n+1} \right|^2 = \left| \frac{1-1.5}{1+1.5} \right|^2 = 0.04$$

We know that $T = \frac{I_T}{I_0} = (1-R)^2 e^{-\gamma x}$

or $x = \frac{1}{\gamma} \ell n \left(\frac{(1-R)^2}{T} \right)$

Thus, we can have $x_1 = \frac{1}{\gamma} \ell n \left(\frac{(1-R)^2}{T_1} \right)$ and $x_2 = \frac{1}{\gamma} \ell n \left(\frac{(1-R)^2}{T_2} \right)$

or $\frac{x_2}{x_1} = \ell n \left(\frac{(1-R)^2}{T_2} \right) \div \ell n \left(\frac{(1-R)^2}{T_1} \right) = 1.944$

or $x_2 = 1.944 \times x_1 = 29.2 \text{ mm}$.

7.7 Atomic Theory of Optical Properties

In the preceding sections, optical properties of materials were discussed by applying Maxwell's electromagnetic wave equations. The expressions obtained for refractive index and reflectivity hold good for far infrared spectrum, i.e., in the lower frequencies range. To understand the optical properties of materials in the visible spectrum, the interaction of electromagnetic waves with the electrons of the materials has to be taken into account. The interaction of the electromagnetic wave on the electrons of dielectric materials and metals is different since dielectrics contain bound electrons whereas metals contains free electrons.

7.7.1 Atomic theory of optical properties of metals

When electromagnetic wave is incident on a metal, the electric field $E = E_0 e^{i\omega t}$ of the electromagnetic wave exerts a varying force on the electrons as a result of which they vibrate. During vibration, the electrons frequently collide with crystal imperfections such as vacancies, interstitial atoms, dislocations, grain boundaries and the like. Therefore, the motion or vibration of the electrons is not free but damped. As usual for low speed, the damping force on the electron F_d is proportional to speed, i.e.,

$$F_d = -bv = -b\frac{dx}{dt} \tag{7.47}$$

The applied force F_{ext} on an electron due to the electric field $E = E_0 e^{i\omega t}$ of the electromagnetic wave is given as

$$F_{ext} = eE = eE_0 e^{i\omega t} \tag{7.48}$$

Applying Newton's laws of motion, the equation of motion of the electron in a damping medium is given by

$$m\frac{d^2 x}{dt^2} + \xi\frac{dx}{dt} = eE \tag{7.49}$$

In Eq. (7.49), we have neglected the magnetic force $e\vec{v} \times \vec{B}$ because the magnetic force $e\vec{v} \times \vec{B}$ in the electromagnetic wave is much smaller than the electric force $e\vec{E}$ in the same electromagnetic wave.

At equilibrium, the drift speed $v_d = \dfrac{dx}{dt}$ of the electron is constant. Therefore, we have

$$\frac{dx}{dt} = v_d$$

and hence, $\dfrac{d^2x}{dt^2} = 0$

Putting these two conditions into Eq. (7.49), we get

$$0 + \xi v_d = eE$$

or $\xi = \dfrac{eE}{v_d}$ (7.50)

However, we know that current density $J = \sigma_{dc} E$ is given as

$$J = Nev_d$$

or $v_d = \dfrac{J}{Ne} = \dfrac{\sigma_{dc} E}{Ne}$ (7.51)

Here σ_{dc} is called the dc conductivity and N is the number of free electrons per unit volume, i.e., concentration of electrons. Putting this equation into Eq. (7.50), we get

$$\xi = \dfrac{Ne^2}{\sigma_{dc}}$$ (7.52)

Equation (7.52) shows that the damping coefficient $\dfrac{\xi}{2m}$ is inversely proportional to the dc conductivity. Substituting Eq. (7.52) into Eq. (7.49), we have

$$m\dfrac{d^2x}{dt^2} + \dfrac{Ne^2}{\sigma_{dc}}\dfrac{dx}{dt} = eE$$ (7.53)

Assuming a solution of this equation as $x = x_0 e^{i\omega t}$, we have

$$-m\omega^2 x + i\dfrac{ne^2\omega}{\sigma_{dc}} x = eE$$

or $x = \dfrac{eE}{i\dfrac{Ne^2\omega}{\sigma_{dc}} - m\omega^2} = \dfrac{E}{i\dfrac{Ne\omega}{\sigma_{dc}} - \dfrac{m\omega^2}{e}}$

or $Nex = \dfrac{NeE}{i\dfrac{Ne\omega}{\sigma_{dc}} - \dfrac{m\omega^2}{e}}$ (7.54)

In Eq. (7.54), ex is the induced dipole moment and N is the number of electrons per unit volume. Therefore Nex will be the polarization P, i.e.,

$$P = Nex$$ (7.55)

The substitution of this equation into Eq. (7.54) gives

$$P = \dfrac{NeE}{i\dfrac{Ne\omega}{\sigma_{dc}} - \dfrac{m\omega^2}{e}}$$ (7.56)

We know that $P = \varepsilon_0(\varepsilon_r - 1)E$. Hence, Eq. (7.56) becomes

$$\varepsilon_0(\varepsilon_r - 1)E = \dfrac{NeE}{i\dfrac{Ne\omega}{\sigma_{dc}} - \dfrac{m\omega^2}{e}}$$

or $\varepsilon_r = 1 + \dfrac{Ne}{i\dfrac{Ne\varepsilon_0\omega}{\sigma_{dc}} - \dfrac{\varepsilon_0 m\omega^2}{e}}$

or $\varepsilon_r = 1 + \dfrac{1}{i\dfrac{\varepsilon_0\omega}{\sigma_{dc}} - \dfrac{\varepsilon_0 m\omega^2}{Ne^2}}$ (7.57)

$\dfrac{Ne^2}{\varepsilon_0 m}$ is the square of the plasma angular frequency ω_p, i.e.,

$$\omega_p^2 = \dfrac{Ne^2}{\varepsilon_0 m}$$ (7.58)

The plasma frequency of metals is of the order of 10^{15} Hz.

Putting $\dfrac{Ne^2}{\varepsilon_0 m} = \omega_p^2$ into Eq. (7.57), we have

$$\varepsilon_r = 1 + \cfrac{1}{i\cfrac{\varepsilon_0 \omega}{\sigma_{dc}} - \cfrac{\omega^2}{\omega_p^2}} = 1 + \cfrac{\omega_p^2}{i\cfrac{\omega \varepsilon_0 \omega_p^2}{\sigma_{dc}} - \omega^2}$$

or $\quad \varepsilon_r = 1 + \cfrac{\omega_p^2}{i\omega_d - \omega^2}$ \hfill (7.59)

where $\dfrac{\varepsilon_0 \omega_0^2}{\sigma_{dc}}$ has the unit of frequency and is called damping frequency ω_d. Thus, we have

$$\omega_d = \frac{\varepsilon_0 \omega_0^2}{\sigma_{dc}}$$ \hfill (7.60)

The damping frequency of metals is of the order of 10^{12} Hertz.

Equation (7.59) shows that the dielectric constant ε_r is complex and hence, we can have

$$\varepsilon_{rc} = 1 + \frac{\omega_p^2}{i\omega_d - \omega^2}$$

or $\quad \varepsilon_{rc} = 1 - \dfrac{\omega_p^2}{\omega_d^2 + \omega^2} - i\dfrac{\omega_p^2 \omega_d}{\omega \omega_d^2 + \omega^3}$

or $\quad \varepsilon_{rc} = 1 - \dfrac{\omega_p^2}{\omega_d^2 + \omega^2} - i\left(\dfrac{\omega_d}{\omega}\right)\dfrac{\omega_p^2 \omega_d}{\omega_d^2 + \omega^2}$ \hfill (7.61)

where ε_{rc} is the complex dielectric constant. The relation between the complex dielectric constant and the complex refractive index n_c is given by

$$\varepsilon_{rc} = n_c^2$$ \hfill (7.62)

Since the refractive index n_c and the dielectric constant ε_{rc} are complex, we can write $n_c = n - ib$ (Hence, $n_c^2 = n^2 - b^2 - 2inb$) and $\varepsilon_{rc} = \varepsilon_r - i\varepsilon_{ri}$. Thus, Eq. (7.62) becomes

$$\varepsilon_r - i\varepsilon_{ri} = n^2 - b^2 - 2inb$$

The imaginary part b of the complex refractive index $n_c = n - ib$ is called the extinction coefficient. It measures the extent of attenuation of the wave.

Putting the previous equation into Eq. (7.61), we have

$$\varepsilon_r - i\varepsilon_{ri} = n^2 - b^2 - i2nb = 1 - \frac{\omega_p^2}{\omega_d^2 + \omega^2} - i\left(\frac{\omega_d}{\omega}\right)\frac{\omega_p^2}{\omega_d^2 + \omega^2}$$

Thus, we have

$$\varepsilon_r = n^2 - b^2 = 1 - \frac{\omega_p^2}{\omega^2 + \omega_d^2} \tag{7.63}$$

$$\varepsilon_{ri} = 2nb = \left(\frac{\omega_d}{\omega}\right)\frac{\omega_p^2}{\omega^2 + \omega_d^2} \tag{7.64}$$

The variation of dielectric constants $\varepsilon_r = n^2 - b^2$ and $\varepsilon_{ri} = 2nb$ with angular frequency of the incident light is shown in Fig. 7.5.

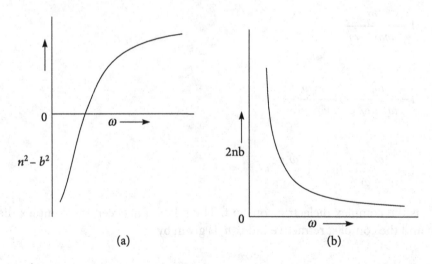

(a)   (b)

Figure 7.5 (a) Plot of the real part of the complex dielectric constant $\varepsilon_r = n^2 - b^2$ versus the angular frequency of the incident light wave. (b) Plot of the imaginary part of the complex dielectric constant $\varepsilon_{ri} = 2nb$ versus the angular frequency of the incident light wave

Case 1: $\omega^2 \gg \omega_d^2$

For visible spectrum $\omega^2 \gg \omega_d^2$. Under this condition, we can neglect ω_d^2 in comparison to ω^2 and Eqs (7.63) and (7.64) reduce to

$$\varepsilon_r = n^2 - b^2 = 1 - \frac{\omega_p^2}{\omega^2} \tag{7.65}$$

and $\varepsilon_{ri} = 2nb = \left(\frac{\omega_d}{\omega}\right)\frac{\omega_p^2}{\omega^2} \tag{7.66}$

Near plasma frequency, i.e., when $\omega \approx \omega_p$, Eqs (7.65) and (7.66) reduce to

$$\varepsilon_r = n^2 - b^2 = 0 \tag{7.67}$$

and $\varepsilon_{ri} = 2nb = \frac{\omega_d}{\omega} \tag{7.68}$

Equation (7.67) shows that at plasma frequency, the imaginary and real parts of the complex refractive indices are equal.

Case 2: $\omega^2 \ll \omega_d^2$

For far infrared spectrum, $\omega^2 \ll \omega_d^2$. Under this condition, we can neglect ω^2 in comparison to ω_d^2 and Eqs (7.63) and (7.64) reduce to

$$\varepsilon_r = n^2 - b^2 = 1 - \frac{\omega_p^2}{\omega_d^2} \tag{7.69}$$

$$\varepsilon_{ri} = 2nb = \left(\frac{\omega_d}{\omega}\right)\frac{\omega_p^2}{\omega_d^2} = \frac{\omega_p^2}{\omega\omega_d}$$

or $\varepsilon_{ri} = 2nb = \frac{\varepsilon_0\omega_p^2}{\varepsilon_0\omega\omega_d} = \frac{\sigma_{ac}}{\varepsilon_0\omega} \tag{7.70}$

where $\dfrac{\varepsilon_0\omega_p^2}{\omega_d}$ has the unit of conductivity and is called ac conductivity σ_{ac}. Thus, we have

$$\sigma_{ac} = \frac{\varepsilon_0\omega_p^2}{\omega_d} \tag{7.71}$$

The comparison of Eqs (7.71) and (7.60) shows that at far infrared region, dc conductivity σ_{dc} and ac conductivity σ_{ac} are equal.

Variation of n and b with ω

Equations (7.63) and (7.64) may be solved for n and b. The variation of n and b with ω is shown in Fig. 7.6. As seen from the figure, the refractive index is less than unity for a wide range of frequencies in the region of plasma frequency. The extinction coefficient b is very large at low frequencies. It decreases monotonically with increasing frequency. The metal thus becomes transparent at high frequencies. The qualitative agreement with these predictions of classical theory is obtained in case of alkali metals and some of the good conductors like silver, gold and copper.

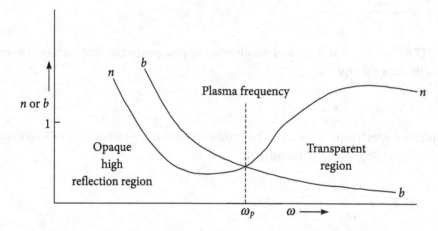

Figure 7.6 | Variation of refractive index n and extinction coefficient b of a metal with angular frequency ω of the incident light wave. The metal is transparent if the angular frequency ω of the incident light wave is more than the plasma angular frequency of the metal; otherwise it is opaque

7.7.2 Atomic theory of optical properties of dielectrics

When electromagnetic wave is incident on dielectrics, the electric field $E = E_0 e^{i\omega t}$ of the electromagnetic wave exerts a varying force on the electrons as a result of which they vibrate. Unlike the metals, the electrons in dielectrics are bound to their respective nuclei as a result of which the nucleus–electron system behave as a spring–body system under the influence of the attractive electrostatic force between the nucleus (positively charged) and the electron (negatively charged). The restoring force on the electron, according to Hooke's law, is given by

$$F_r = -ax \tag{7.72}$$

During vibration, the electrons frequently collide with crystal imperfections such as vacancies, interstitial atoms, dislocations and grain boundaries. Therefore, the motion/vibration of the electrons is also damped. As usual for low speed, the damping force on the electron F_d is proportional to speed, i.e.,

$$F_d = -\xi v = -\xi \frac{dx}{dt} \tag{7.73}$$

The applied force F_{ext} on an electron due to the electric field $E = E_0 e^{i\omega t}$ of the electromagnetic wave is given as

$$F_{ext} = eE = eE_0 e^{i\omega t} \tag{7.74}$$

Applying Newton's laws of motion, the equation of motion of the electron in a damping medium is given by

$$m \frac{d^2 x}{dt^2} = eE - \xi \frac{dx}{dt} - ax$$

or $$m \frac{d^2 x}{dt^2} + \xi \frac{dx}{dt} + ax = eE \tag{7.75}$$

In Eq. (7.75), we have neglected the magnetic force $e\vec{v} \times \vec{B}$ because the magnetic force $e\vec{v} \times \vec{B}$ in the electromagnetic wave is much smaller than the electric force $e\vec{E}$ in the same electromagnetic wave.

Assuming a solution of this equation as $x = x_0 e^{i\omega t}$, we have

$$-m\omega^2 x + i\xi \omega x + ax = eE$$

or $$x = \frac{eE}{-m\omega^2 + a + i\xi\omega} = \frac{\dfrac{e}{m}E}{-\omega^2 + \dfrac{a}{m} + i\dfrac{\xi}{m}\omega} = \frac{\dfrac{e}{m}E}{-\omega^2 + \omega_0^2 + i\dfrac{\xi}{m}\omega} \tag{7.76}$$

or $$Nex = \frac{\dfrac{Ne^2}{m}E}{-\omega^2 + \omega_0^2 + i\dfrac{\xi}{m}\omega} \tag{7.77}$$

Here

$$\omega_0^2 = \frac{a}{m} \tag{7.78}$$

In Eq. (7.78), ex is the induced dipole moment and N is the number of electrons per unit volume. Therefore, Nex will be the polarization P, i.e.,

$$P = Nex \tag{7.79}$$

The substitution of this equation into Eq. (7.77) gives

$$P = \frac{\dfrac{Ne^2}{m}E}{-\omega^2 + \omega_0^2 + i\dfrac{\xi}{m}\omega} \tag{7.80}$$

We know that $P = \varepsilon_0(\varepsilon_r - 1)E$. Hence, Eq. (7.80) becomes

$$\varepsilon_0(\varepsilon_r - 1)E = \frac{\dfrac{Ne^2}{m}E}{-\omega^2 + \omega_0^2 + i\dfrac{\xi}{m}\omega}$$

or

$$\varepsilon_r = 1 + \frac{\dfrac{Ne^2}{\varepsilon_0 m}}{\omega_0^2 - \omega^2 + i\dfrac{\xi\omega}{m}} \tag{7.81}$$

Equation (7.81) shows that the dielectric constant ε_r is complex and hence, we can have

$$\varepsilon_{rc} = 1 + \frac{\dfrac{Ne^2}{\varepsilon_0 m}}{\omega_0^2 - \omega^2 + i\dfrac{\xi\omega}{m}}$$

or

$$\varepsilon_{rc} = 1 + \frac{\dfrac{Ne^2}{\varepsilon_0 m}\left(\omega_0^2 - \omega^2\right)}{\left(\omega_0^2 - \omega^2\right)^2 + \dfrac{\xi^2\omega^2}{m^2}} - i\frac{\dfrac{Ne^2}{\varepsilon_0 m}\dfrac{b\omega}{m}}{\left(\omega_0^2 - \omega^2\right)^2 + \dfrac{\xi^2\omega^2}{m^2}}$$

or $\quad \varepsilon_{rc} = 1 + \dfrac{\dfrac{Ne^2 m}{\varepsilon_0}\left(\omega_0^2 - \omega^2\right)}{m^2\left(\omega_0^2 - \omega^2\right)^2 + \xi^2\omega^2} - i\,\dfrac{\dfrac{Ne^2 b\omega}{\varepsilon_0}}{m^2\left(\omega_0^2 - \omega^2\right)^2 + \xi^2\omega^2}$

or $\quad \varepsilon_{rc} = 1 + \dfrac{Ne^2 m}{\varepsilon_0}\,\dfrac{\omega_0^2 - \omega^2}{m^2\left(\omega_0^2 - \omega^2\right)^2 + \xi^2\omega^2} - i\,\dfrac{Ne^2}{\varepsilon_0}\,\dfrac{\xi\omega}{m^2\left(\omega_0^2 - \omega^2\right)^2 + \xi^2\omega^2}$ $\qquad$ (7.82)

where ε_{rc} is the complex dielectric constant. The relation between the complex dielectric constant and the complex refractive index n_c is given by

$$\varepsilon_{rc} = n_c^2 \qquad (7.83)$$

Since the refractive index n_c and the dielectric constant ε_{rc} are complex, we can write $n_c = n - ib$ (Hence, $n_c^2 = n^2 - b^2 - 2inb$) and $\varepsilon_{rc} = \varepsilon_r - i\varepsilon_{ri}$. Thus, Eq. (7.83) becomes

$$\varepsilon_r - i\varepsilon_{ri} = n^2 - b^2 - 2inb$$

The imaginary part b of the complex refractive index $n_c = n - ib$ is called the extinction coefficient.

Putting this equation into Eq. (7.82), we have

$$\varepsilon_r - i\varepsilon_{ri} = n^2 - b^2 - i2nb$$

$$= 1 + \dfrac{ne^2 m}{\varepsilon_0}\,\dfrac{\omega_0^2 - \omega^2}{m^2\left(\omega_0^2 - \omega^2\right)^2 + \xi^2\omega^2} - i\,\dfrac{ne^2}{\varepsilon_0}\,\dfrac{\xi\omega}{m^2\left(\omega_0^2 - \omega^2\right)^2 + \xi^2\omega^2}$$

Thus, we have

$$\varepsilon_r = n^2 - b^2 = 1 + \dfrac{ne^2 m}{\varepsilon_0}\,\dfrac{\omega_0^2 - \omega^2}{m^2\left(\omega_0^2 - \omega^2\right)^2 + \xi^2\omega^2} \qquad (7.84)$$

and $\quad \varepsilon_{ri} = 2nb = \dfrac{ne^2}{\varepsilon_0}\,\dfrac{\xi\omega}{m^2\left(\omega_0^2 - \omega^2\right)^2 + \xi^2\omega^2}$ $\qquad$ (7.85)

The variation of dielectric constants $\varepsilon_r = n^2 - b^2$ and $\varepsilon_{ri} = 2nb$ with the angular frequency of the incident light is shown in Fig. 7.7.

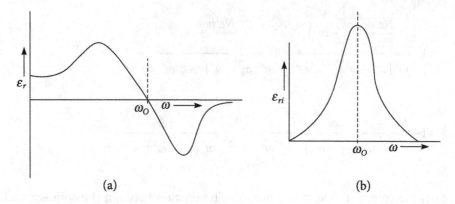

(a)   (b)

Figure 7.7 (a) Plot of the real part of the complex dielectric constant $\varepsilon_r = n^2 - b^2$ versus the angular frequency of the incident light wave. (b) Plot of the imaginary part of the complex dielectric constant $\varepsilon_{ri} = 2nb$ versus the angular frequency of the incident light wave

Equations (7.84) and (7.85) may be solved for n and b to show their variation with the frequency of the incident light wave. The variation of n and b with the frequency of the incident light wave is shown in Fig. 7.8. The absorption of the light wave by the dielectric material is greatest at the natural frequency ω_0. The index of refraction is greater than unity for small frequencies and increases with frequency as the natural frequency is approached.

This is the case of normal dispersion and is exhibited by the most transparent material over the visible range of the spectrum – the natural frequency being in the ultraviolet region. At or near the natural frequency, however, the dispersion [the variation of refractive index with frequency] becomes anomalous in the sense that the index of refraction decreases with increasing frequencies.

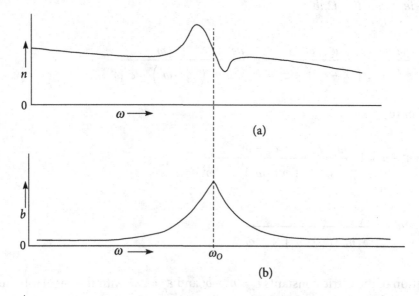

(a)

(b)

Figure 7.8 Variation of n and b with the frequency of the incident light wave

Questions

7.1   What do you mean by optical property of a material?

7.2   What are the possible phenomena that may occur when light falls on a material?

7.3   Explain the colours of the horizon during sunrise and sunset.

7.4   Why does fog look white?

7.5   What is light vector?

7.6   Explain how an electron scatters light.

7.7   Explain the blue colour of the sky.

7.8   What is reflectivity?

7.9   Derive an expression for reflectivity of a dielectric material in terms of the refractive index.

7.10 Derive an expression for reflectivity of a dielectric material in terms of the dielectric constant.

7.11 Prove that for a good conductor, reflectivity is nearly one.

7.12 Explain the statement "good conductors are good reflectors as well as good absorbers".

7.13 What is refraction coefficient?

7.14 What are the factors on which reflectivity of a metal depends?

7.15 Derive an expression for refractivity of a dielectric material.

7.16 What are the factors on which refractivity of a metal depends?

7.17 What do you mean by absorption?

7.18 Derive an expression for the intensity of refracted light inside a material.

7.19 Derive an expression for absorption coefficient of a material.

7.20 Explain, according to quantum theory, how light is absorbed by matter?

7.21 What is a phonon?

7.22 Explain why visible light cannot be absorbed by a non-metallic material having a forbidden energy band more than 3.1 eV.

7.23 Explain why visible light cannot be absorbed by a non-metallic material having a forbidden energy band less than 1.8 eV.

7.24 What is direct inter-band transition?

7.25 What is indirect inter-band transition?

7.26 Distinguish between direct inter-band transition and indirect inter-band transition?

7.27 What do you mean by intra-band transition?

7.28 Explain how impurities affect absorption of light by materials.

7.29 Derive the equation $I_T = I_0 \left(1 - R\right)^2 \left(1 - e^{-\gamma \ell}\right)$.

7.30 Show that at plasma frequency, in case of metals, the imaginary and real parts of complex refractive indices are equal.

7.31   Show that at far infrared region, in case of metals, the dc conductivity σ_{dc} and ac conductivity σ_{ac} are equal.

7.32   Show that the dielectric constant of a dielectric material is generally complex.

7.33   What is anomalous dispersion?

Problems

7.1   Diamond has a band gap of 5.6 eV. Find the maximum wavelength for which diamond is opaque.   [Ans   2200 nm]

7.2   In a material, transition occurs between a metastable state and an energy level of 0.5 eV and the wavelength of the radiation emitted is 2200 nm. Calculate the energy of the metastable state.   [Ans   1.07 eV]

7.3   In a step index optical fiber, the refractive indices of the core and the cladding are 1.45 and 1.35 respectively. Calculate the numerical aperture of the fiber.   [Ans   0.529]

7.4   Zinc selenide has a band gap of 2.58 eV. Over what range of wavelength of the visible light is it transparent?   [Ans   Up to 484 nm]

7.5   The dielectric constant of quartz is 1.55. What is its refractive index?
                                                    [Ans $\sqrt{\varepsilon_r} = 1.25$]

7.6   The fraction of the non-reflected radiation that is transmitted through a 10 mm thick transparent material is 0.95. If the thickness is increased to 20 mm, what fraction of light will be transmitted?   [Ans   0.90]

7.7   The transmissivity of a dielectric material of 15 mm thickness to normally incident light is 0.80. If the refractive index of the material is 1.55, compute the thickness of the material that will yield a transmissivity of 0.70. All the reflection losses may be included.   [Ans   20.4 mm]

7.8   The transmissivity of a dielectric material of 20.4 mm thickness to normally incident light is 0.70. If the refractive index of the material is 1.55, compute the absorption coefficient of the material.   [Ans   0.013 mm^{-1}]

Multiple Choice Questions

1.   All the possible optical phenomena occur always simultaneously.
        (i)   true   (ii)   false

2.   Which of the following relations are correct?

        (i) $c = \dfrac{1}{\sqrt{\varepsilon_0 \mu_0}}$ (ii) $c = \sqrt{v \varepsilon_r}$

(iii)　$v = \dfrac{c}{\varepsilon_r}$

(iv)　$n = \sqrt{\mu_r \varepsilon_r}$

(v)　$c = v\sqrt{\varepsilon_r}$

3.　In indirect inter-band transition, the electron under goes
   (i)　no change in both energy and momentum
   (ii)　change in energy but no change in momentum
   (iii)　change in both energy and momentum
   (iv)　change in momentum but no change in energy

4.　What is the range of visible radiation?
   (i)　7×10^{-5} cm to 4×10^{-5} cm
   (ii)　6×10^{-5} cm to 4×10^{-5} cm
   (iii)　7×10^{-5} cm to 3×10^{-5} cm
   (iv)　6×10^{-5} cm to 3×10^{-5} cm

5.　Which of the following in the light wave is responsible for almost all optical phenomena?
   (i)　electric vector
   (ii)　magnetic vector
   (iii)　propagation vector
   (iv)　none of the above

6.　Which of the following phenomenon makes the object visible?
   (i)　absorption
   (ii)　refraction
   (iii)　interference
   (iv)　scattering

7.　The blue colour of the sky is due to
   (i)　refraction
   (ii)　diffraction
   (iii)　interference
   (iv)　scattering

8.　For a good conductor, the value of reflection coefficient is
   (i)　1
   (ii)　0
   (iii)　100
   (iv)　∞

9.　For a good conductor, the value of refraction coefficient is
   (i)　1
   (ii)　0
   (iii)　100
   (iv)　∞

10.　For a good conductor, the value of skin depth in cm is
   (i)　1
   (ii)　0
   (iii)　100
   (iv)　∞

11.　For a perfect insulator, the value of absorption coefficient is
   (i)　1
   (ii)　0
   (iii)　100
   (iv)　∞

12. Phonons are the particles of
    (i) electromagnetic waves
    (ii) gravitational waves
    (iii) acoustic/sound waves
    (iv) microwaves

13. Photons are the particles of
    (i) electromagnetic waves
    (ii) gravitational waves
    (iii) acoustic/sound waves
    (iv) microwaves

14. At far infrared region
    (i) dc conductivity = ac conductivity
    (ii) dc conductivity > ac conductivity
    (iii) dc conductivity < ac conductivity
    (iv) dc conductivity = two times the ac conductivity

15. The component of the visible light wave responsible to induce electronic polarization in atoms in materials is
    (i) Electric vector
    (ii) Light vector
    (iii) Magnetic vector
    (iv) Propagation vector

Answers

1 (ii)	2 (i & iv)	3 (iii)	4 (i)	5 (i)	6 (iv)	7 (iv)	8 (i)
9 (ii)	10 (ii)	11 (ii)	12 (iii)	13 (i & iv)	14 (i)	15 (i) & (ii)	

8 Optoelectronic Devices

8.1 Introduction

Optoelectronic devices are transducers for electrical-to-optical or optical-to-electrical signals. In these devices, photon, the basic unit of light plays the most vital role. Such devices can be divided into four groups: (1) photo-detectors and solar cells that convert photons into electrical current, (2) light-emitting diodes (LED) and semiconductor lasers (or diode lasers) that convert an electric energy to light energy, (3) optical fibers that guide light within a small plastic or glass fiber between a light source and a detector, and (4) optical-fiber amplifiers that convert the energy of an optical pump source to photons identical to an optical signal. A useful expression when dealing with optoelectronics is the conversion between the wavelength of the light and energy of a photon by the relation

$$\lambda = \frac{12400}{E} \text{Å} \tag{8.1}$$

where the energy of the photon, E, is in electron volts.

8.2 Laser

The absorption and emission of electromagnetic radiation by materials has been very ingeniously and skillfully exploited in making a device that amplifies electromagnetic radiation and generates extremely intense, coherent, monochromatic radiation. The device is called LASER, an acronym for light amplification by stimulated emission of radiation.

8.2.1 Metastable state

An atom is said to be in an excited state when one or more of the revolving electrons of the atom jump to a higher orbit or higher energy level by absorbing energy from an external source. Atoms generally cannot stay in the excited state for infinite time. Within a certain finite time Δt, all the excited atoms drop to ground state. Energy width ΔE ($E_1 - E_2$) and life time are related inversely to each other by Heisenberg's uncertainty principle involving energy and time, i.e.,

$$\Delta E \Delta t = \frac{\hbar}{2} \tag{8.2}$$

Equation 8.2 indicates that the atoms in the ground state have the longest life time. Normally, the excited atom drops to ground state in 10^{-8} second, i.e., the life time of an excited atom is 10^{-8} second. The key to laser is the presence, in many atoms, of one or more energy levels whose life time may be 10^{-3} second or more, instead of the usual 10^{-8} second. Such long-lived excited states are called metastable states. The metastable state lies in between the ground state and the excited state; more close to the excited state.

8.2.2 Electronic transition

When electromagnetic radiation of a suitable frequency to match the energy difference between the higher level and the lower level is incident on a material, electrons at the lower energy level are excited to the higher energy level by absorbing the incident photon energy or electrons at the higher energy level drop to the lower energy level by emitting photons. Three kinds of atomic transition involving electromagnetic radiation are possible between two energy levels E_0 and E_1 of an atom. They are (i) spontaneous emission, (ii) induced or stimulated absorption and (iii) induced or stimulated emission.

Spontaneous emission

The atoms in the excited state at the higher energy level E_1 may drop to the lower energy level spontaneously without external provocation by emitting a photon of energy $E_1 - E_0 = h\nu$. This process is called spontaneous emission. The spontaneous emission can be represented mathematically as

Atom* $\rightarrow$ atom + photon

Atom* represents the atom in an excited state.

Stimulated absorption

The atoms in the ground state at the lower energy level E_0 may be excited to the higher energy level E_1 by absorbing a photon of energy $E_1 - E_0 = h\nu$ from the incident electromagnetic radiation. This process is called induced or stimulated absorption. The stimulated absorption can be represented mathematically as

Atom + photon $\rightarrow$ atom*

Induced or stimulated absorption is caused by externally incident electromagnetic waves. An atom in the ground state absorbs a photon of proper energy and makes a transition to the excited state.

Stimulated emission

The atoms in the excited state at the higher energy level E_1 may be forced to drop to the lower energy level with external provocation by the incidence of a photon of energy $E_1 - E_0 = h\nu$ This process is called induced or stimulated emission. Stimulated emission was first envisioned by Albert Einstein in the year 1917 and later on confirmed by quantum physics. In induced or stimulated emission two photons each of energy $h\nu$ are emitted. The stimulated emission can be represented mathematically as

Atom* + photon → atom + 2 photons

In induced or stimulated emission, the two photons emitted per event travels exactly in the same direction to conserve momentum; they have exactly the same energy (monochromatic) rendering the emitted electromagnetic waves to be in the same phase, i.e., the emitted electromagnetic waves are coherent. These electromagnetic waves constitute the laser beam. Thus, key to laser production is the enhancement of stimulated emission.

Spontaneous emission, stimulated absorption and stimulated emission are illustrated in Fig. 8.1.

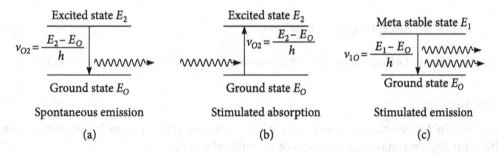

Figure 8.1 Electronic transitions between two energy levels in an atom can occur by (a) spontaneous emission, (b) stimulated absorption, and (c) stimulated emission

8.2.3 Spontaneous and stimulated emission probabilities

Let us consider an assembly of atoms in thermal equilibrium at temperature T with radiation frequency v and energy density $u(v)$. Let N_1 be the number of atoms per unit volume in lower energy states and N_2 the number of atoms per unit volume in higher energy states. At any instant of time, all the three transition processes, namely, stimulated emission, stimulated absorption and spontaneous emission can occur simultaneously.

Stimulated absorption (Upward transition)

Let us consider two energy states of a particular atom. If the atom is initially in lower energy state 1 having energy E_1, it can go up to higher energy state 2 having energy E_2 by absorbing a photon of light whose frequency is

$$v = \frac{E_2 - E_1}{h} \qquad (8.3)$$

The probability that atoms will actually undergo the transition from lower energy state 1 to higher energy state 2 by absorbing photons of light is proportional to the rate at which photons of frequency v fall on it and therefore, to the energy density $u(v)$, i.e.,

Stimulated absorption rate $\propto u(v)$.

Again as obvious, the stimulated absorption rate is directly proportional to the number of atoms present per unit volume N_1 in the lower energy states, i.e.,

Stimulated absorption rate $\propto N_1$.

Combining these two ideas, we have

Stimulated absorption rate $\propto N_1 u(v)$.

or   Stimulated absorption rate $= B_{12} N_1 u(v)$. $\qquad (8.4)$

Here the constant of proportionality B_{12} is known as Einstein's absorption coefficient.

Emission

An atom in higher energy states can make a radiative transition to lower energy states either through spontaneous emission or stimulated emission.

Spontaneous emission (Downward transition)

The probability of spontaneous emission is independent of the energy density of the radiation field and is only proportional to the number of atoms present per unit volume N_2 in the higher energy states, i.e.,

Spontaneous emission rate $\propto N_2$.

or   Spontaneous emission rate $= A_{21} N_2$. $\qquad (8.5)$

Here the constant of proportionality A_{21} is known as Einstein's spontaneous emission coefficient.

Stimulated emission (Downward transition)

The probability that atoms will actually undergo transition from higher energy state 2 to lower energy state 1 due to the incidence of certain photons of light is proportional to the rate at which photons of frequency v fall on it and therefore, to the energy density $u(v)$, i.e.,

Stimulated emission rate $\propto u(v)$

Again as obvious, the stimulated emission rate is directly proportional to the number of atoms present per unit volume in the higher energy states, i.e.,

Stimulated emission rate $\propto N_2$.

Combining these two ideas, we have

Stimulated emission rate $\propto N_2 u(v)$.

or   Stimulated emission rate $= B_{21} N_2 u(v)$.   (8.6)

Here the constant of proportionality B_{21} is known as Einstein's stimulated emission coefficient. The values of B_{21}, A_{21}, and B_{21} are determined from the atomic systems. Thus,

Total emission rate $= B_{21} N_2 u(v) + A_{21} N_2$ (8.7)

At thermal equilibrium, at any instant of time, the number of upward transitions must be equal to the number of total downward transitions. Thus, at thermal equilibrium, we have

$$B_{12} N_1 u(v) = B_{21} N_2 u(v) + A_{21} N_2$$ (8.8)

or $$u(v) = \frac{N_2 A_{21}}{N_1 B_{12} - N_2 B_{21}} = \frac{\dfrac{A_{21}}{B_{21}}}{\left(\dfrac{B_{12}}{B_{21}}\right)\left(\dfrac{N_1}{N_2}\right) - 1}$$ (8.9)

The population density N_1 (number of atoms per unit volume) in the lower energy state 1 in thermal equilibrium is given by the Maxwell–Boltzmann distribution law as

$$N_1 = N_0 e^{-\frac{E_1}{kT}}$$ (8.10)

Similarly, $$N_2 = N_0 e^{-\frac{E_2}{kT}}$$ (8.11)

In these equations, N_o is the population density of the ground state. From Eqs (8.10) and (8.11), we have

$$\frac{N_1}{N_2} = e^{\frac{E_2-E_1}{kT}}$$
(8.12)

Putting N_1/N_2 from Eq. (8.12) into Eq. (8.9), we get

$$u(v) = \frac{\frac{A_{21}}{B_{21}}}{\left(\frac{B_{12}}{B_{21}}\right)e^{\frac{E_2-E_1}{kT}}-1} = \frac{A_{21}}{B_{21}}\frac{1}{\left(\frac{B_{12}}{B_{21}}\right)e^{\frac{hv}{kT}}-1}$$
(8.13)

This expression gives the energy density of photons of frequency v in equilibrium with atoms in the two energy states E_1 and E_2 at temperature T. Now according to Planck's law, the energy density of radiation is given by

$$u(v) = \frac{8\pi h v^3}{c^3}\frac{1}{e^{\frac{hv}{kT}}-1}$$
(8.14)

Equation (8.14) is consistent with Planck's law if we set

$$\frac{A_{21}}{B_{21}} = \frac{8\pi h v^3}{c^3}$$
(8.15)

and $B_{12} = B_{21} = B$
(8.16)

These two equations are referred to as Einstein's relations.

Observe that if we had not assumed the existence of stimulated emission, we would not have been able to arrive at an expression for $u(v)$ that is similar to Planck's law. In order to obtain the correct form for $u(v)$, Einstein predicted the existence of stimulated emission, which was later confirmed by quantum theory.

Putting Eq. (8.16) into Eq. (8.13), we get

$$u(v) = \frac{A_{21}}{B_{21}}\frac{1}{e^{\frac{hv}{kT}}-1}$$
(8.17)

or $$\frac{A_{21}}{B_{21}u(v)} = e^{\frac{hv}{kT}}-1$$
(8.18)

This equation gives the ratio of the number of spontaneous emissions to stimulated emissions. For a normal optical source, $T \approx 10^3 \, \text{K}$ and with $\lambda = 6000 \, \text{Å}$,

$$\frac{A_{21}}{B_{21} u(v)} \approx 10^{10}.$$

Thus, at optical frequencies, the emission is predominantly spontaneous. Therefore, the emission from a usual light source is incoherent.

8.2.4 Basic principle of lasers

Suppose a group of atoms are in the same excited state. Out of all the atoms in the excited state, one atom undergoes stimulated emission by the action of an incident photon resulting in two unidirectional secondary photons of equal frequency. These two secondary photons will cause stimulated emission in two other excited atoms producing four unidirectional secondary photons of equal frequency. These secondary photons create further stimulated emission causing an avalanche of secondary photons till an intense beam of unidirectional coherent photons are obtained. This is the basis of operation of a laser.

Population inversion

The simplest model explained earlier has the following practical difficulties.

i.   The group of excited atoms cannot be kept in an excited state for longer periods of time without spontaneous emission.

ii. The secondary photons produce stimulated absorption in the atoms that are just rendered to ground state, hampering the characteristics of the laser.

Under normal conditions, at thermal equilibrium, the population of atoms in the ground state is more than that of those in the excited state. However, in population inversion, the population of atoms in the excited state is more than that of those in the ground state. Population inversion is the inverse of the number of atoms in the ground state to the in the excited state. This phenomenon is unnatural and is achieved by artificial means. It is essential for the operation of the laser. A system in which population inversion is achieved is called an active system.

Mathematical analysis of population inversion

Equation (8.12) gives the ratio of the population density N_1 in the lower energy state to the population density N_2 in the higher energy state at thermal equilibrium. From this equation, we can have

$$N_2 - N_1 = N_1 \left(e^{\frac{E_2 - E_1}{kT}} - 1 \right) \tag{8.19}$$

Since the RHS of Eq. (8.19) is negative (why?), we have $N_1 > N_2$ and therefore, the photons of appropriate energy incident on the system have more probability to interact with atoms at lower energy state than at higher energy state. Further, the probability for stimulated absorption is more than that for stimulated emission. If somehow the population in the higher energy state is more than the lower energy state, the photons of appropriate energy incident on the system will have more probability to interact with atoms at the higher energy state causing more stimulated emission and stimulated emission will dominate over stimulated absorption causing lasing action. Therefore for lasing action to occur, the population density of the higher energy state has to be higher than that of the lower energy state. The situation in which the population density of the higher energy state is higher than that of the lower energy state is called population inversion, because it is opposite to the normal situation where population density of the lower energy state is higher than that of the higher energy state. The process of achieving population inversion is called pumping. To achieve population inversion by optical pumping requires appropriate energy.

Equation (8.19) shows that if absolute temperature T becomes negative (which is impossible!), the RHS of Eq. (8.19) becomes positive implying that $N_2 > N_1$, i.e., population inversion. Thus, at negative absolute temperature, population inversion is a normal case. Therefore, population inversion is sometimes referred to as negative temperature state. It should be noted that negative absolute temperature is not a physical reality, but only a mathematical concept to represent population inversion.

Pumping

As mentioned before, the method of achieving population inversion (increasing the number of atoms in the excited state to be more than that in the ground state) by an external source of energy is called pumping. Population inversion cannot be achieved between two energy levels by just pumping between the same two levels. Minimum of three energy levels are involved in population inversion. If possible, let there be two states, namely, the metastable state and the ground state. Atoms in the ground state are excited to the metastable state by absorbing energy of the incident photons. When half of the atoms are in each state, the rate of stimulated emission and stimulated absorption are equal as a result of which the number of atoms in the metastable state cannot be more than half of the total atoms in the system. In this situation, laser action cannot occur. Population inversion is only possible when stimulated absorption is to a higher energy level E_2 than the metastable level E_1 from which stimulated emission takes place. This prevents the pumping from depopulating the metastable state. Thus, more than half the atoms must be in the metastable state for stimulated emission to predominate. Pumping is done in ruby lasers by irradiating the ground state with highly intense radiation from a xenon flash lamp[*].

Pumping Mechanism

Excitation of an active material in lasers is often called pumping. There are several methods of optical pumping of which a few are cited here.

---

[*] The xenon flash lamp consists of a transparent glass or quartz tube filled with xenon (or, occasionally, other noble gases) and fitted with electrodes. High voltage from a capacitor charges the electrodes and causes the gas to ionize; when an ionization path is complete, a pulse of current passes between the electrodes, causing the gas to flash and discharging the capacitor. The duration of the flash can be as short as one microsecond, and the circuitry can be arranged to cause the lamp to operate several thousand times per second.

i.   *Optical pumping*   The most widely spread and simplest pumping method is the optical pumping. Most conventional sources of optical pumping are flash discharge tubes, mainly xenon ones. Beside flash discharge tubes, LED and other lasers can be used as optical pumping sources.

ii.   *Solar pumping*   In this method of pumping, sunlight is focused on a sapphire sphere adjoined to one end of a ruby rod by means of a parabolic mirror. Laser radiation emerges from the other end of the rod.

iii.   *Electric pumping*   The light energy of an electrically exploding wire is utilized in exciting an active material in lasers. An exploding wire is a very powerful source of light. The wire is placed in one of the focus lines of an elliptical reflector made from stainless steel. The ruby rod is disposed in the other focus line of the reflector. The wire is exploded by high current pulses. This method of pumping is promising for the creation of super powerful lasers.

iv.   *Cathodoluminescent pumping*   Cathodoluminescent pumping is based on the use of cathode ray tubes. The cylindrical surface of these tubes is coated with phosphor. The ruby rod is placed axially inside the cathode ray tube. The phosphor is luminescent when bombarded with electrons. The radiant energy of such luminescence causes population inversion.

v.   *Laser pumping*   A very interesting and evidently promising method of exciting one laser by the beam of another laser.

vi.   Chemical pumping: Here certain chemical reactions result in excitation and thus create population inversion.

vii.   *Nuclear pumping*   Beside the aforementioned pumping techniques, experiments were reported on pumping active materials by means of nuclear sources. With this method, the laser radiation obtained will be very close to X-ray radiation.

Optical resonator

When an active material is placed in between two perfectly parallel laser mirrors out of which one is partially reflecting, the reflected laser light passing through the active material induces more lasing action in the active material. Such a system of two mirrors and active material represents an optical cavity or optical resonator. Due to reflections from both the mirrors, optical standing waves are formed in the active material. If L is the length of the active material (distance between the two mirrors) and λ is the wavelength of the light, we have

$$\frac{m\lambda}{2} = L \, , m = 1, 2, 3, \dots \tag{8.20}$$

(Actually m is a very large number. See Example 8.1)

or $\quad v = \dfrac{mv}{2L} = \dfrac{mc}{2L\mu} \; \text{ since } v = \lambda v \tag{8.21}$

Here μ is the refractive index of the active material. Equation (8.21) gives the discrete frequencies of the oscillation of the modes. Different values of m lead to different oscillation frequencies, which constitute the longitudinal modes of the optical cavity. For an optical cavity, m is a very large number. From Eq. (8.21), it is clear that the output radiation from the laser consists of several closely spaced resonant lines lying within the bandwidth of the radiation.

The frequency difference Δv between adjacent longitudinal modes is obtained as

$$\Delta v = (m+1)\frac{c}{2L\mu} - m\frac{c}{2L\mu} = \frac{c}{2L\mu} \tag{8.22}$$

The number of optical modes N within the bandwidth can be obtained by dividing the emission line width of the beam by Δv, i.e.,

$$N = \frac{\text{Bandwidth}}{\Delta v} \tag{8.23}$$

Example 8.1

For a fully transparent active material of length 60 cm, calculate the value of m in Eq. (8.21) for a light of wavelength 5000 Å.

Solution

The data given are $\mu = 1$, $L = 60$ cm, $\lambda = 5000$ Å $= 5000 \times 10^{-8}$ cm We have

$$m = \frac{2L}{\lambda} = \frac{2 \times 60}{5000 \times 10^{-8}} = 2.4 \times 10^6; \text{ a very large number.}$$

Example 8.2

Determine the frequency difference between the cavity modes of a laser that is 60 cm long and contains gas of refractive index 1.0204. If the bandwidth of the laser beam is 1.5×10^9 Hz, find the number of optical modes in the output radiation.

Solution

The data given are $\mu = 1.0204$, $L = 60$ cm. The frequency difference is calculated as

$$\Delta v = \frac{c}{2L\mu} = \frac{3 \times 10^{10}}{2 \times 60 \times 1.0204} = 2.45 \times 10^8 \text{ Hz.}$$

The number of optical modes $N = \frac{\text{Bandwidth}}{\Delta v} = 6.$

There are 6 optical modes in the output.

Diffraction loss

In deriving Eq. (8.21) for various oscillating frequencies, it is assumed that plane light wave propagates to and fro unmodified inside the optical cavity. However, practically, the mirrors of any practical optical resonator have finite transverse dimension and hence, only that portion of the light wave that strikes the mirror would get reflected; the portion of the wave lying outside the transverse dimension of the mirror would be lost from the cavity due to diffraction effects. This loss constitutes a basic loss called diffraction loss. For the diffraction loss to be minimum in a resonator cavity employing plane mirrors, we should have

$$\frac{\lambda}{D} << \frac{D}{L} \quad \text{or} \quad \frac{D^2}{\lambda L} >> 1 \tag{8.24}$$

The quantity $D^2/\lambda L$ is known as the Fresnel number. For the diffraction loss to be minimum, the Fresnel number should be much more than unity. The diffraction loss can be minimized by using concave mirrors in place of plane mirrors, by grinding one end of the active material to a prism shape and so on. For the diffraction loss to be minimum in a resonator cavity employing spherical mirrors, we should have

$$0 \leq \left(1 - \frac{L}{R_1}\right)\left(1 - \frac{L}{R_2}\right) \leq 1 \tag{8.25}$$

R_1 and R_2 are the radius of curvature of the two spherical mirrors.

Example 8.3

Prove that for a resonator cavity consisting of a plane circular mirror of diameter 1 cm and an active material of length 60 cm employing light of wavelength 6000 Å, the diffraction loss will be minimum.

Solution

The data given are $D = 1$ cm, $L = 60$ cm, $\lambda = 6000$ Å 6000×10^{-8} cm. The Fresnel number will be

$$\frac{D^2}{\lambda L} = \frac{1^2}{6000 \times 10^{-8} \times 60} = 278,$$

which is much more than 1 implying that the diffraction loss is the minimum.

Laser action

The key to laser action is the presence of a number of energy levels of different widths in atoms. For a three-level laser system, there must be (i) ground state of energy E_0, (ii) excited state of energy E_2 and (iii) in between E_0 and E_2 metastable state of energy E_1 nearer to E_2.

For a four-level laser system, there must be (i) ground state of energy E_0, (ii) excited state of energy E_2, (iii) in between E_0 and E_2 metastable state of energy E_1 nearer to E_2 and (iv) in between E_0 and E_2 intermediate state of energy E' nearer to E_0.

8.2.5 Three-level laser systems (ruby laser)

As discussed earlier, population inversion takes place when there are at least three energy states, namely, the ground state of energy E_0, the metastable state of energy E_1 and the excited state of energy E_2. The life time of an atom in the ground state E_0 is nearly infinity; in the excited state, it is of the order of 10^{-8} second; in the metastable state, it is of the order of 10^{-3} second. Atomic transitions to ground state, excited state and metastable state are given by the following relations and are illustrated in Fig. 8.2

$$\text{Ground state} \xrightarrow{\text{stimulated absorption}} \text{Excited state}$$

$$\text{Excited state} \xrightarrow[10^{-8}\text{ second}]{\text{spontaneous emission}} \text{metastable state} + (E_2 - E_1) \text{ energy}$$

$$\text{Metastable state} \xrightarrow[10^{-3}\text{second}]{\text{Stimulated emission}} \text{ground state} + \text{photons}$$

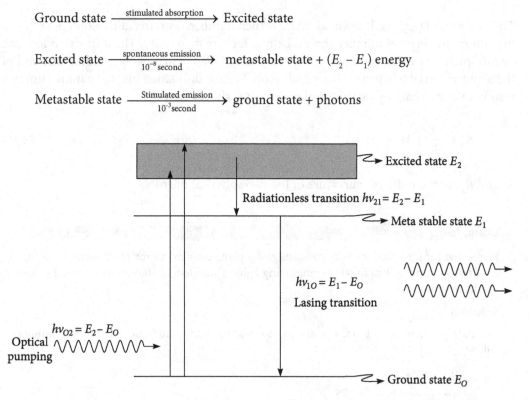

Figure 8.2 Explanation of laser action in a three-level laser system. Population inversion is achieved between ground state E_0 and metastable state E_1. Photons of frequency $v_{02} = (E_2 - E_0)/h$) from an external source strike the atoms in the ground state to excite it to the excited state E_2. Within around 10^{-8} second, atoms from the excited state E_2 drops to the metastable state E_1 by emission of energy $E_2 - E_1$ and energy $E_2 - E_1$ is absorbed by the phonons. Within around 10^{-3} second, the photons emitted by the spontaneous emission from this metastable state initiate stimulated emission as a result of which atoms from the metastable state E_1 drop to the ground state E_0 by emission of laser photons of frequency $v_{01} = (E_1 - E_0)/h$)

As mentioned earlier, in normal conditions, at thermal equilibrium, the ground state is highly populated and the other excited states are sparsely populated. (That means the number of atoms in the ground state is much more than that in the excited states.). Optical pumping is done by irradiating the ground state with highly intense radiation from a xenon flash lamp. Due to the short life time of atoms in the excited state E_2, the energy band of this state has to be broad in accordance with Heisenberg's uncertainty principle. Therefore, the incident radiation from the xenon flash lamp has a narrow range of frequencies instead of a single frequency. The transition of atoms from the excited state E_2 to the metastable state E_1 is non-radiative. The energy released during this transition is absorbed by the phonons of the system and the system gets heated up. As the life time of atoms in the excited state E_2 (10^{-8} second) is many times less than the lifetime of atoms in the metastable state E_1 (10^{-3} second), with continuous optical pumping, the number of atoms in the metastable state keeps on increasing and a stage will come when population inversion is achieved between the ground state and the metastable state.

Now to invoke the stimulated emissions of atoms in the metastable state, we require photons of frequency $v_{01} = (E_1 - E_0)/h)$. These photons are obtained by the spontaneous emission from the metastable state E_1 to the ground state E_0. The photons emitted by the spontaneous emission from the metastable state E_1 to the ground state E_0 stimulate a lasing transition from the metastable state E_1 to the ground state E_0 emitting coherent unidirectional photons profusely. All the photons moving coherently along the axis of the optical cavity will be emitted out as a highly intense, coherent and monochromatic radiation of frequency $v_{01} = (E_1 - E_0)/h)$.

For continuous emission of laser beam of constant intensity, optical pumping should be maintained constantly at such a rate that decrease in the number of atoms in the metastable state is compensated by the optical pumping. This is made possible with the help of a xenon flash lamp. Ruby laser is an example of three-level lasers.

8.2.6 Four-level laser systems (He–Ne laser)

In the previous section, we described in detail the concepts behind a three-level laser. The difficulty in the three-level laser is that as the lasing transition occurs, the population of the ground state is increased, upsetting the population inversion. As a result of this, excess population in the ground state allows absorption of lasing transition, thereby removing the photons that might contribute to the lasing action. This difficulty is removed in four-level laser systems. A four-level laser system is depicted in Fig. 8.3.

The ground state is pumped to an excited state that decays rapidly to the metastable state as with the three-level laser systems. The lasing transition proceeds from the metastable state to yet another lower energy state called the intermediate state which in turn decays rapidly to the ground state. The atom in its ground state thus cannot absorb the energy of the lasing transition. The lower short-lived intermediate state decays rapidly, its population is always smaller than that of the metastable state, which maintains the population inversion. Helium–neon laser works on the principle of four-level laser systems.

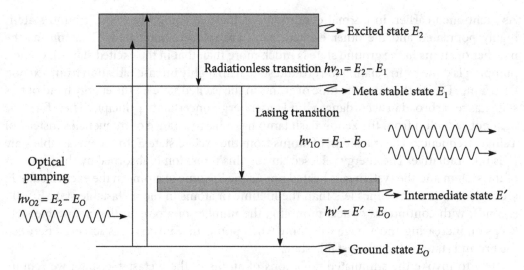

Figure 8.3 | Explanation of laser action in a four-level laser system. Population inversion is achieved between ground state E_0 and metastable state E_1. Photons of frequency $v_{02} = (E_2 - E_0)/h)$ from an external source strike the atoms in the ground state to excite it to the excited state E_2. Within around 10^{-8} second, atoms from the excited state E_2 drops to the metastable state E_1 by emission of energy $E_2 - E_1$ and the energy $E_2 - E_1$ is absorbed by the phonons. Within around 10^{-3} second, the photons emitted by the spontaneous emission from this metastable state initiate the stimulated emission as a result of which atoms from the metastable state E_1 drop to the intermediate state E' by emission of laser photons of frequency $v_{01} = (E_1 - E')/h)$. The atoms from the excited intermediate state E' drop to the ground state very quickly by emission of energy $hv' = E' - E_0$

8.2.7 Broadening of laser radiation

Laser radiation does not have a single frequency but a very small range of frequencies. The spreading of frequencies over a small range is attributed mostly to (i) Heisenberg's uncertainty principle

$$\Delta E \Delta t = \frac{\hbar}{2}.$$

The other reasons for the broadening of laser radiation are (ii) collision of excited atoms with other atoms/ molecules/ electrons and (iii) Doppler effect.

Heisenberg's uncertainty principle

Heisenberg's uncertainty principle states that

$$\Delta E \Delta t = \frac{\hbar}{2}.$$

Here, ΔE is the width of an energy level and Δt is the life time of an excited atom in this energy level. We know that the ground state of an atom is most stable, i.e., the life time is nearly infinite ($\Delta t \to \infty$) implying that the energy level is very sharp in atoms in the ground state. However, the life time of an atom in the excited state is at best 10^{-3} second as a result of which the energy level cannot be sharp and must have a certain width called the width of energy level of excited atoms. Therefore, energy of all the photons emitted during electronic transition from the excited state to the ground state cannot have the same energy but energies spread over a small energy range. Since all the photons have different energies spread over a small range, they have different frequencies spread over a small range.

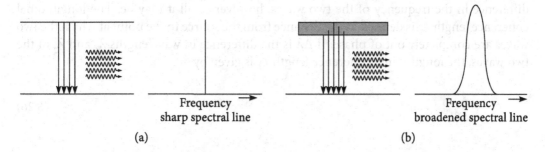

Frequency →
sharp spectral line
(a)

Frequency →
broadened spectral line
(b)

Figure 8.4 Broadening of laser radiation. (a) Transition of electrons from one sharp energy level to another sharp energy level, produces sharp spectral lines. (b) Transition of electron from one broadened energy level of a certain width to another sharp energy level produces spectral lines of a certain width

As Heisenberg's uncertainty principle is inherent in nature, width of the energy level is inherent of excited atoms.

Collisional broadening

The excited atoms may drop to the ground state at an earlier time than its natural time (i.e., Δt decreases) when they collide with other atoms/ molecules/ electrons. This causes the uncertainty ΔE to increase, making the range of frequency to increase.

Doppler effect broadening

In the laser medium, the atoms are in random motion. Due to the motion of the excited atoms, when they drop to the ground state, there will be a shift in the frequency of the emitted radiation. Moreover, there will be a small distribution of frequency in the emitted radiation about the actual frequency v_0. This is called the Doppler effect broadening and is more pronounced in gas laser.

Total broadening due to all these effects has been illustrated in Fig. 8.4.

8.2.8 Coherence

As all the photons coming out of the laser system are in phase, the laser is a coherent source of radiation. However, it is not possible to maintain coherence throughout the propagation distance because all the photons of a laser beam do not have exactly the same frequency but frequencies spread over a small range as described earlier. Depending upon the positions up to which laser beam maintains coherence, coherence are of two types: (i) longitudinal coherence and (ii) transverse coherence.

Longitudinal coherence

Let us consider two waves emitted from a single source. Very close to the source, the two waves are almost in phase, i.e., they have zero or nearly zero phase difference. As the length of propagation increases, the two waves become more and more out of phase and at a certain distance, the two waves become completely out of phase. This is because of the difference in the frequency of the two waves, however small it may be. The longitudinal coherence length ℓ_L is defined as the distance from the source to the point at which the two waves are completely out of phase. If $\Delta\lambda$ is the difference in wavelengths λ_1 and λ_2 of the two waves, the longitudinal coherence length ℓ_L is given by

$$\ell_L = \frac{\lambda^2}{\Delta\lambda} \qquad\qquad (8.26)$$

where $\lambda = \dfrac{\lambda_1 + \lambda_2}{2}$ and $\Delta\lambda = \lambda_1 \sim \lambda_2$.

Equation (8.26) shows that the closer is the two waves in wavelengths, greater will be the longitudinal coherence length ℓ_L. The longitudinal coherence is also called temporal coherence. This has been depicted in Fig. 8.5(a).

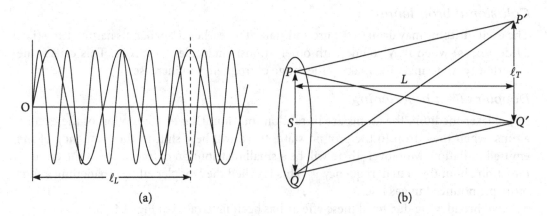

(a)   (b)

Figure 8.5 | Illustration of (a) longitudinal coherence or temporal coherence and (b) transverse coherence or spatial coherence

Transverse coherence

Let us consider two waves emitted from two points P and Q of a single source as shown in Fig. 8.5(b). The points P and Q are situated on a single plane transverse to the laser beam. Very close to the source the two waves at the points P and Q are almost in phase, i.e., they have zero or nearly zero phase difference. As the length of propagation increases, the two waves loses their parallelism and become more and more out of phase laterally and after travelling a certain distance L, the two waves become completely out of phase laterally, i.e., the two waves at P′ and Q′ are completely out of phase. This is because of the difference in the frequency of the two waves, however small it may be. The transversal coherence length ℓ_T is defined as the lateral distance between two points (P′ and Q′) situated on a plane transverse to the laser beam at which the two waves are completely out of phase. If the longitudinal distance of the transversal plane passing through these two points P′ and Q′ is L and s is the distance between the two points P and Q, then the transversal coherence length ℓ_T is given by

$$\ell_T = \frac{L\lambda}{s} \tag{8.27}$$

where λ is the wavelength of the laser source.

Example 8.4

A laser beam of wavelength 7400 Å has coherence time 4×10^{-5} s Determine the temporal coherence length, spectral width and purity factor.

Solution

The data given are $\lambda = 7400$ Å $= 7.4 \times 10^{-7}$ m, $\tau = 4 \times 10^{-5}$ s, $c = 3 \times 10^8$ m/s.

The temporal coherence length $\ell = \tau c = 4 \times 10^{-5} \times 3 \times 10^8 = 12 \times 10^3$ m $= 12$ km.

The spectral width $\Delta\lambda = \dfrac{\lambda^2}{\ell} = \dfrac{(7.4 \times 10^{-7})^2}{12000} = 4.56 \times 10^{-17}$ m.

The purity factor $Q = \dfrac{\lambda}{\Delta\lambda} = \dfrac{7.4 \times 10^{-7}}{4.56 \times 10^{-17}} = 1.6 \times 10^{10}$

8.3 Practical Lasers

8.3.1 Ruby laser

The first successful laser was the ruby laser based on the three energy level concept. It is a solid state laser device. The different parts of the ruby laser are shown in Fig. 8.6. It consists of three main parts:

a.   *Active working material*   The working material here is a rod of ruby crystal.
b.   *Resonant cavity*   The resonant cavity is made up of two optically plane mirrors (one fully reflecting and the other partially reflecting) attached to the two ends of the ruby rod.
c.   *Exciting system*   In case of a ruby laser, it is a xenon flash tube.

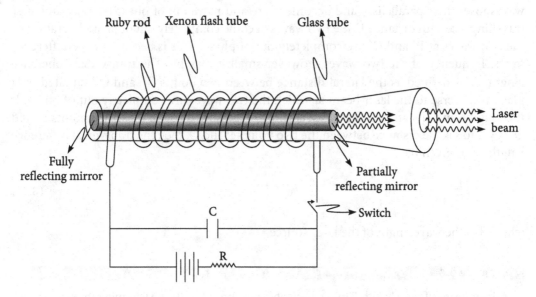

Figure 8.6 │ Different parts of a ruby laser

Ruby is simply a single crystal of Al_2O_3 (specifically, α-alumina, a stable form of aluminum oxide) to which has been added an order of 0.05% Cr^+. These ions impart to ruby its characteristic red colour. As previously explained, these ions provide electron states that are essential for the laser to function. The ruby laser is in the form of rods, the ends of which are optically flat, parallel and highly polished. Both ends are silvered such that one end is highly (~96%) reflecting and other end is partially transmitting (~50%). The ruby rod is surrounded by a helical xenon flash tube. The ruby rod is illuminated with light from the xenon flash tube. Before this exposure, virtually all the Cr^+ ions are in their ground state. However, photons of wavelength 5600 Å from xenon flash tube excite electrons from the Cr^+ ions into higher energy states. These electrons can decay back into their ground state by two different paths. Some fall back directly and the associated photon emissions are not part of the laser beam. Other electrons decay into a metastable state where they may reside up to 0.003 second before spontaneous emission. The energy levels of a chromium atom are shown in Fig. 8.7.

The initial spontaneous photon emission by a few of these electrons is the stimulus that triggers an avalanche of emissions from the remaining electrons in the metastable state. Of the photons directed parallel to the long axis of the ruby rod, some are transmitted through the partially silvered end; others incident on the highly reflected end are reflected. The photons that are not emitted in this axis direction are lost. The photon beam repeatedly

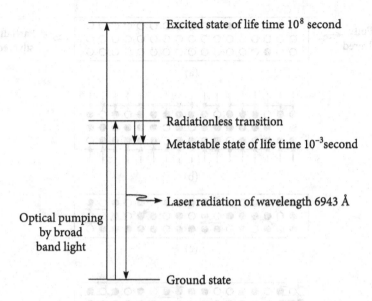

Excited state of life time 10^8 second

Radiationless transition

Metastable state of life time 10^{-3} second

Laser radiation of wavelength 6943 Å

Optical pumping
by broad
band light

Ground state

Figure 8.7 | Illustration of the energy levels of chromium atom in ruby laser. Laser action takes place between the metastable state and the ground state of chromium atom

travels back and forth along the rod length, and its intensity increases as more and more emissions are stimulated. The stimulated emission and light amplification for a ruby laser are shown schematically in Fig. 8.8.

Ultimately, a high intensity, coherent and highly collimated laser light beam of wavelength 6943Å of short duration is transmitted through the partially silvered end of the ruby rod.

8.3.2 He–Ne gas laser

The main drawback of ruby laser is that though the output beam is very intense, it is not continuous. For continuous laser beam, gas lasers are in use. The He–Ne laser is a four-level gas laser. A mixture of He–Ne (10:1) is enclosed in a glass tube at a pressure of 13.33 Pa. One end of the glass tube is highly reflecting while the other end is partially reflecting so that a resonator cavity is formed. The laser action takes place in the energy level of the neon atom. Helium atoms help to achieve population inversion by imparting their energy to the neon atoms. The electric discharge produced in the tube excites the He atoms from the ground state to the excited state at 20.61 eV. The Ne atom has a metastable state at 20.66 eV, which is very close to the excited state of He atom at 20.61 eV. Due to random motion, the excited He atoms collides with Ne atoms in the ground state as a result of which Ne atoms jump to the metastable state by absorbing energy from the excited He atoms during collision and He atoms returns to the ground state. Again the electric discharge produced in the tube excites the He atoms from the ground state to excited state at 20.61 eV and the transition of Ne atoms from the ground state to the excited state goes on due to collision with excited He atoms. Very soon population inverse is achieved in Ne atoms between the metastable state

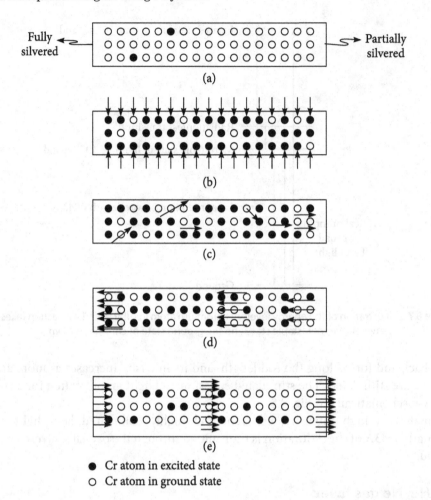

Fully silvered

Partially silvered

(a)

(b)

(c)

(d)

(e)

● Cr atom in excited state
○ Cr atom in ground state

Figure 8.8 Schematic representation of the stimulated emission and light amplification for a ruby laser. (a) The chromium ions before excitation. (b) Electrons in some chromium atoms are excited into higher energy states by the xenon light flash. (c) Emissions from metastable electron states are stimulated by the photons that are spontaneously emitted. (d) Upon reflection from the silvered ends, the photons continue to stimulate emissions as they traverse the rod length. (e) The coherent and intense beam is finally emitted through the partially silvered end

at 20.66 eV and the lower energy intermediate state at 18.70 eV. The laser action takes place during the transition of Ne atoms from the metastable state at 20.66 eV to the intermediate state at 18.70 eV. In the He–Ne laser, other stimulated emissions like $3s \rightarrow 3p$ and $2s \rightarrow 2p$ also takes place whose spectrum lies in the infrared region. The relevant energy level diagram of helium and neon atoms is shown in Fig. 8.9.

The main advantages of gas lasers are exceptionally high monochromaticity, pure spectrum, and high stability of frequency.

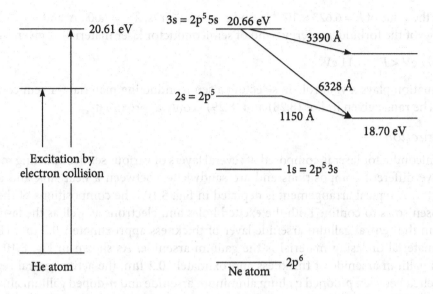

Figure 8.9 Relevant energy level diagram of helium and neon atoms. The notations 1s, 2s, 3s, 2p, and 3p are used to denote the excited state of the Ne atoms

8.3.3 Semiconductor lasers

A light emitting diode (LED), basically a p-n junction, emits incoherent light due to spontaneous emission. By suitably exploiting a p-n-junction, population inversion and stimulated emission can be achieved in it; laser action will be possible in a semiconductor p-n junction. This device is called a semiconductor laser or diode laser or laser diode. A typical semiconductor laser is made of gallium arsenide.

Principle

In radiative transition of excited electrons, from the conduction band to the valence band, to combine with holes, photons are emitted in the visible range. The forbidden energy gap (band gap) E_g between the valence band and the conduction band in the materials used for the semiconductor laser is that the wavelength λ associated with band gap energy

$$E_g \left(\lambda = \frac{hc}{E_g} \right)$$

must be within the visible range, i.e., $4000 \text{ Å} \leq \lambda \leq 7000 \text{ Å}$. The forbidden energy gap (band gap) E_g in these materials will be given approximately by

$$E_g = \frac{hc}{\lambda}, \text{ with } 4000 \text{ Å} \leq \lambda \leq 7000 \text{ Å}. \tag{8.28}$$

Putting the value of $h = 6.623 \times 10^{-34}$ Js, and $c = 3 \times 10^8$ m/s, $\lambda_{min} = 4000$ Å and $\lambda_{max} = 7000$ Å, the range of the forbidden energy band of semiconductor laser materials is given by

$$1.77 \text{ eV} < E_g < 3.11 \text{ eV} \tag{8.29}$$

This equation plays a vital role in selecting a semiconducting material for semiconductor lasers. The range given in Eqs (8.28) and (8.29) is only approximate.

Construction

The semiconductor laser is composed of several layers of various semiconducting materials that have different compositions and are sandwiched between a heat sink and a metal conductor. A typical arrangement is depicted in Fig. 8.10. The compositions of the layers are chosen so as to confine both the excited holes and electrons as well as the laser beam to within the central gallium arsenide layer of thickness approximately 0.2 μm. Here the active material or lasing material is the gallium arsenide. As shown in Fig. 8.10, a thin layer of gallium arsenide of thickness approximately 0.2 μm, the active central region, is sandwiched between p-doped gallium aluminum arsenide and n-doped gallium aluminum arsenide. The p-doped gallium aluminum arsenide and n-doped gallium aluminum arsenide are called confining regions. The confining regions are different from the active central region in the percentage of aluminum. This is done to control the refractive index and the forbidden energy band of the materials. The confining regions should have slightly higher forbidden energy band than the active material (here gallium arsenide), as shown in Fig. 8.11, to ensure population inversion. The variation of refractive index ensures confinement of emitted photons to the central active region.

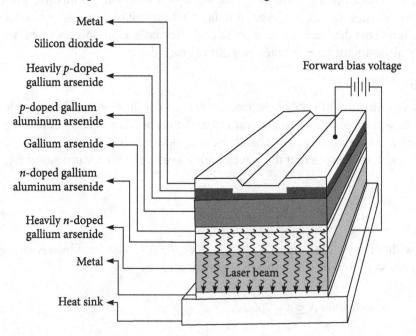

Figure 8.10 | Construction of a semiconductor laser

One end of the semiconductor laser is fully reflecting and other end is partially reflecting. The highly coherent laser beam comes out through the partially reflecting end.

Working principle

A voltage is applied to the arrangement as shown in Fig. 8.11 in forward bias to inject electrons into the n-type material and holes into the p-type material. The electrons injected into the n-type material from the external circuit diffuse into the active region. The electrons are prevented from diffusing into the p-type material by a potential barrier. Hence, they are concentrated in the active region. Similarly, the holes injected into the p-type material from the external circuit diffuse into the active region. The holes are prevented from diffusing into the n-type material by the potential barrier. Hence, they are concentrated in the active region. This creates the population inversion discussed earlier.

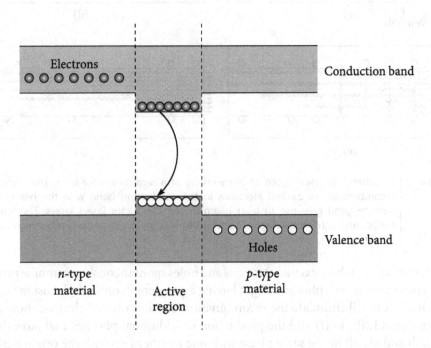

Figure 8.11 | Energy bands in a semiconductor laser. The active region has a smaller energy gap than the n-type and p-type regions (confining region) on either side

Working

A voltage is applied to the arrangement shown in Fig. 8.10. The applied voltage excites the electrons from the valence band to the conduction band across the forbidden energy band resulting in the creation of holes in the conduction band as shown in Fig. 8.12(a). Figure 8.12(a) shows the energy band diagram of the semiconducting material along with several holes and excited electrons.

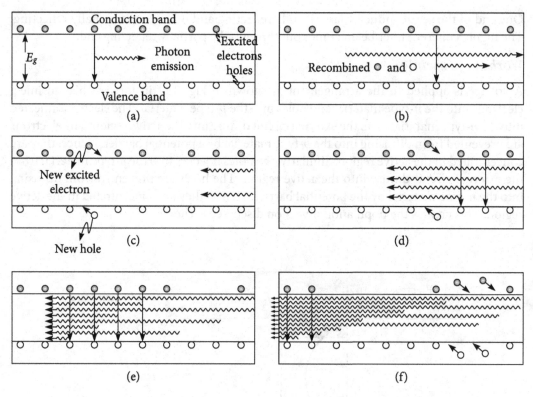

Figure 8.12 | Schematic representations of the working of a semiconductor laser. The stimulated recombination of excited electrons in the conduction band with the holes in the valence band give rise to laser beam in semiconductor lasers lasers. The left side surface and right side surface of each figure are respectively partially reflecting and fully reflecting

Subsequently, a few of these excited electrons and holes spontaneously recombine producing spontaneous emission of a photon of light having a wavelength obtained by using Eq. (8.28). One such photon will stimulate the recombination of other excited electron–hole pairs as shown in Figs 8.12(b) to (f) and the production of additional photons that have the same wavelength and are all in the same phase with one another and with the original photons. Thus, a monochromatic and coherent beam is produced. In a semiconductor laser, one end is fully reflecting and other end is partially reflecting. At the fully reflecting end of the semiconductor laser, this light beam is reflected back into the material so that additional recombination will be stimulated. The other end of the laser is partially reflecting, which allows for some of the photons to escape. With this type of laser, a continuous laser beam is produced inasmuch as a constant applied voltage ensures that there is always a steady source of holes and excited electrons.

Detailed explanation of Fig. 8.12

The working of a semiconductor laser is figuratively explained in detail in Fig. 8.12. The stimulated recombination of excited electrons in the conduction band with the holes in the valence band gives rise to a laser beam in semiconductor lasers. Each part of Fig. 8.12 is explained as follows.

Figure 8.12(a)   One excited electron recombines with a hole and the energy associated with this recombination is emitted as a photon of light of wavelength $\lambda = \dfrac{hc}{E_g}$.

Figure 8.12(b)   The photons emitted by (a) stimulates the recombination of another excited electron and a hole resulting in the emission of another photon of light having $\lambda = \dfrac{hc}{E_g}$.

Figure 8.12(c)   The two photons emitted in (a) and (b) having the same wavelength and being in phase with one another are reflected by the fully reflected end back into the active central region. In addition, new excited electrons and new holes are generated by a current that passes through the semiconductor.

Figure 8.12(d) and (e)   In proceeding through the semiconductor, more excited electron–hole combinations are stimulated, which give rise to additional photons of light that also become part of the monochromatic coherent laser beam.

Figure 8.12(f)   Some of the laser beam escapes through the partially reflecting end of the semiconductor laser.

Diode lasers are small in size and consume very less power (typically 10mW, compared with standard He–Ne laser that may consume several watts). As a result, diode lasers can be powered by batteries. Efficiency of the order 20% are possible (i.e., 20% of the electrical power supplied to the device is converted to laser light), compared with 0.1% in the He–Ne laser. The light signal can be turned off and on in switching times that are characteristic of semiconductors ($< 10^{-10}$ s), and thus, we have a device that can rapidly modulate the beam.

Applications of semiconductor lasers

Semiconductor lasers have wide applications from shopping complexes to the research laboratory.

i.   Semiconductor lasers are used in bar code reading devices.

ii.   They are used in computer CD and DVD drives to read and write.

iii.   Semiconductor lasers are used in fiber optical communication systems.

Example 8.5

In a material, transition occurs between a metastable state and an energy level of 0.25 eV. The wavelength of the radiation emitted is 1100nm. Calculate the energy of the metastable state.

Solution

The data given are $E_e = 0.25$ eV $= 0.4 \times 10^{-19}$ J and $\lambda = 1100 \times 10^{-9}$ m.

We know

$$E_m - E_e = h\nu = \frac{hc}{\lambda} = \frac{6.63 \times 10^{-34} \times 3 \times 10^8}{1100 \times 10^{-9}} \text{ J}$$

or $E_m = 0.4 \times 10^{-19}\text{J} + \dfrac{6.63 \times 10^{-34} \times 3 \times 10^8}{1100 \times 10^{-9}} \text{ J} = 2.2 \times 10^{-19}\text{J} = 1.375 \text{ eV}$

The energy of the metastable state is 1.375 eV.

8.4 Applications of Lasers

The light produced by lasers is in general far more monochromatic, directional, powerful, and coherent than that from any other light sources. Nevertheless, individual kinds of lasers differ greatly in these properties as well as in wavelength, size, and efficiency. There is no single laser suitable for all purposes, but combinations of properties can do things that were difficult or impossible before lasers were developed.

i.   *Installation of large structures*   A continuous visible beam of laser provides an ideal straight line for all kinds of alignment applications. The beam from such a laser typically diverges by less than one part in a thousand, approaching the theoretical limit. The beam's divergence can be reduced by passing it backward through a telescope, although fluctuations in the atmosphere then limit the sharpness of the beam over a long path. Lasers have come to be widely used for alignment in large constructions. For example, to guide machines for drilling tunnels and for laying pipelines. The alignments of nuclear accelerators or particle accelerators like the large hadron collider are done using laser beams.

ii. *Measurement of long distances*   A pulsed laser can be used in light radar, sometimes called LIDAR (LIght Detection And Ranging). The narrowness of its beam permits sharp definition of targets. As with radar, the distance to an object is measured by the time taken for the light to reach and return from it, since the speed of light is known. LIDAR echoes have been returned from the moon, facilitated by a multiprism reflector that was placed there by the first astronauts to land there. Distances can be measured from an observatory on the earth to the lunar mirror with an accuracy of several centimeters. Simultaneous measurements of the mirror's distance and direction from two observatories on different parts of the earth could give an accurate value for the distance between the two observatories. A series of such measurements can tell the rate at which continents are drifting relative to each other.

iii. *Ultra fine mappings*   Vertically directed laser radar in an airplane can serve as a fast, high-resolution device for mapping fine details, such as the contours of steps in a stadium or the shape of the roof of a house. With pulsed laser radar, returns can be obtained from dust particles and even from air molecules at higher altitudes. Thus, air densities can be measured and air currents can sometimes be traced.

iv. *Laser interferometers* The high coherence of a laser's output is very helpful in measurement and other applications involving interference of light beams. If a light beam is divided into two parts that travel different paths, when the beams come together again they may be either in phase so that they reinforce each other or out of phase so that they cancel one another. Thus, the brightness of the recombined wave changes from light to dark, producing interference fringes, when the difference in path lengths is changed by one-half of a wavelength. Such devices are called laser interferometers. Very small displacements can be detected, and larger distances can be measured with precision. With lasers, these measurements can be carried out over extremely long distances. Laser interferometers are used to monitor small displacements in the Earth's crust across geological faults. In manufacturing, such devices are employed to gauge fine wires, to monitor the products of automated machine tools, and to test optical components.

v. *Doppler effects* Lasers can be so monochromatic that a small shift in the light frequency can be detected. Light reflected from an object that is moving toward the laser is raised in frequency by an amount depending on the velocity of the object (Doppler effect). For a receding object, the frequency is lowered. Small velocities can be measured by recording the change of frequency.

vi. *Holography* The brightness and coherence of laser light make it especially suitable for holography, the three dimensional imaging. The holography could not have been possible without lasers.

vii. *Diamond processing* The light from many lasers is relatively powerful and can be focused by a conventional lens system to a small spot of great intensity. Thus, even a moderately small pulsed laser can vaporize a small amount of any substance and drill narrow holes in the hardest materials. Ruby lasers, for example, are used to drill holes in diamonds for wire drawing dies and in sapphires for watch bearings.

viii. *Biology* For biological research, a finely focused laser can vaporize parts of a single cell, thus permitting microsurgery of chromosomes.

ix. *Retina operation* Strong heating can be produced by a laser at a place where no mechanical contact is possible. One of the earliest applications of lasers was for surgery on the retina of the eye.

x. *Electronics* Lasers are also used for small-scale cutting and welding. They can trim resistors to exact values by removing material and can alter connections within integrated arrays of microcircuit elements. A pulse of light from a laser can vaporize a sample of a substance for analysis by suitable instruments. By this method, an extremely small sample can be analyzed without introducing contaminants.

xi. *Raman effect* The high brightness, pure colour, and directionality of laser light make it ideally suitable for experiments on light scattering. Even a small amount of light that is scattered with a change of wavelength or direction can be readily identified. Laser light Raman effect produces characteristic wavelength shifts by which molecular species can be identified. With laser sources and sensitive spectrography, small samples of transparent liquids, gases, or solids can be analyzed. It is even possible to measure contaminants in the atmosphere at a considerable distance by the Raman scattering of light from a laser beam.

xii. *Communications* Laser beams can be used for communications. Because the light frequency is so high ($\sim 10^{15}$ Hz for visible light), the intensity can be rapidly altered to encode very complex signals. In principle, one laser beam could carry as much information as all existing radio channels.

xiii. *Computers* Laser technology is integral to optical disc recording and storage systems. In such a system, digital data are recorded by burning a series of microscopic holes, commonly referred to as pits, with a laser beam into thin metallic film on the surface of a small disc. In the read mode, laser light of low intensity is reflected off the disc surface and is "read" by light-sensitive diodes. The amount of light received by the diodes varies according to the presence or the absence of the pits, and this input is digitized by the diode circuits. The digital signals are subsequently converted into analog information on a video screen.

Lasers are also used in computer printers. Laser printers employ a laser beam and a system of optical devices to etch images on a photoconductor drum. The images are carried from the drum to paper by means of electrostatic photocopying.

Defence US–British forces devastated Iraq in 1991 and 2003 by using laser guided missiles, smart bombs warheads and the like. Most modern day warheads are laser guided as a result of which casualties of human lives is kept minimum.

8.5 Light Emitting Diodes

A light emitting diode (LED) is a forward biased p-n junction diode, emitting incoherent light due to spontaneous emission generated by the annihilation of injected holes and electrons in the depletion layer. The injected holes and electrons annihilate each other across the junction, thus generating photons if the quantum efficiency[**] is unity. LEDs can radiate ultra violet, visible or infrared rays depending upon the type of compound semiconductors used. They convert electrical energy to light energy. The first visible LED with extensive applications was based on $GaAs_{1-x}P_x$[***] grown on GaAs substrates. The composition for maximum brightness depends on both the quantum efficiency of the LED and the sensitivity of the eye. The maximum brightness occurs near $x = 0.4$, which

---

[**] The ratio of the number of emitted photons to the number of electrons crossing the p-n junction is the quantum efficiency.

[***] In the notation used for ternary solid solutions, for example, the x in $Al_xGa_{1-x}As$ means that x percent of the group III elements are aluminum and $(1 - x)$ percent of the group-III elements are gallium. For example, with x = 0.3, the $Al_{0.3}Ga_{0.7}As$ would have 30 percent aluminum and 70 percent gallium. For quaternary solid solutions such as $Ga_xIn_{1-x}As_yP_{1-y}$, the x represents the percentage of group III elements, while the y represents the percent of group V elements. The energy gap and therefore, the wavelength of the emitted light, changes with the composition of x or x and y. The wavelength of the emitted light can be changed by varying the semiconductor composition.

corresponds to an energy gap of approximately 1.9 eV. LED emission is generally in the visible part of the spectrum with wavelengths from 0.4 μm to 0.7 μm.

8.5.1 Principle

In radiative recombination, the excited electrons in the conduction band spontaneously jump to the valence band to combine with holes and are accompanied by the emission of photons in the visible range. When a *p-n* junction is forward biased, the electrons and holes, injected into the *p*-region and *n*-regions respectively through the depletion layer are recombined to give out photons of energy equal to the energy gap between the valence band and the conduction band. This process is called injection electroluminescence.

The forbidden energy gap, i.e., the band gap E_g between the valence band and the conduction band in materials used for LED, is that region where the wavelength λ associated with band gap energy

$$E_g\left(\lambda = \frac{hc}{E_g}\right)$$

must be within visible range, i.e., 4000 Å $\leq \lambda \leq$ 7000 Å. The forbidden energy gap (band gap) E_g in these materials will be given by

$$E_g = \frac{hc}{\lambda}, \text{ with 4000 Å} < \lambda < 7000 \text{ Å}$$

Putting the value of $h = 6.626 \times 10^{-34}$ Js and $c = 3 \times 10^8$ m/s, $\lambda_{min} = 4000$ Å and $\lambda_{max} = 7000$ Å, the range of forbidden energy band of semiconductor laser materials is given by

$$1.77 \text{ eV} < E_g < 3.11 \text{ eV}$$

This equation plays a vital role in selecting a semiconducting material for visible LED. The materials InP, GaAs, GaAs$_{1-x}$P$_x$ are suitable for visible LED.

Two types of radiative recombination mechanisms are commonly encountered in LEDs depending upon the band gap and characteristics of the semiconductor material used. These are (i) direct recombination and (ii) indirect recombination.

i.   **Direct recombination**   The emission of photons as a result of recombination of electrons and holes is possible only when both energy and momentum are conserved. A photon can have considerable energy but its momentum $h\nu/c$ is very small. Therefore, the simplest and most probable recombination process will be that where electrons and holes have the same value of momentum. This situation exists in many III–V compound semiconductors called direct band gap semiconductors (e.g., GaAs) where the conduction band minimum and valence band maximum occurs at zero momentum position as shown in Fig. 8.13. In this case, recombination is called direct recombination.

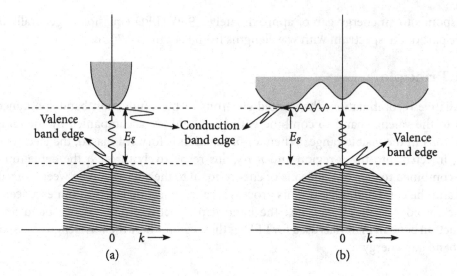

Figure 8.13 | (a) Band structure of direct band gap semiconductor with photon emission and (b) Band structure of indirect band gap semiconductor with photon emission

ii. **Indirect recombination**   In indirect band gap semiconductors, the conduction band minimum and valence band maximum occurs at different values of momentum and therefore, in indirect recombination, to conserve momentum, third particles such as phonons or chemical impurities has to be involved. The probability of indirect recombination in indirect band gap semiconductors [Fig. 8.13(b)] is thus obviously much less than direct recombination in direct band gap semiconductors.

8.5.2 Construction

Silicon, the most commonly used semiconductor for electronic devices and integrated circuits, is not suitable for LEDs. However, in compound semiconductors like gallium arsenide (GaAs), gallium arsenide phosphide ($GaAs_{1-x}P_x$), aluminum gallium arsenide ($Al_xGa_{1-x}As$), aluminum gallium indium phosphide [$(Al_xGa_{1-x})_yIn_{1-y}P$] and gallium indium arsenide phosphide ($Ga_xIn_{1-x}As_yP_{1-y}$), radiative recombination can occur with ease. The peak intensity of the emission spectrum occurs at a photon energy slightly less than the semiconductor energy gap. The radiative recombination must compete with various non-radiative recombination processes, because of undesirable impurities and crystal defects, including precipitations and dislocations. Careful material processing is necessary to obtain useful LEDs.

In a typical LED, the clear epoxy dome serves as a structural element to hold the lead frame together, as a lens to focus the light, and as a refractive index match to permit more light to escape from the chip. The LED chip, typically 250 μm $\times$ 250 μm $\times$ 250 μm, is mounted in a reflecting cup formed in the lead frame. The n- and p-type GaP : N layers represent nitrogen added to GaP to give green emission, the n- and p-type GaAsP : N layers represent nitrogen added to $GaAs_xP_{1-x}$ to give orange and yellow emission, and the p-type GaP: ZnO layer represents zinc and oxygen added to GaP to give red emission. Another useful material is $Al_xGa_{1-x}As$ for red LEDs, although a more complex structure is required than for GaP: ZnO. The $Al_xGa_{1-x}As$ light-emitting layer is confined between two larger energy gap $Al_yGa_{1-y}As$ layers $(x < y)$; this permits the more efficient injection of carriers into the light-emitting layers and allows the confining layers to be transparent to the generated light, permitting a high light-extraction efficiency.

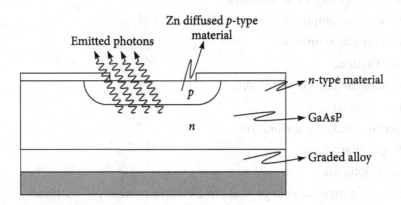

Figure 8.14 | Construction of a typical LED

In the early 1990s, two further enhancements to LEDs were developed. One was the introduction of LEDs, based on $(Al_xGa_{1-x})_yIn_{1-y}P$, which has a larger energy gap than $Al_xGa_{1-x}As$ and permits higher-efficiency LEDs throughout the spectral region from green to red-orange. The other was the commercialization of blue-emitting LEDs, based on silicon carbide (SiC), although these blue LEDs are at least an order of magnitude less efficient than the other LEDs and more costly. However, blue LEDs can be combined on a cluster with other colour LEDs to give all colours, including white, for full-colour moving-message panel applications. In 1994, blue LEDs, based on the III–V compound GaN were demonstrated with 2.0 percent efficiencies.

8.5.3 Applications of LED

More than 20 billion LEDs are produced each year. This shows the utility of LEDs. LEDs have a wide variety of applications starting from commercial establishments to scientific work, few of which are cited here.

i. LEDs in conjunction with a photodiode or other photosensitive device can be used as light sources in optical fiber communication systems.

ii. Visible LEDs are used as numeric displays or indicator lamps and are sufficiently bright that a row of red LEDs are used in an automobile spoiler to replace the conventional rear-window brake light.

iii. Infrared LEDs are employed in optoisolators and in television remote controls.

iv. LEDs are used in digital displays in all modern day electronic devices.

8.5.4 Merits of LED over conventional incandescent lamps

The following are the merits of LED over other conventional incandescent and other types of lamps.

i. Low working voltages and currents.

ii. Less power consumption.

iii. No need for warm up time.

iv. Very first action.

v. Emission of monochromatic light.

vi. Small size and weight.

vii. No effect of mechanical vibration

viii. Less fragile than glass.

ix. Extremely long life.

Typical LED uses a forward voltage of about 2V and a current of 5 to 10mA.

Gallium arsenide (GaAs) produces infra red light while red, green and orange lights are produced by gallium arsenide phosphide (GaAsP) and gallium phosphide (GaP).

8.6 Optical Fibers

The optical frequencies are extremely high ($\sim 10^{15}$Hz) in comparison to radio waves ($\sim 10^{6}$Hz) or microwaves ($\sim 10^{10}$Hz) and therefore, a light beam acting as a carrier wave can carry far more information than radio or microwaves. The attempts in this direction gave birth to optical fibers. Optical fibers are wave guides that carry light through long distances with very low losses. They are made from certain materials by using very high precision techniques. Signal transmission through metallic wires is electronic, whereas in optical fibers, it is photonic. Using optical fibers, speed of transmission, information density and transmission density have increased many times with decrease in error rate. Regarding information density, two small optical fibers can transmit the equivalent of 24000 telephone calls simultaneously. 33 tonnes of copper wires and 100 gm of optical fibers transmit the same amount of information! The special features of optical fibers include their tremendous information carrying capacity. They are inexpensive, light weight, have high reliability exceedingly small attenuation and are immune to electromagnetic radiation.

8.6.1 Structure of optical fibers

Most optical fibers are made of silica glass. They are generally thinner than human hair. Fiber core diameters range between 1 μm and 200 μm, while cladding diameters are between 100 μm and 400 μm.

The cross-sectional view of an optical fiber is shown in Fig. 8.15. The fiber components are core, cladding and jacket.

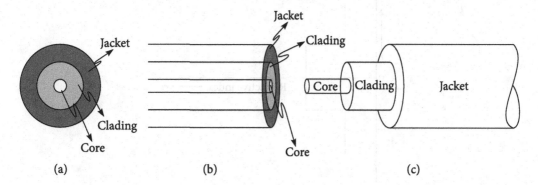

Figure 8.15 | (a) The cross-sectional view of an optical fiber. The central part of the optical fiber is called core, its covering is called cladding and the outer most coating is called jacket. (b) and (c) show the core, cladding and jacket longitudinally

The cylindrical central part of an optical fiber is called core. The core is made up of pure silica. The diameter of the core is of the order of few μm. The cylindrical coating over the core is called cladding. The diameter of the cladding is of the order of 100–400 μm. The cladding is made of silica doped with suitable amounts of germanium and fluorine to control the refractive index. The refractive index of the cladding is less than that of the core so that occurrence of internal reflection is possible. The outer-most coating, of thickness of the order of 60 μm and made up of polymeric material, is called the jacket. The jacket protects the core and cladding from external damages that might result from abrasion and external pressures.

8.6.2 Classification of optical fibers

There are two methods of classification of optical fibers; one method is based on the variation of the refractive index of the core and the other method is based on the core diameter or modes of transmission of the signal.

Classification on the basis of variation of the refractive index of the core

Depending upon the variation of the refractive index of the core from its axis to the surface (or up to the inner surface of the cladding), optical fibers are of two types: (i) step index fiber and (ii) graded index fiber.

Step index fiber

In step index fiber, the refractive index of the core is constant throughout it. The refractive index of the core changes step-wise with respect to the radius of the core as shown graphically in Fig. 8.16.

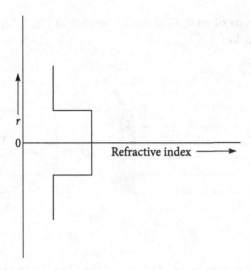

Figure 8.16 | Variation of refractive index of the core of a step index optical fiber with its radius

Light rays entering into the step index fiber at different angles of incidence travel different optical paths before they arrive at the other end of the fiber as illustrated in Fig. 8.17. As can be seen from the figure, smaller is the angle of incidence of a ray, larger is its optical path. The shortest optical path travelled by a light ray is equal to the length of the optical fiber. Since the refractive index of the core is the same for all the rays entering into the fiber, the rays travel with the same speed but covers different optical paths inside the core. Therefore, all the rays reach the other end at different times

$$\left[\because t = \frac{\text{optical paths}}{\text{speed}} \right]$$

resulting in slight distortion in the output.

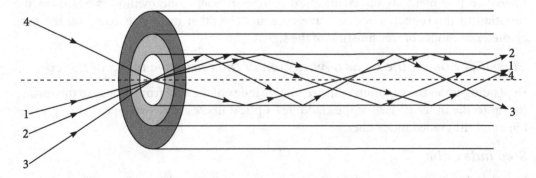

Figure 8.17 | All the light rays entering into the step index fiber at different angles of incidence travels different optical paths. Out of the four rays, ray 3 travels the longest optical path in the step index fiber as illustrated in the figure

Numerical aperture of a step index fiber

The numerical aperture of an optical fiber is a measure of the ability of the optical fiber to confine the optical signals within the core of the optical fiber. Only those light rays that strike the core–cladding interface at an angle more than the critical angle can undergo internal reflections.

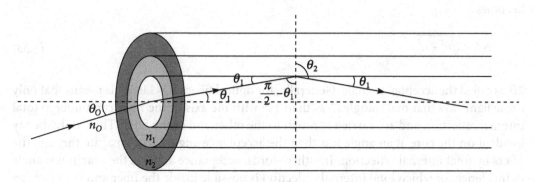

Figure 8.18 | Calculation of numerical aperture

The numerical aperture NA of a step index optical fiber is defined by

$$NA = n_0 \sin \theta_0 \qquad (8.30)$$

where

n_0 = Refractive index of the medium in which the fiber is immersed, generally air.

θ_0 = Angle of incidence of the light ray as shown in Fig. 8.18.

Applying Snell's law to the cross-sectional area, we have

$$\frac{\sin \theta_0}{\sin \theta_1} = \frac{n_1}{n_0} \quad \text{or} \quad n_0 \sin \theta_0 = n_1 \sin \theta_1 \qquad (8.31)$$

Similarly, applying Snell's law to the core–cladding interface, we have

$$n_2 \sin \theta_2 = n_1 \cos \theta_1 \qquad (8.32)$$

Sum of the squares of Eqs (8.31) and (8.32) gives

$$n_1^2 = n_0^2 \sin^2 \theta_0 + n_2^2 \sin^2 \theta_2$$

or $n_1^2 = NA^2 + n_2^2 \sin^2 \theta_2$

or $NA^2 = n_1^2 - n_2^2 \sin^2 \theta_2$

For total internal reflection to occur, θ_2 should be at least 90° for which this equation becomes

$$NA = \sqrt{n_1^2 - n_2^2} \qquad (8.33)$$

$2\theta_0$ is called the acceptance angle. Generally, for optical fibers $\theta_0 \approx 11°$. This means that only those light rays that make angles less than 11° with the axis of the fiber can undergo total internal reflection and are carried through to the other end of the fiber. Thus, only the ray incident on the core at an angle less than the acceptance angle can propagate through the fibers by total internal reflection. In other words, acceptance angle is the maximum angle of incidence for which total internal reflection is possible inside the fiber and the fiber can transmit electromagnetic radiation to the other end efficiently. The value of the acceptance angle can be calculated for an optical fiber by using the following relation:

$$n_0 \sin \theta_a = \sqrt{n_1^2 - n_2^2} \qquad (8.34)$$

Equation (8.33) can be rewritten in the form

$$NA = n_1 \left(2\Delta\right)^{\frac{1}{2}} \qquad (8.35)$$

where

$\Delta = \dfrac{n_1^2 - n_2^2}{2n_1^2}$ is called the relative refractive indices difference. $\qquad (8.36)$

Example 8.6

In a step index optical fiber, the refractive indices of core and cladding are 1.55 and 1.53 respectively. Calculate the numerical aperture of the fiber.

Solution

The data given are $n_1 = 1.55$ and $n_2 = 1.53$
    The numerical aperture is given by

$$NA_{step} = \sqrt{n_1^2 - n_2^2} = \sqrt{1.55^2 - 1.53^2} = 0.248$$

Graded index fiber

In graded index fiber the refractive index of the core is not constant throughout it. The refractive index of the core varies parabolically with respect to the radius of the core as shown graphically in Fig. 8.19. The refractive index of the core is maximum at its core and minimum at its surface region as illustrated in Fig. 8.19. Therefore, in graded index fiber, the speed of light is certain maximum at the surface and certain minimum along the axis of the core as $v = c/n$.

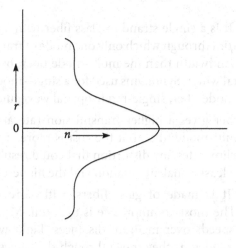

Figure 8.19 The variation of refractive index of the core of a graded index optical fiber with its radius. Refractive index of the core of a graded index optical fiber varies parabolically along its radius

The refractive index is so accurately varied along the diameter that all the light rays entering into the graded index optical fiber arrive at the other end of the optical fiber, at the same time minimizing the distortion produced in step index optical fiber. The path of a light ray in graded index optical fiber is shown in Fig. 8.20.

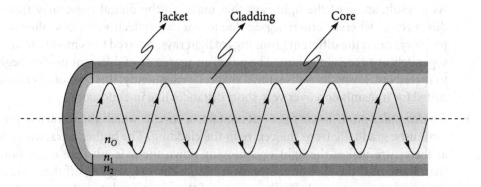

Figure 8.20 The path of a light ray in graded index optical fiber. Here the longitudinal section of a fiber is shown

Classification on the basis of core diameter

Again, depending upon the core diameters, optical fibers are of two types; (i) single-mode fiber and (ii) multi-mode fibers. Fibers with a larger core diameter (>50 μm) are called multimode fibers, because more than one electromagnetic field configuration can propagate through such a fiber. A single-mode fiber has a small core diameter, and the difference in refractive index between the core and the cladding is smaller than that for the multi-mode fiber. Only one electromagnetic field configuration propagates through a single-mode fiber.

i.   *Single-mode cable*   It is a single strand of glass fiber with a typical core diameter of around 8 μm to 10.5 μm through which only one mode of transmission can propagate. It can carry higher bandwidth than the multi-mode fiber, but requires a light source with a narrow spectral width. Synonyms used for a single-mode cable are mono-mode optical fiber, single-mode fiber, single-mode optical waveguide, uni-mode fiber.

A single-mode fiber gives a higher transmission rate and up to 50 times more distance than the multi-mode fiber, but the cost is more. The small core and single light wave virtually eliminates any distortion that could result from overlapping light pulses, providing the least signal attenuation and the highest transmission speeds.

ii. *Multi-mode cable*   It is made of glass fibers, with core diameters in the range 50 μm to 200 μm. The most common size is 62.5 μm. Multimode fiber gives high bandwidth at high speeds over medium distances. Light waves are dispersed into numerous paths, or modes, as they travel through the cables core. However, in long cables, multiple paths of light can cause signal distortion at the receiving end, resulting in an unclear and incomplete data transmission. Therefore, multi-mode optical fibers are mostly used for communication over short distances, such as within a building or on a campus. It is more cost effective than single mode fibers.

Multi-mode fibers can be classified into two types, namely, (a) step index multimode fibers and (b) graded index multi-mode fibers, depending upon whether the fiber is step index or graded index.

a.   *Step index multi-mode fiber*   It has a large core, up to 200 microns in diameter. As a result, some of the light rays that make up the digital pulse may travel a direct route, whereas others zigzag due to internal reflections. These alternative pathways cause the different groupings of light rays, referred to as modes, to arrive separately at a receiving point. The pulse, an aggregate of different modes, begins to spread out, losing its well-defined shape. Consequently, this type of fiber is best suited for transmission over very short distances like in endoscopy.

b.   *Graded index multi-mode fiber*   It contains a core in which the refractive index diminishe gradually from the centre to the cladding. The higher refractive index at the centre makes the light rays moving down the axis advance more slowly than those near the cladding. Moreover, rather than zigzagging off the cladding, light in the core curves helically because of the graded index, reducing its travel distance. The shortened path and the higher speed allow light at the periphery to arrive at a receiver at about the same time as the slow but straight rays in the core axis. The result: a digital pulse suffers less dispersion.

Figure 8.21 gives the comparison of single-mode fiber, step index multi-mode fiber and graded index multi-mode fiber.

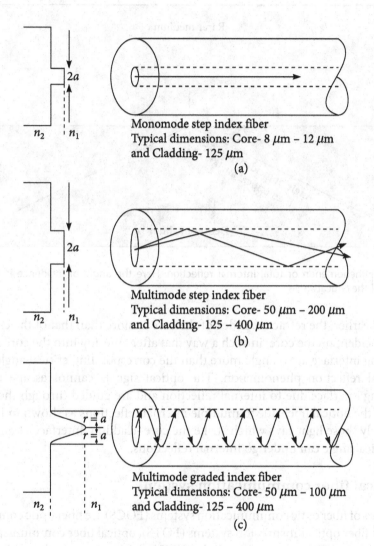

Monomode step index fiber
Typical dimensions: Core- 8 μm – 12 μm
and Cladding- 125 μm
(a)

Multimode step index fiber
Typical dimensions: Core- 50 μm – 200 μm
and Cladding- 125 – 400 μm
(b)

Multimode graded index fiber
Typical dimensions: Core- 50 μm – 100 μm
and Cladding- 125 – 400 μm
(c)

Figure 8.21 | Comparison of (a) single-mode fiber, (b) step index multi-mode fiber, and (c) graded index multi-mode fiber

8.6.3 Principle of optical fiber communication

The operation of simple optical fibers is based on the phenomenon of total internal reflection. In total internal reflection, if light is refracting from a denser medium to a rarer medium and the angle of incidence is more than the critical angle, there will be no refraction and the refracted ray will be reflected back into the first medium obeying the laws of reflection. This is depicted in Fig. 8.22.

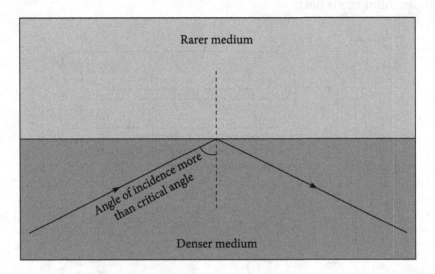

Figure 8.22 | Phenomenon of total internal reflection. Here the angle of incidence is more than the critical angle

As discussed earlier, the refractive index of the core is more than that of the cladding. The light ray is incident on the core, in such a way that after entering into the core it strikes the core–cladding interface at an angle more than the corresponding critical angle producing total internal reflection phenomenon. The optical signals cannot escape through the core–cladding interface due to internal reflection and are guided through the core to the other end of the optical fiber due to multiple internal reflections as shown in Fig. 8.17 and Fig. 8.20. Only those light rays which strike the core–cladding interface at an angle more than the critical angle can under go internal reflections.

8.6.4 Optical fiber communication system

The synonyms of fiber optic communication systems (FOCS) are fiber optic communication link (FOCL), fiber optic transmission systems (FOTS), optical fiber communication system (OFCS), and optical fiber transmission link (OFTL). A fiber optic communication systems (FOCS) is shown schematically in Fig. 8.23. The basic components are the transmitter, optical fiber and receiver. Additional elements include optical splice, repeaters, beam splitters and optical amplifiers.

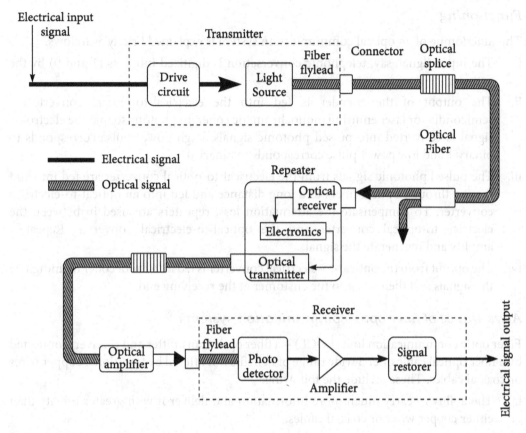

Figure 8.23 Schematic diagram showing the major components of the optical fiber communication system

Optical transmitter

The transmitter converts an electrical analog or digital signal into its corresponding optical signal. The source of the optical signal can be either a light emitting diode, or a solid state laser diode. The most popular wavelengths of operation for optical transmitters are 850 nm, 1310 nm, or 1550 nm.

Fiber optic cable

The cable consists of one or more glass fibers, which act as waveguides for the optical signal. Fiber optical cable is similar to electrical cable in its construction, but provides special protection for the optical fiber within. For systems requiring transmission over distances of many kilometers, or where two or more fiber optical cables must be joined together, an optical splice is commonly used.

Optical receiver

The receiver converts the optical signal back into a replica of the original electrical signal. The detector of the optical signal is either a PIN-type photodiode or an avalanche-type photodiode.

Functioning

The functioning of an optical communication system is explained briefly as follows.

i.   The input signal, say, telephonic conversation is digitized into bits (1 and 0) by the encoder.

ii. The output of the encoder is fed into the electrical-to-optical converter, a semiconductor laser emitting monochromatic coherent photons so that the electronic signal is converted into pulsed photonic signals. High power pulse corresponds to binary 1 and low power pulse corresponds to binary 0.

iii. The pulsed photonic signals from the electrical-to-optical converter are fed into and carried through optical fibers to a long distance and fed into an optical-to-electrical converter. To compensate the attenuation loss, repeaters are used in between the electrical-to-optical converter and the optical-to-electrical converter. Repeaters amplify and regenerate the signals.

iv. The output from the optical-to-electrical converter is fed into the decoder to undigitize the signals and then given to the customer at the receiving end.

Advantages of fiber optic communication systems

Fiber optic communication link (FOCL) – a fiber optic transmitter and receiver, connected by a fiber optical cable offer a wide range of benefits not offered by traditional copper wires or coaxial cables. These include the following:

i.   The ability to carry much more information and deliver it with greater fidelity than either copper wires or coaxial cables.

ii. Fiber optical cables can support much higher data rates, and at greater distances, than coaxial cables, making them ideal for transmission of serial digital data.

iii. The fiber is totally immune to virtually all kinds of interference, including lightning, and will not conduct electricity. It can therefore come in direct contact with high voltage electrical equipment and power lines without any damage. It will also not create ground loops of any kind.

iv. As the basic fiber is made of glass, it will not corrode and is unaffected by most chemicals. It can be buried directly in most kinds of soil or exposed to most corrosive atmospheres in chemical plants without significant concern.

v.   Since the only carrier in the fiber is light, there is no possibility of a spark from a broken fiber. Even in the most explosive of atmospheres, there is no fire hazard, and no danger of electrical shock to personnel repairing broken fibers.

vi. Fiber optical cables are virtually unaffected by outdoor atmospheric conditions, allowing them to be lashed directly to telephone poles or existing electrical cables without concern for extraneous signal pickup. Fiber optical cables have greater resistance to electromagnetic noise such as radios, motors or other nearby cables.

vii. A fiber optical cable, even one that contains many fibers, is usually much smaller and lighter in weight than a wire or coaxial cable with similar information carrying capacity. It is easier to handle and install, and uses less duct space. It can frequently be installed without ducts.

viii. By using an optical fiber, a high degree of data security in data transmission is afforded, since an optical signal is well confined within the wave guide. This makes the fibers attractive in applications where information security is important, such as military establishments, business establishments, computer networks, banking and the like.

ix. The functioning of optical fiber is not affected either by low temperatures or by high temperatures.

x. Fiber optical cables cost much less for maintenances.

xi. Silica, the principal material of which optical fibers are made, is abundantly and cheaply available on the earth.

Today's low-loss glass fiber optical cable offers almost unlimited bandwidth and unique advantages over all previously developed transmission media. Useful light energy can be taken from one point to another point without loss.

8.6.5 Characteristics of light source

The light source or the optical source is the main component of the transmitter. This has to meet a number of requirements. These are delineated as follows:

i. The physical dimensions of the light source must be compatible with the size of the fiber optical cable being used. This means that it must emit light in a cone with cross-sectional diameter $8\ \mu$ –$100\ \mu$. Otherwise, it cannot be coupled into the fiber optical cable.

ii. Second, the optical source must be able to generate enough optical power so that the desired BER (bit error rate; the fraction of bits transmitted that are received incorrectly) can be met.

iii. Third, there should be high efficiency in coupling the light generated by the optical source into the fiber optical cable.

iv. Fourth, the optical source should have sufficient linearity to prevent the generation of harmonics and inter-modulation distortion. If such interference is generated, it is extremely difficult to remove. This would cancel the interference resistance benefits of the fiber optical cable.

v. Fifth, the optical source must be easily modulated with an electrical signal and must be capable of high-speed modulation. Otherwise, the bandwidth benefits of the fiber optical cable are lost.

vi. Finally, there are the usual requirements of small size, low weight, low cost and high reliability. The light emitting junction diode stands out as matching these requirements. It can be modulated at the needed speeds. The proper selection of semiconductor materials and processing techniques results in high optical power and efficient coupling of it to the fiber optical cable. These optical sources are easily manufactured using standard integrated circuit processing. This leads to low cost and high reliability.

There are two types of light emitting junction diodes that can be used as the optical source of the transmitter. These are the light emitting diode (LED) and the laser diode (LD). LEDs are simpler and generate incoherent, lower power light. LD's are more complex and generate coherent, higher power light. Figure 8.24 illustrates the optical power output P of these devices as a function of the electrical input current I. As the figure indicates, LED has a relatively linear P-I characteristic while the LD has a strong non-linearity or threshold effect. The LD may also be prone to kinks where the power actually decreases with increasing bandwidth. With minor exceptions, LDs have advantages over LEDs in the following ways.

i.   They can be modulated at very high speeds.

ii.   They produce greater optical power.

iii. They have higher coupling efficiency to the fiber optical cable.

LEDs have advantages over LD's because they have

i.   higher reliability

ii.   better linearity

iii. lower cost

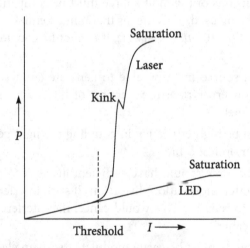

Figure 8.24 | Power (P) and current (I) characteristics of LED and laser diodes

Both LED and LD generate an optical beam with such dimensions that it can be coupled into a fiber optical cable. However, the LD produces an output beam with much less spatial width than an LED. This gives it greater coupling efficiency. Each can be modulated with a digital electrical signal. For very high-speed data rates, the link architect is generally driven to a transmitter having an LD. When cost is a major issue, the link architect is generally driven to a transmitter having an LED.

A key difference in the optical output of an LED and an LD is the wavelength spread over which the optical power is distributed. The spectral width impacts the effective transmitted signal bandwidth. A larger spectral width takes up a larger portion of the fiber optical cable link bandwidth. Figure 8.25 illustrates the spectral width of the two devices. The optical power generated by each device is the area under the curve. The spectral width is the half-power spread. An LD will always have a smaller spectral width than an LED. The specific value of the spectral width depends on the details of the diode structure and the semiconductor material. However, typical values for an LED are around 40 nm for operation at 850 nm wavelength and 80 nm at 1310 nm wavelength.

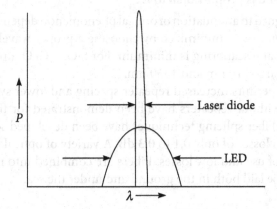

Figure 8.25 | LED and laser spectral widths

Once a transmitter is selected on the basis of being either an LED or an LD, additional concerns should be considered in reviewing the specifications. These concerns include packaging, environmental sensitivity of device characteristics, heat sinking and reliability.

With either an LED or an LD, the transmitter unit must have a transparent window to transmit light into the fiber optical cable. In reality, there is a small hemispherical lens to help focus the light into the fiber optical cable with the transmitter units.

8.6.6 Attenuation in optical fibers

Exceptionally pure and high quality fibers are fabricated using advanced and sophisticated techniques. In spite of all the possible precautions, due to the inherent properties of the optical fiber material, light signals get attenuated during their propagation in the core. Impurities and other defects in the materials absorb and scatter light to attenuate the light signals. The sources of attenuations are as follows.

i.   The energy absorption by lattice vibrations of the ions of the silica glass.

ii.   The energy absorption by the impurities of the optical fiber material. The presence of copper, iron and vanadium is especially detrimental for light signal propagation. Their concentrations must be reduced to the order of several parts per billion. Impurities like hydroxyl and water contaminants attenuate the light signal to a great extent. Their concentrations are made extremely low. At longer wavelengths, absorption by the atomic vibrations in the silicon–oxygen atoms rapidly increases the loss.

iii.   Uniformity of fiber cross-sectional dimension and core roundedness are critical for light signal propagation. Tolerances of these parameters to within $1.0\ \mu m$ over one kilometer of length are permissible.

iv.   Microscopic variations of the refractive index of the fiber optical material scatter light signals. Unintentional microscopic variations arise due to the disordered structure of silica glass. Rayleigh scattering is caused by microscopic variations in the refractive index of a fiber and is proportional to λ^4.

All these reasons assigned to attenuation of optical phenomenon depend on the wavelengths. Therefore, attenuation can be minimized by choosing a proper wavelength of light signal for which absorption and scattering is minimum. For $SiO_2 - GeO_2$ optical fibers, the most suitable wavelengths are 1310 nm and 1550 nm.

    The low fiber loss permits increased repeater spacing and lower system cost. High-bit-rate digital systems without repeaters have been demonstrated for fiber lengths of more than 100 kilometers. Fiber splicing techniques have been developed so that repairs can be made in the field with losses of only 0.1 to 0.3 dB. A variety of optical connectors are used, providing both ease of use and low losses. Fibers are combined into many different kinds of cables, which can be laid both in the ground and under the sea.

8.6.7 Applications of optical fibers

The simplest application of optical fibers is the transmission of information to other locations. The use of fiber-optics was generally not available until 1970. An optical fiber with a loss of 20dB/km was then constructed. It was recognized that optical fiber would be feasible for telecommunication transmission only if fibers could be developed so pure that attenuation would be much less than 20dB/km. After all, attenuation limits are based on intended applications.

Application of optical fiber in data transmission

The main application of optical fiber is in the field of transmission of information or data to different locations. In the following, few applications of optical fibers in data transmission are cited. er

i.   *Communications*   The applications of optical fiber communications have increased at a rapid rate since the first commercial installation of a fiber optical system in 1977. Telephone companies began early on, replacing their old copper wire systems with optical fiber lines. Today's telephone companies use optical fibers throughout their system as well as in long distance connections between city phone systems.

ii.   *Cable TV*   Cable television companies have also begun integrating fiber optics into their cable systems. The trunk lines that connect central offices have generally been replaced with optical fiber. Some providers have begun experimenting with fiber using a fiber/coaxial hybrid. Such a hybrid allows for the integration of fiber and coaxial at a neighbourhood location. This location, called a node, would provide the optical receiver that converts the light impulses back to electronic signals. The signals could then be fed to individual homes via coaxial cable.

iii. *Networking* Local area network (LAN) is a collective group of computers, or computer systems, connected to each other allowing for shared program software or data bases. Colleges, universities, office buildings, and industrial plants, just to name a few, all make use of optical fiber within their LAN systems.

iii. *Power companies* Power companies are emerging groups that have begun to utilize fiber optics in their communication systems. Most power utilities already have fiber optical systems

iv. *Imaging* Image transmission by optical fibers is also widely used in photocopiers, in phototypesetting, in computer graphics, and in other imaging applications.

Other applications of optical fibers

The simplest application of optical fibers is the transmission of light to locations that are otherwise difficult to illuminate.

Medical field

Optical fibers have been used in modern day medical equipments extensively.

i. *Dental surgery* Dentists' drills, for example, often incorporate a fiber optical cable that lights up the insides of patients' mouths.

ii. *Endoscopy* Optical fibers are used in some medical instruments to transmit images of the inside of the human body. Physicians use an instrument called an endoscope to view these inaccessible regions. The endoscope sends a beam of light into a body cavity, such as the inside of the stomach, via a fiber. A bundle of fibers returns a reflection of the inside of the cavity. The bundle consists of several thousands of very thin fibers assembled precisely side by side and optically polished at their ends. Each individual fiber carries a tiny bit of the final image, which is reconstituted and observed through a magnifier or a television camera.

iii. *Surgery* Fiber optical lasers are sometimes used for surgery. In laser surgery, optical fibers are used to transmit the laser beam to the point of interest where surgery is to be carried out.

As sensors

Optical fibers are used in a wide variety of sensing devices, ranging from thermometers to gyroscopes. The potential in this field is nearly unlimited because transmitted light is not sensitive to many environmental parameters, including pressure, sound waves, structural strain, heat, and motion. The fibers are especially useful where electrical effects make ordinary sensors or wiring useless, less accurate, or even hazardous.

Material processing

Fibers have also been developed to carry high power laser beams for cutting and drilling. In laser processing of materials like drilling, cutting, welding, polishing and so on, the high power laser is located at one place and the laser beam can be carried to different locations through optical fibers.

Current research

The development of new optical techniques will expand the capability of fiber optical systems. Newly developed optical fiber amplifiers, for example, can directly amplify optical signals without first converting them to an electrical signal, speeding up transmission and lowering power requirements. Dense wave division multiplexing (DWDM), another new fiber optical technique, puts many colours of light into a single strand of fiber optical cable. Each colour carries a separate data stream. Using DWDM, a single strand of fiber optical cable can carry up to 3 trillion bits of information per second. At this rate, downloading of the entire contents of the big data banks can be performed within a few minutes! With a broadband dial-up modem, it would take few hundred years.

Questions

8.1   What do you mean by optoelectronic devices?

8.2   What are the different electronic transitions in an atom?

8.3   Explain, applying Heisenberg's uncertainty principle, that atoms in the ground state have the longest life time.

8.4   Define spontaneous emission.

8.5   Define stimulated absorption

8.6   Define stimulated emission

8.7   What is the basic principle of laser?

8.8   What do you mean by population inversion?

8.9   What do you mean by optical pumping?

8.10   Explain how the emission from a usual light source is incoherent.

8.11   Justify the name population inversion.

8.12   Why is population inversion called negative temperature state?

8.13   Distinguish between active material, population inversion and pumping in laser.

8.14   What are the different methods of pumping in laser?

8.15   Explain how a three-level laser system works.

8.16   Explain how a four-level laser system works.

8.17   What are the possible reasons of broadening of laser radiation?

8.18   What do you mean by longitudinal coherence?

8.19   What do you mean by spatial coherence?

8.20   What is laser light?

8.21   What are the properties of laser light?

8.22   Describe the construction and working of an optical cavity.

8.23   Explain with a neat diagram the construction and action of a ruby laser.

8.24   Explain with a neat diagram the construction and action of a gas laser.

8.25   What is LED?

8.26   What is a direct band gap semiconductor? Give the band diagram.

8.27   What is direct recombination in LEDs?

8.28   What is a indirect band gap semiconductor? Give the band diagram.

8.29   What is indirect recombination in LEDs?

8.30   Discuss the principle of LED.

8.31   Describe the construction of a LED.

8.32   What are the merits of LED over conventional incandescent lamps?

8.33   Mention a few materials with which LEDs are made of.

8.34   What is quantum efficiency of LED?

8.35   Mention a few materials with which diode lasers are made of.

8.36   Discus the principle of semiconductor lasers.

8.37   Describe the construction of semiconductor lasers.

8.38   Describe the working principle of a semiconductor laser.

8.39   Describe the working of a semiconductor laser.

8.40   Mention a few applications of semiconductor lasers.

8.41   Distinguish between LED and diode lasers.

8.42   Explain why the semiconductors for visible LEDs must have band gaps between 1.77eV and 3.1eV?

8.43   Explain how very small distances are measured by laser light.

8.44   Explain how very large distances are measured by laser light.

8.45   Explain how very large structures are erected perfectly with the help of laser light.

8.46   What are the roles of laser in computer systems?

8.47   What do you mean by optical fibers?

8.48   What are the merits of optical fiber communication system over conventional communication systems?

8.49   Explain the structure of optical fibers.

8.50   Distinguish between step index optical fibers and graded index optical fibers.

8.51   Derive an equation to calculate the number of optical modes N within a bandwidth.

8.52   What do you mean by bandwidth?

8.53   What do you mean by diffraction loss?

8.54 Distinguish between single mode and multiple mode optical fibers?

8.55 Distinguish between step index multiple mode and graded index multiple mode optical fibers?

8.56 What are the characteristics of step index multiple mode optical fibers?

8.57 What are the characteristics of graded index multiple mode optical fibers?

8.58 Explain the principle of optical fibers communication.

8.59 Explain with a block diagram the optical fibers communication system.

8.60 Derive the expression for numerical aperture of a step index optical fiber in terms of refractive indices of core and cladding.

8.61 What do you mean by acceptance angle?

8.62 Derive an expression for acceptance angle.

8.63 What are the advantages of fiber optic communication systems over the traditional coaxial or copper cable communication?

8.64 What are the characteristics of light sources used in FOCL?

8.65 What are the advantages of LD over LED in OFTL?

8.66 What are the advantages of LED over LD in OFCS?

8.67 What is the difference between light emitted from LED and light emitted from diode lasers.

8.68 Mention a few applications of optical fiber in data transmission.

8.69 Mention a few applications of optical fiber in medical fields.

8.70 Explain how optical fibers are used in material processing.

8.71 What are the future perspectives of fiber optics?

8.72 What are the sources of attenuation in optical fiber communication?

8.73 How are optical fibers in use to improve the living conditions on the earth?

Problems

8.1 For a fully transparent active material of length 60 cm, calculate the value of m in $v = \dfrac{mv}{2L} = \dfrac{mc}{2L\mu}$ (optical cavity) for a light of wavelength 6000 Å. [Ans 2×10^6]

8.2 Determine the frequency difference between the cavity modes of a laser that is 60 cm long with active material having refractive index 1.2. If the bandwidth of the laser beam is 5.3×10^9 Hz, find the number of optical modes in the output radiation.

[Ans $2.08 \times 10^8, 7$]

8.3 Prove that for a resonator cavity consisting of a plane circular mirror of diameter 1cm and active material of length 60 cm employing light of wavelength 5000 Å, the diffraction loss will be minimum.

[Ans $\dfrac{D^2}{\lambda L} \approx 333 \gg 1$]

8.4 In a material, transition occurs between a metastable state and an energy level of 0.5 eV and the wavelength of the radiation emitted is 2200 nm. Calculate the energy of the metastable state. [Ans 1.07 eV]

8.5 A laser beam of wavelength 1.15 μm has coherence time 26.7×10^{-9} s. Determine the temporal coherence length, spectral width and purity factor.
[Ans 8.01 m, 1.65×10^{-13} m, 6.96×10^{6}]

8.6 The wavelengths of sodium yellow light are 5896 Å and 5890 Å. Calculate the longitudinal coherence length for these wavelengths. [Ans 5.79 cm]

8.7 A particular green LED emits light of wavelength 5490 Å. Calculate the energy band gap of the semiconductor material used in making green LED. [Ans 2.26 eV]

8.8 In a step index optical fiber, the refractive indices of core and cladding are 1.45 and 1.35 respectively. Calculate the numerical aperture of the fiber. [Ans 0.529]

8.9 What is the critical angle for a ray in a step index optical fiber for which the refractive index of the core is 1.53 and that of the cladding is 2.5% less of the core.
[Ans 77.2°]

8.10 In a step index optical fiber, the refractive index of the core and the required numerical aperture of the fiber are 1.55 and 0.25 respectively. Calculate the refractive index of the cladding. [Ans 1.53]

Multiple Choice Questions

1. The full form of LASER is
   (i) Light Amplification of Stimulated Emission of Radiation
   (ii) Light Amplification by Stimulated Emission of Radiation
   (iii) Light Acceleration of Stimulated Emission of Radiation
   (iv) Light Acceleration by Stimulated Emission of Radiation

2. Which of the following atomic states has longest life time?
   (i) Excited state   (ii) Metastable state
   (iii) Intermediate state   (iv) Ground state

3. Which of the following is correct?
   (i) The metastable state lies in between the ground state and the excited state
   (ii) The excited state lies in between the ground state and the metastable state
   (iii) The ground state lies in between the metastable state and the excited state
   (iv) None of the above

4. Which of the following conclusion is incorrect?
   (i) Atom + photon → atom* implies stimulated absorption.
   (ii) Atom* → atom + photon implies spontaneous emission.
   (iii) Atom* + photon → atom + 2 photons implies stimulated emission
   (iv) None of the above

5. The range of wavelengths for visible light is
   (i) 0.4 μm to 0.7 μm   (ii) 0.3 μm to 0.6 μm
   (iii) 0.2 μm to 0.8 μm   (iv) 0.1 μm to 0.9 μm

6. Which of the following devices convert light energy to electric energy?
   (i) LED   (ii) Semiconductor laser
   (iii) Solar cells   (iv) Optical fibers

7. Which of the following devices convert electric energy to light energy?
   (i) LED   (ii) Semiconductor laser
   (iii) Solar cells   (iv) optical fibers

8. The light incident on the core of an optical fiber at an angle more than the acceptance angle can reach the other end of the fiber.
   (i) True   (ii) False

9. At any time, stimulated emission, stimulated absorption and spontaneous emission cannot occur simultaneously.
   (i) True   (ii) False

10. Which of the following is correct? In active material
    (i) number of atoms in the ground state is more than that in the excited state
    (ii) number of atoms in the excited state is more than that in the ground state
    (iii) number of atoms in the ground state and the excited state are equal
    (iv) none of the above

Answers

1 (ii)	2 (iv)	3 (i)	4 (iv)	5 (i)	6 (iii)	7 (i & ii)	8 (ii)
9 (ii)	10 (ii)						

9   Dielectric Materials

9.1 Introduction

Dielectric materials are materials that do not conduct electricity, but can sustain an electric field. All dielectric materials are insulators. Insulators resist electric current when electric potential difference is applied. Dielectric materials can store electric energy. The molecules of a dielectric material are either polar or non-polar. Polar molecules are defined as molecules in which the centre of gravity of positive nuclei and negative electrons are not at the same point. Non-symmetrical molecules such as H_2O, N_2O and the like are polar molecules. Non-polar molecules are defined as molecules in which the centre of gravity of positive nuclei and negative electrons are at the same point. Symmetrical molecules such as H_2, N_2, O_2 and the like are non-polar molecules.

9.2 An Overview of Dielectric Polarization

The phenomenon of slight relative shift of positive and negative electric charges in opposite directions within a dielectric, induced by an external stimuli like electric field, temperature, pressure etc is called dielectric polarization. Under the influence of an external applied electric field, non-polar molecules in dielectrics are polarized and become induced dipoles. This is depicted in Fig. 9.1.

Under the influence of an externally applied electric field, the polar molecules in a dielectric [polar molecules are itself electric dipoles having inherent dipole moments] become aligned in the direction of the applied electric field. The stronger the applied electric field, the greater is the aligning effect. This is depicted in Fig. 9.2.

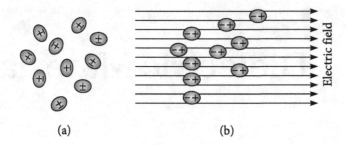

(a)   (b)

Figure 9.1 (a) Behaviour of non-polar molecules in the absence of an externally applied electric field. (b) Behaviour of non-polar molecules in the presence of an externally applied uniform electric field

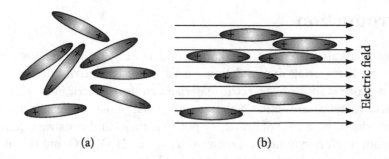

(a)   (b)

Figure 9.2 (a) Behaviour of polar molecules in the absence of an externally applied electric field. (b) Behaviour of polar molecules in the presence of an externally applied uniform electric field

The net effect of applying an electric field to either polar or non-polar dielectrics in the form of a slab is more or less the same. When an electric field is applied to a slab of dielectric in a direction perpendicular to one of its surfaces, negative charges are developed on one surface and positive charges developed on the opposite surface as shown in Fig. 9.3. Thus, the dielectric as a whole becomes polarized. These positive and negative charges are called induced charges as these charges are induced by applied electric field. These induced charges are not free but bound to the molecules lying just inside the surfaces.

Therefore, these positive and negative charges are also called bound charges. The net (sum of positive and negative) charges per unit volume on these two opposite surfaces are not zero, whereas deep inside the dielectric, the net charges per unit volume is zero.

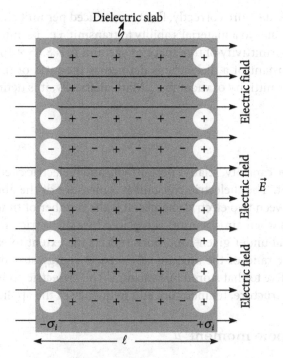

Dielectric slab

$-\sigma_i$ ℓ $+\sigma_i$

Electric field

$\vec{E}$

Figure 9.3 | Polarization of a dielectric under the influence of an externally applied uniform electric field. Thin layers of induced charges are developed on two opposite surfaces

9.3 Dielectric Parameters

Following are the dielectric parameters of a dielectric.

i.   Dielectric constant ε_r.
ii. Electric dipole moment $\vec{\mu}_e$
iii. Dielectric polarization $\vec{P}$
iv. Polarizability α

9.3.1 Dielectric constant ε_r

We know that permittivity ε first appears in the mathematical formulation of Coulomb's law $\left(F = \dfrac{q_1 q_2}{4\pi\varepsilon r^2} \right)$ or electric field intensity $\left(E = \dfrac{q}{4\pi\varepsilon r^2} \right)$. It also appears in Gauss's law in electrostatics $\left(\oint_S E \cdot ds = \dfrac{q}{\varepsilon} \right)$.

It is the measure of the resistance that is encountered when creating an electric field in a dielectric medium. In other words, permittivity is a measure of how an electric field affects, and is affected by a dielectric medium. The permittivity of a medium describes how much electric field (more correctly, flux) is produced per unit charge in that medium. Thus, permittivity relates to a material's ability to transmit, i.e., permit an electric field. The numerical value of permittivity of free space or vacuum ε_0 is $8.85 \times 10^{-12}\,C^2/Nm^2$.

The dielectric constant of a dielectric is defined as the ratio of the permittivity of the dielectric ε to the permittivity of vacuum ε_0. Mathematically, it is defined by

$$\varepsilon_r = \frac{\varepsilon}{\varepsilon_0} \tag{9.1}$$

The specific inductive capacity, relative permittivity and dielectric coefficient are synonyms of dielectric constant. The dielectric constant is a measure of the ability of a medium to reduce the force between two electric charges. It is the number of times the capacitance of a capacitor increases when air (vacuum) is replaced by a dielectric material. The dielectric constant of a material under given conditions reflects the extent to which it concentrates electric flux. It is the ratio of the amount of electrical energy stored in a material by an applied voltage, relative to that stored in a vacuum. The dielectric constant depends on the composition, microstructure, temperature and frequency of the applied field.

9.3.2 Electric dipole moment $\vec{\mu}_e$

The dipole moment of a dipole is defined as the product of magnitude of charge of one pole and the distance between the two opposite charges constituting the electric dipole. It is a vector quantity with direction from the negative to the positive charge of the dipole (Fig. 9.4). Mathematically, it is defined by

$$\vec{\mu}_e = q\vec{r} \tag{9.2}$$

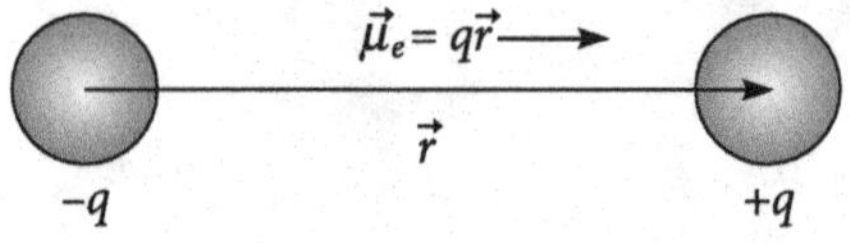

Figure 9.4 | The electric dipole moment $\vec{\mu}_e$. It is a vector quantity with direction from the negative to the positive charge of the dipole

9.3.3 Dielectric polarization

Polarization vector $\vec{P}$ in its quantitative meaning is defined as the amount of induced dipole moments per unit volume of a polarized dielectric material. Mathematically, polarization vector $\vec{P}$ is defined as

$$\vec{P} = N\vec{\mu}_e \tag{9.3}$$

where

$\bar{\mu}_e$ = dipole moment of a single polarized molecule.

N = number of polarized molecules per unit volume of the polarized dielectric.

The direction of the polarization vector is the same as the direction of the dipole moment. The SI unit of the polarization vector $\bar{P}$ is Cm^{-2}. The extent of polarization of a dielectric is described by the polarization vector $\bar{P}$. As shown in Fig. 9.3, the opposite surfaces of the polarized dielectric slab contains equal and opposite charges and hence, the dielectric slab as a whole may be treated as a electric dipole. Figure 9.5 shows a polarized elemental dielectric slab of width ℓ.

Let

dq_i = induced charges on either surfaces of the dielectric on an elemental area $\bar{ds}$.

ℓ = distance between two opposite surfaces containing induced charges of the opposite sign
= width of the polarized elemental dielectric slab.

Therefore, induced surface charge per unit area σ_i will be given by

$$\sigma_i = \frac{dq_i}{ds} \tag{9.4}$$

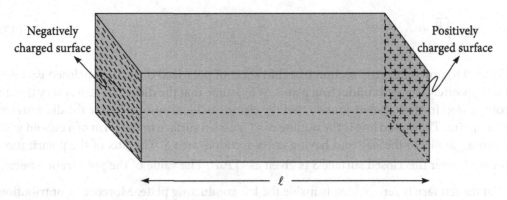

Figure 9.5 | Dipole moment of a polarized elemental dielectric slab of width ℓ

The magnitude of the dipole moment vector of the polarized elemental dielectric slab from the definition of dipole moment is given by

$$\mu_e = \ell dq_i \tag{9.5}$$

The volume of the polarized elemental dielectric slab is $= \ell ds$.

From the definition of the polarization vector P, we have

$$\vec{P} = \frac{\text{Total dipole moments of the polarized elemental dielectric slab}}{\text{Volume of the polarized elemental dielectric slab}}$$

or $\vec{P} = \dfrac{\vec{\mu}_e}{\ell ds}$ (9.6)

Putting the value of dipole moment from Eq. (9.5) into Eq. (9.6), we have

$$P = \frac{\ell dq_i}{\ell ds}$$

or $P = \dfrac{dq_i}{ds} = \sigma_i =$ surface charge density.   (9.7)

Thus polarization is numerically equal to the surface charge density of induced charges. From Eq. (9.7), in general, we have

$$dq_i = \vec{P} \cdot \vec{ds}$$

or $q_i = \oint_S \vec{P} \cdot \vec{ds}$ (9.8)

Figure 9.6 shows the cross-section of a thin sheet of polarized dielectric enclosed between two oppositely charged conducting plates. We assume that the dielectric slab is very thin in comparison to other dimensions so that the electric field is uniform inside the dielectric at every point. The dotted line is the outline of a Gaussian surface in the form of a cuboid with its ends parallel to the slab and having cross-sectional area S. The flux of the polarization vector $\vec{P}$ over this closed surface S is given as $\oint_S \vec{P} . \vec{ds}$. The value of the polarization vector $\vec{P}$ at the left face is zero, which is inside the left conducting plate. Moreover, contribution to this integral from the end surfaces is zero as the angle between $\vec{P}$ and $\vec{ds}$ is 90°. Only the right face perpendicular to $\vec{P}$ contributes to this integral. From Eq. (9.8), we have

$\oint_S \vec{P} . \vec{ds} = q_i =$ the total induced charge enclosed by the Gaussian surface. In general, we have

$$\oint_S \vec{P} . \vec{ds} = -q_i$$ (9.9)

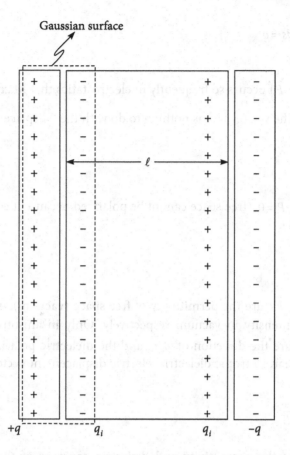

Gaussian surface

ℓ

$+q$ q_i q_i $-q$

Figure 9.6 | Cross-section of a thin sheet of polarized dielectric enclosed between two oppositely charged conducting plates

The flux of the polarization vector $\vec{P}$ over a closed surface is outward, whereas the enclosed induced charge is negative and vice versa. Therefore, there is the negative sign in the RHS of Eq. (9.9). This equation is the Gauss' law for polarization vector $\vec{P}$. Gauss's law for polarization vector $\vec{P}$ can be stated as, "the flux of the polarization vector $\vec{P}$ over a closed surface is equal to the negative of the total induced charge within the surface". Mathematically, this statement is translated as $\oint_S \vec{P}.\overrightarrow{ds} = -q_i$.

In the Gaussian surface shown in Fig. 9.6, Gauss' law in the presence of both free and induced charge is

$$\oint_S \vec{E} \cdot \overrightarrow{ds} = \frac{1}{\varepsilon_0}\left(q + q_i\right)$$

or $$\oint_S \varepsilon_0 \vec{E} \cdot \overrightarrow{ds} = q - \oint_S \vec{P} \cdot \overrightarrow{ds}$$

or $\oint\limits_{s}\left(\vec{P}+\varepsilon_{0}\vec{E}\right)\cdot\vec{ds}=q$ (9.10)

The quantity $\left(\vec{P}+\varepsilon_{0}\vec{E}\right)$ occurs so frequently in electrostatics that Maxwell called it electric displacement $\vec{D}$ (The vector $\vec{D}$ has nothing to do with the "displacement vector"). Thus, we have

$$\vec{D}=\vec{P}+\varepsilon_{0}\vec{E}.$$ (9.11)

In free space, since $\vec{P}=0$ (free space cannot be polarized because it contains nothing), we have from Eq. (9.11),

$$\vec{D}_{0}=\varepsilon_{0}\vec{E}_{0}$$ (9.12)

where ε_{0}, $\vec{D}_{0}$, and $\vec{E}_{0}$ are the permittivity of free space (vacuum), electric displacement and electric field intensity in vacuum respectively. Only in anisotropic dielectrics, like crystalline quartz, are the direction of $\varepsilon_{0}\vec{E}$ and the dielectric polarization vector $\vec{P}$ not the same. For a linear, isotropic dielectric, electric displacement vector from Eq. (9.11) is given by

$$\vec{D}=\varepsilon\vec{E}=\varepsilon_{0}\varepsilon_{r}\vec{E}$$ (9.13)

where ε and ε_{r} are the permittivity and dielectric constant of the dielectric material respectively.

Dielectric polarization exists in the material as long as there is an externally applied electric field. By putting the value of $\vec{D}$ from Eq. (9.13) into Eq. (9.11), we have

$$\varepsilon_{0}\varepsilon_{r}\vec{E}=\vec{P}+\varepsilon_{0}\vec{E}$$

or $\vec{P}=\varepsilon_{0}(\varepsilon_{r}-1)\vec{E}=\chi\varepsilon_{0}\vec{E}$ (9.14)

This equation is applicable to linear dielectrics such as benzene, water, diamond and the like where polarization is directly proportional to the applied electric field intensity.

Electric susceptibility χ is defined by

$$\chi=\varepsilon_{r}-1$$ (9.15)

χ is a dimensionless proportionality constant that indicates the degree of polarization of a dielectric material in response to an applied electric field. It is a measure of how easily a dielectric is polarized in response to an electric field. The greater is the electric susceptibility, greater is the ability of a material to be polarized in response to the field, and thereby reduce the total electric field inside the material (and stored energy). The higher is the dielectric polarization, higher is the dielectric constant of the material. It is in this way that the electric susceptibility influences the permittivity of the material and thus, influences many other phenomena in that medium, from the capacitance of capacitors to the speed of light. For a linear, isotropic dielectric material, dielectric polarization vector is determined by Eq. (9.14).

The role of the electric field intensity vector $\vec{E}$ and the electric displacement vector $\vec{D}$ are the same except that the vector $\vec{E}$ is associated with all type of charges including polarization charges, whereas vector $\vec{D}$ is associated only with free charges. The electric susceptibility χ defined by $\chi = \varepsilon_r - 1$ [= departure of the dielectric constant of the medium by unity] for an isotropic media has the same value for any direction of the applied electric field. For a non-isotropic media such as most of crystals, the magnitude of polarization varies with the direction of applied electric field and consequently, electric susceptibility χ must be expressed as a tensor. The χ-tensor of a crystal summarizes most of its optical properties.

9.3.4 Polarizability α

Since the polarization of the medium, i.e., the alignment of dipole moments $\vec{\mu}_e$ is produced by the electric field, it is plausible to assume that the dipole moment $\vec{\mu}_e$ is proportional to the field. Thus, we have

$$\vec{\mu}_e \propto \vec{E}$$

or $\vec{\mu}_e = \alpha \vec{E}$ (9.16)

The proportionality constant α is called polarizability of the molecule. Polarizability is an atomic parameter and is characteristic of that particular atom/molecule. It is a measure of the easiness with which the atoms get polarized under an electric field. According to Eq. (9.16), the polarizability of a molecule/atom can be defined as the induced dipole moment per unit applied electric field at the site of the atom. It is the coefficient of the applied electric field in Eq. (9.16). The unit of polarizability is Fm^2 (Farad.meter^2). Expression (9.16) gives the average value of polarizability. In general, it is given by

$$\alpha = \sum_j w_j \alpha_j$$ (9.17)

where w_j is the weight fraction of α_j.

The polarization of the medium $\vec{P}$ according to Eq. (9.16) can be defined as

$$\vec{P} = N\alpha\vec{E}$$ (9.18)

where N is the number of molecules per unit volume of the polarized dielectric. From Eq. (9.11), we have

$$\vec{D} = \vec{P} + \varepsilon_0 \vec{E} = N\alpha\vec{E} + \varepsilon_0 \vec{E} = \varepsilon_0 \left(1 + \frac{N\alpha}{\varepsilon_0}\right)\vec{E}$$

or $$\varepsilon\vec{E} = \varepsilon_0 \left(1 + \frac{N\alpha}{\varepsilon_0}\right)\vec{E}$$

or $$\varepsilon_r = 1 + \frac{N\alpha}{\varepsilon_0} \qquad (9.19)$$

This equation expresses the macroscopic parameter dielectric constant ε_r in terms of polarizability α, a microscopic parameter. From Eq. (9.19), we can have

$$\chi = \frac{N\alpha}{\varepsilon_0} \qquad (9.20)$$

where N is the number of atoms/molecules per unit volume. If m is the mass of each molecule, the density of the medium ρ will be $\rho = Nm$ [since density is the mass per unit volume] and this equation becomes

$$\chi = \frac{\rho\alpha}{m\varepsilon_0} \qquad (9.21)$$

If M is the molar mass and N_A is Avogadro's number ($= 6.023 \times 10^{23}$), the mass of each molecule 'm' will be

$$m = \frac{M}{N_A} \qquad (9.22)$$

Thus, in terms of density and molar mass, the electric susceptibility is given by

$$\chi = \frac{\rho N_A \alpha}{M\varepsilon_0}. \qquad (9.23)$$

Moreover, in terms of density and molar mass, the dielectric constant ε_r is given by

$$\varepsilon_r = 1 + \frac{\rho N_A}{M\varepsilon_0}\alpha \qquad (9.24)$$

This equation shows that the dielectric constant increases linearly with density.

Example 9.1

The polarizability of argon is 1.8×10^{-40} C^2m/N. Calculate the dielectric constant and electric susceptibility of argon at NTP.

Solution

The data given is

$$\alpha = 1.8 \times 10^{-40} \, C^2 m/N$$

The dielectric constant is given by $\varepsilon_r = 1 + \dfrac{N\alpha}{\varepsilon_0}$

At NTP, $22.4 \times 10^{-3} m^3$ contains 6.023×10^{23} atoms. Therefore, we have

$$N = \frac{6.023 \times 10^{23}}{22.4 \times 10^{-3}} = 26.888 \times 10^{24} \, atoms/m^3$$

Hence, we have $\varepsilon_r = 1 + \dfrac{N\alpha}{\varepsilon_0} = 1 + \dfrac{26.888 \times 10^{24} \times 1.8 \times 10^{-40}}{8.85 \times 10^{-12}} = 1.0005469$

Example 9.2

When an electric field of 300 V/cm is applied to the benzene C_6H_6, calculate the contribution of one benzene molecule to the polarization of benzene. Data given are: density = 0.8 gm/cc, static dielectric constant = 2.28

Solution

The data given are

$$E = 300 \, V/cm = 3 \times 10^4 \, V/m, \; \rho = 0.8 \, \frac{gm}{cm^3} = 800 \, kg/m^3, \; \varepsilon_r = 2.28.$$

The polarization P is given by

$$P = \varepsilon_0 (\varepsilon_r - 1)E = (8.85 \times 10^{12}) \times (3 \times 10^4)(2.28 - 1) = 3.398 \times 10^{-7} \, C/m^2$$

The number of molecules per unit volume N is given by $N = \dfrac{\rho N_A}{M}$. The molecular mass M of C_6H_6 is calculated as

$$M = 12 \times 6 + 1 \times 6 = 78$$

$$N = \frac{800 \times 6.023 \times 10^{26}}{78} = 6.64 \times 10^{27} \, molecules/m^3$$

6.64×10^{27} molecules/m^3 contribute 3.398×10^{-7} C/m^2. Then one molecule contributes

3.398×10^{-7} C/m$^2 \div 6.64 \times 10^{27}$ molecules/m$^3 = 5.12 \times 10^{-35}$ Cm

This is nothing but the dipole moment of a single benzene molecule!

9.4 Microscopic Field $\vec{E}_L$

Experimental observations show that the equations

$$\varepsilon_r = 1 + \frac{N\alpha}{\varepsilon_0} \quad \text{and} \quad \varepsilon_r = 1 + \frac{\rho N_A}{M\varepsilon_0}\alpha$$

does not hold good in case of solids and liquids! The reason is that we have assumed that the electric field affecting the molecules/atoms in the materials is equal to the externally applied field E_0, which is true only for gases. Under the action of an externally applied field E_0, the molecules in the solid dielectric are under the influence of the applied field E_0 as well as other dipoles present in the vicinity of the molecules. The central problem in the theory of solid and liquid dielectrics is to calculate the total field called internal field or local field $\vec{E}_L$ at the position of a given atom. For condensed matter (solids and liquids), Eq. (9.16) becomes

$$\vec{\mu}_e = \alpha \vec{E}_L \tag{9.25}$$

The electric field $\vec{E}_L$ is also called microscopic field. To evaluate $\vec{E}_L$, we must calculate the total field acting on a certain typical dipole inside a solid or liquid dielectric subjected to an external electric field. The interaction of the typical dipole with other dipoles present inside the cavity is however to be treated microscopically, which is necessary since the discrete nature of the medium very close to the dipoles should be taken into account.

The local field acting on the central dipole [see Fig. 9.7] is thus given by the vector sum

$$\vec{E}_L = \vec{E}_0 + \vec{E}_1 + \vec{E}_2 + \vec{E}_3 \tag{9.26}$$

where

$\vec{E}_0$ = Externally applied electric field.

$\vec{E}_1$ = The electric field produced due to polarization charges residing on the external surface of the specimen.

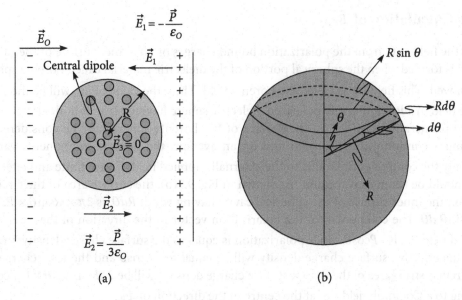

Figure 9.7 | (a) The procedure for calculating the local field $\vec{E}_L$. (b) The procedure for calculating field $\vec{E}_2$, the electric field produced due to polarization charges residing on the surface of the Lorentz sphere inside the specimen. The length (perimeter) and width of the shaded strip is $2\pi R \sin\theta$ and $Rd\theta$ respectively giving the area of the strip to be $2\pi R \sin\theta \times Rd\theta$

$\vec{E}_2$ = The electric field produced due to polarization charges residing on the surface of the Lorentz sphere inside the specimen.

$\vec{E}_3$ = The electric field produced due to other dipoles inside the Lorentz sphere.

9.4.1 Calculation of $\vec{E}_1$

$\vec{E}_1$ is the electric field produced due to polarization charges residing on the external surface of the specimen. As shown in Fig. 9.7(a), the direction of the externally applied electric field $\vec{E}_0$ and $\vec{E}_1$ are opposite and $\vec{E}_1$ opposes $\vec{E}_0$. Therefore, $\vec{E}_1$ is called the depolarizing field. As proved earlier, the magnitude of the polarization vector $\vec{P}$ is numerically equal to the surface charge density. Thus, we have

$$E_1 = -\frac{\sigma_i}{\varepsilon_0} = -\frac{P}{\varepsilon_0}$$

or $\quad \vec{E}_1 = -\dfrac{\vec{P}}{\varepsilon_0}$ (9.27)

9.4.2 Calculation of $\vec{E}_2$

$\vec{E}_2$ is the field at O from the polarization bound charges on the inner surface of the cavity, which is formed when the spherical portion of the dielectric material called Lorentz sphere is removed. This field $\vec{E}_2$ is called the Lorentz field. Thus, the value of $\vec{E}_2$ will be the same as the field inside a uniformly polarized dielectric sphere having polarization $-\vec{P}$.

The polarization charges on the surface of the Lorentz sphere may be considered as forming a continuous distribution and on an average, the normal components vanish and only the components parallel to the externally applied field $\vec{E}_0$ are finite and effective and should be taken into account. As shown in Fig. 9.7(b), the surface area of the circular strip on the inner surface of the spherical cavity having width $Rd\theta$ is $2\pi \times R\sin\theta \times Rd\theta = 2\pi R^2 \sin\theta\, d\theta$. The component of the polarization vector in the direction of the externally applied field $\vec{E}_0$ is $-P\cos\theta$. Since polarization is equal to the surface charge density, in this case, the effective surface charge density will be equal to $-P\cos\theta$ and the total charge on the circular strip [area of the strip × surface charge density] will be $2\pi R^2 \sin\theta\, d\theta \times (-P\cos\theta)$ leading to a Coulomb field dE_2 at the centre in the direction of $\vec{E}_0$.

$$dE_2 = \frac{2\pi R^2 \sin\theta d\theta \left(-P\cos\theta\right)\left(-\cos\theta\right)}{4\pi\varepsilon_0 R^2}$$

or

$$E_2 = \frac{P}{2\varepsilon_0} \int_0^{\pi} \cos^2\theta \sin\theta d\theta$$

or

$$E_2 = \frac{P}{2\varepsilon_0}\frac{2}{3} = \frac{P}{3\varepsilon_0}$$

Vectorially, $\vec{E}_2 = \dfrac{\vec{P}}{3\varepsilon_0}$ $\qquad\qquad\qquad\qquad\qquad$ (9.28)

9.4.3 Calculation of $\vec{E}_3$

The last term in Eq. (9.26), i.e., $\vec{E}_3$ is the field at O due to the electric dipoles inside the Lorentz sphere. There are a number of important cases for which $\vec{E}_3 = 0$. If there are a great many dipoles in the cavity and if these are oriented parallel but randomly distributed in position and also if there are no correlation among the positions of the dipoles, then $\vec{E}_3 = 0$. This is the situation that might prevail in a gas or liquid. Similarly, if the dipoles in the cavity are located at the regular atomic positions (lattice points) of a cubic crystal, then again $\vec{E}_3 = 0$. In general, $\vec{E}_3 \neq 0$ and if the dielectric contains several species of molecules,

$\vec{E}_3$ may differ at the various molecular positions. However, we restrict our discussions to the rather large class of materials in which

$$\vec{E}_3 = 0 \tag{9.29}$$

Combining Eqs (9.26) to (9.29), we have

$$\vec{E}_L = \vec{E}_0 - \frac{\vec{P}}{\varepsilon_{01}} + \frac{\vec{P}}{3\varepsilon_{02}} + 0 \tag{9.30}$$

or $\quad \vec{E}_L = \vec{E}_0 - \dfrac{2\vec{P}}{3\varepsilon_0} \tag{9.31}$

From Eq. (9.11), we have

$$\vec{E}_0 \varepsilon_0 = \vec{P} + \varepsilon_0 \vec{E}$$

or $\quad \vec{E}_0 = \dfrac{\vec{P}}{\varepsilon_0} + \vec{E} \tag{9.32}$

In this equation, $\vec{E}$ is the average applied electric field called macroscopic field/Maxwell's field. Putting Eq. (9.32) into Eq. (9.31), we get

$$\vec{E}_L = \frac{\vec{P}}{\varepsilon_0} + \vec{E} - \frac{2\vec{P}}{3\varepsilon_0}$$

or $\quad \vec{E}_L = \vec{E} + \dfrac{\vec{P}}{3\varepsilon_0} \tag{9.33}$

Equation (9.33) is called the Lorentz relation. It is a relation between microscopic field $\vec{E}_L$ and macroscopic field $\vec{E}$. Comparing Eqs (9.33) and (9.30), the mathematical expression for the macroscopic field is obtained as

$$\vec{E} = \vec{E}_0 - \frac{\vec{P}}{\varepsilon_0} \tag{9.34}$$

The difference between the microscopic field $\vec{E}_L$ and the macroscopic field $\vec{E}$ may be examined as follows. The field $\vec{E}$ is a macroscopic quantity and as such is an average field, the average being taken over a large number of dipoles. It is this field that enters into

Maxwell's equations, which are used for the macroscopic description of a dielectric media. As plotted in Fig. 9.8, this field is shown as a constant field.

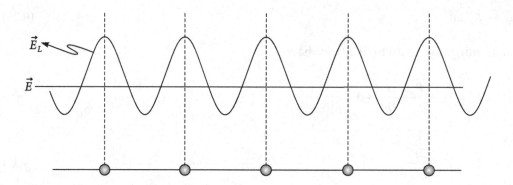

Figure 9.8 | The difference between the microscopic field $\vec{E}_L$ and the macroscopic field $\vec{E}$. The solid spheres represent atoms/molecules

On the other hand, the microscopic field $\vec{E}_L$ fluctuates within the dielectric with respect to the position of points. As Fig. 9.8 indicates, this field $\vec{E}_L$ is quite large at the dipole sites themselves and hence, the atoms/molecules are more effectively polarized than they would be by the macroscopic field $\vec{E}$.

Example 9.3

When an electric field 300 Volt/cm is applied to benzene, it is polarized to an extent of 3.398×10^{-7} C/m^2. Calculate the local field acting on the benzene molecule.

Solution

The data given are $E_0 = 300$ V/cm $= 3 \times 10^4$ V/m, $P = 30398 \times 10^{-7}$ C/m^2
   The local field acting on a molecule/atom of a polarized medium is given by

$$E_L = E_0 - \frac{2P}{3\varepsilon_0} = 3 \times 10^4 \ \text{V/m} - \frac{2 \times 3.398 \times 10^{-7}}{3 \times 8.85 \times 10^{-12}} \text{V/m} = 4403 \ \text{V/m}$$

Example 9.4

When an electric field 300 Volt/cm is applied to benzene, it is polarized to an extent of 3.398×10^{-7} C/m^2. Calculate the polarizability of benzene if there are 6.64×10^{27} molecule/m^3.

Solution

The data given are $E_0 = 300$ V/cm $= 3 \times 10^4$ V/m, and $N = 6.64 \times 10^{27}$ molecules/m^3

The local field acting on a molecule/atom of a polarized medium is given by

$$E_L = E_0 - \frac{2P}{3\varepsilon_0} = 3 \times 10^4 \text{ V/m} - \frac{2 \times 3.398 \times 10^{-7}}{3 \times 8.85 \times 10^{-12}} \text{V/m} = 4403 \text{ V/m}$$

The dipole moment μ_e of a single benzene molecule is given by

$$\mu_e = \frac{P}{N} = \frac{3.398}{6.64 \times 10^{27}} \text{Cm} = 5.12 \times 10^{-35} \text{Cm}$$

The polarizability of a benzene molecule is given by

$$\alpha = \frac{\mu_e}{E_L} = \frac{5.12 \times 10^{-35}}{4403} \frac{\text{Cm}^2}{\text{V}} = 1.16 \times 10^{-38} \frac{\text{Cm}^2}{\text{V}} = 1.16 \times 10^{-38} \text{Fm}^2$$

9.5 Polarization Mechanisms

As discussed previously atoms/molecules in a material get polarized under the influence of a microscopic field at the molecular site. There are three different types of polarizations as listed here.

i.   Electronic polarization
ii. Ionic polarization
iii. Dipolar (orientation) polarization

9.5.1 Electronic polarization

In dielectrics such as diamond, phosphorous and inert gases in which there is no ionic bonding or permanent dipole, the polarization is completely due to electronic polarization. Electronic polarization occurs due to the response of individual atoms in the dielectric to the local field at the atoms. Local field is produced inside the material whenever there is an external field. Under the action of local fields, the negatively charged electron cloud is displaced in the opposite direction to that of the local field and the positively charged nucleus is displaced in the direction of the local field. Thus, the local field tries to separate the electron cloud and the nucleus from each other and the atom becomes polarized. The polarization produced due to the displacements of the electron cloud and the nucleus is called electronic polarization. Soon, due to the attractive Coulomb force between the electron cloud and the nucleus, equilibrium is established. Figure 9.9 illustrates electronic polarization.

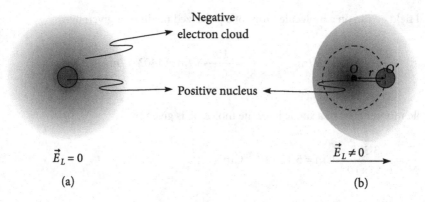

Negative
electron cloud

Positive nucleus

$\vec{E}_L = 0$

(a)

$\vec{E}_L \neq 0$

(b)

Figure 9.9 | (a) In the absence of a local field, the centres of the nucleus and the electron cloud coincide due to spherical symmetry. (b) In the presence of a local field, the centres of the nucleus and the electron cloud are separated through a displacement $\vec{r}$

Theory of electronic polarization

At equilibrium, under the influence of a local field, let $\vec{r}$ be the displacement between the centre of the electron cloud and the centre of the nucleus (i.e., $\vec{r}$ = displacement of the electron cloud). The induced dipole moment acquired by the atom under the action of a local field is given according to the definition by

$$\vec{\mu}_e = Ze\vec{r}, \ Z \text{ being the atomic number} \tag{9.35}$$

It can be seen that the direction of the induced dipole moment is the same as that of local fields. Using Eq. (9.25), we can have

$$\alpha_e \vec{E}_L = Ze\vec{r} \tag{9.36}$$

The polarizability in case of electronic polarization is called electronic polarizability. The magnitude of electronic polarizability is given by Eq. (9.36) as

$$\alpha_e = \frac{Zer}{E_L} \tag{9.37}$$

Let us calculate the electronic polarizability in terms of the atomic radius R. Under the action of the local field, the nucleus of charge Ze is supposed at O' and the centre of the electron cloud of effective charge q' is supposed at O as shown in Fig. 9.9. The nucleus is immersed in the electron cloud. The effective charge of the electron cloud for the nucleus is the amount of charge inside a sphere of radius r, r being the distance between the centre of the electron cloud and the centre of the nucleus. [See Fig. 9.9]. If ρ is the negative charge density of the electron cloud, the amount of charge inside a sphere of radius r will be

$$q' = \rho \times \frac{4\pi}{3} r^3 \tag{9.38}$$

Neglecting the volume of the nucleus, the negative charge density ρ is given by

$$\rho = -\frac{Ze}{\dfrac{4\pi}{3} R^3}, \ R \text{ being the radius of the atom.} \tag{9.39}$$

Therefore,

$$q' = -\frac{Ze}{\dfrac{4\pi}{3} R^3} \times \frac{4\pi}{3} r^3 = -Ze \frac{r^3}{R^3} \tag{9.40}$$

Thus, the magnitude of the force of attraction F_C between the nucleus and the electron cloud according to Coulomb's law is given by

$$F_C = \frac{kZeq'}{r^2} = \frac{Ze}{4\pi\varepsilon_0 r^2} Ze \frac{r^3}{R^3} = \frac{Z^2 e^2 r}{4\pi\varepsilon_0 R^3} \tag{9.41}$$

The force $\vec{F}_L$ that separates the centre of the electron cloud and the centre of the nucleus is due to local field $\vec{E}_L$. The magnitude of this force is given by the definition of the electric field intensity as

$$F_L = E_L Ze \tag{9.42}$$

At equilibrium, the magnitudes of these two forces F_C and F_L are equal, i.e.,

$$F_C = F_L$$

or $\quad \dfrac{Zer}{E_L} = 4\pi\varepsilon_0 R^3 \tag{9.43}$

Using Eq. (9.37), we obtain

$$\alpha_e = 4\pi\varepsilon_0 R^3 \tag{9.44}$$

The electronic polarizability is thus found to be proportional to the cube of the atomic radius. According to Eq. (9.18), the electronic polarization $\vec{P}_e$ will be given by

$$\vec{P}_e = N\alpha_e\vec{E}_L = 4\pi\varepsilon_0 NR^3\vec{E}_L$$

Replacing $\vec{E}$ by $\vec{E}_L$ in Eq. (9.14), susceptibility χ is defined as

$$\chi = \frac{P_e}{\varepsilon_0 E_L} = 4\pi NR^3 \tag{9.45}$$

From Eq. (9.15), the dielectric constant is given by

$$\varepsilon_r = 1 + \chi$$

or $\quad \varepsilon_r = 1 + 4\pi NR^3 = 1 + \dfrac{N\alpha_e}{\varepsilon_0}$

$$\tag{9.46}$$

For dielectrics such as diamond, phosphorous and inert gases in which there is no ionic bonding or permanent dipole, the polarization is completely due to electronic polarization. Susceptibility and dielectric constant can be calculated by using Eqs (9.45) and (9.46) respectively. The displacement of the electron cloud r i.e., the displacement between the centre of the electron cloud and the centre of the nucleus as obtained from Eq. (9.43) is

$$r = \frac{4\pi\varepsilon_0 E_L R^3}{Ze} = \frac{4\pi NR^3 \varepsilon_0 E}{NZe} = \frac{(\varepsilon_r - 1)\varepsilon_0 E}{NZe} \tag{9.47}$$

Example 9.5

The lattice parameter of KCl (a simple cubic) is 6.29×10^{-10} m containing four atoms per unit cell. The electronic polarizability of K^+ is 1.264×10^{-40} Fm^2 and that of Cl$^-$ is 3.408×10^{-40} Fm^2. Calculate the dielectric constant of KCl.

Solution

The electronic polarizability α_e of KCl is given by

$$\alpha_e = \alpha_{eK^+} + \alpha_{eCl^-} = \left(1.264 \times 10^{-40} + 3.408 \times 10^{-40}\right)\text{Fm}^2 = 4.672 \times 10^{-40}\,\text{Fm}^2\,.$$

The volume of the unit cell of KCl = $(6.29 \times 10^{-10}\text{ m})^3 = 2.49 \times 10^{-32}$ m^3
According to the question, 2.49×10^{-32} m^3 contains 4 atoms. Hence, the unit volume of KCl contains

$$N = \frac{4\text{atoms}}{2.49 \times 10^{-32}\,\text{m}^3} = 1.6 \times 10^{28}\,\text{atoms/m}^2$$

The dielectric constant of KCl is given by

$$\varepsilon_r = 1 + \frac{N\alpha_e}{\varepsilon_0} = 1 + \frac{1.607 \times 10^{28} \times 4.672 \times 10^{-40}}{8.85 \times 10^{-12}} = 1.848$$

Example 9.6

The dielectric constant of helium measured at 0K and at one atmosphere is 1.0000681. Under these conditions, gas contains 2.7×10^{23} atoms/m³. Calculate the radius of the electron cloud (= atomic radius). Also calculate the displacement of the electron clouds in the helium atom.

Solution

The data given are

$\varepsilon_r = 1.0000681$ and $N = 2.7 \times 10^{25}$ atoms/m³.

We know that

$$\varepsilon_r = 1 + 4\pi N R^3$$

or $$R = \left(\frac{\varepsilon_r - 1}{4\pi N}\right)^{\frac{1}{3}}$$

Putting the given data into the equation, we get

$$R = \left(\frac{1.0000684 - 1}{4\pi \times 2.7 \times 10^{25}}\right)^{\frac{1}{3}} \text{m} = 5.864 \times 10^{-11} \text{m}$$

The displacement of the electron clouds r in the helium atom is given by

$$r = \frac{(\varepsilon_r - 1)\varepsilon_0 E}{NZe} = \frac{(1.0000864 - 1) \times 8.85 \times 10^{-12} \times 10^6}{2.7 \times 10^{25} \times 2 \times 1.6 \times 10^{-19}} \text{m}$$

$$= 7.01 \times 10^{-17} \text{m}$$

9.5.2 Ionic polarization

When an electric field is applied to ionic solids (NaCl, KBr, KCl, LiBr, etc) having ionic bonding, the positive charges will be displaced in the direction of the field and the negative charges will be displaced in the opposite direction due to the influence of local field. Thus, the positive and negative charges are displaced from each other as shown in Fig. 9.10, resulting in net dipole moment in the material. The polarization arising due to displacement of positive charges in the field direction and negative charges in the direction opposite to the field in ionic solids is called ionic polarization $\vec{P}_i$ and the corresponding dipole moment per unit local field is called ionic polarizability α_i. In addition to ionic polarization in ionic solids there will be electronic polarization. In general, alkali halides exhibit ionic polarization.

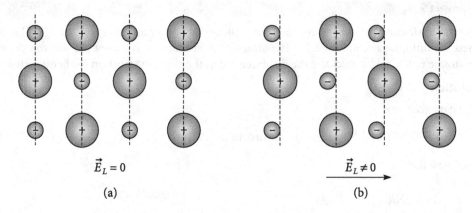

$$\vec{E}_L = 0 \qquad\qquad \vec{E}_L \neq 0 \longrightarrow$$

(a)   (b)

Figure 9.10 | Ionic polarization. (a) Ionic dielectric in the absence of an external field. (b) Ionic dielectric under the influence of an external field. The positive ions get displaced in the direction of the field and the negative ions get displaced in the opposite direction of the field

Theory of ionic polarization

Consider the arrangements of ionic molecules as shown in the Fig. 9.10. In the absence of an electric field as shown in Fig. 9.10(a), there is no displacement of ions. However, as shown in Fig. 9.10(b), when an electric field is applied, there is a displacement of positive and negative ions resulting in an induced dipole moment $\vec{\mu}_i$. The polarization produced in a crystal having N number of dipole moments per unit volume is given by

$$\vec{P}_i = N\alpha_i \vec{E}_L , N = \text{number of ions per unit volume.}$$

Let $\vec{x}_1$ and $\vec{x}_2$ be the displacements of positive and negative ions (singly ionized) from their equilibrium positions due to local field $\vec{E}_L$. The induced dipole moment $\vec{\mu}_i$ is given by

$$\vec{\mu}_i = e\left(\vec{x}_1 + \vec{x}_2\right) \tag{9.48}$$

Let $\vec{F}$ be is the force experienced by the ions due to local field $\vec{E}_L$. The magnitude of the restoring force acting on the positive and negative ions according to Hooke's law, are respectively given by

$$F_+ = \beta_1 x_1 \tag{9.49}$$

and

$$F_- = \beta_2 x_2 \tag{9.50}$$

At equilibrium, they are equal, i.e.,

$$\beta_1 x_1 = \beta_2 x_2$$

The proportionality constants β_1 and β_2 are functions of mass of the respective ions and natural angular frequency ω_0. If m_+ and m_- are the mass of positive and negative ions respectively, then β_1 and β_2 are given by

$$\beta_1 = m_+ \omega_0^2 \text{ and } \beta_2 = m_- \omega_0^2$$

Putting the values of β_1 and β_2 into Eqs (9.49) and (9.50), we have

$$F_+ = m_+ \omega_0^2 x_1 \tag{9.51}$$

and

$$F_- = m_- \omega_0^2 x_2 \tag{9.52}$$

Since both ions are singly ionized, we have

$$F_+ = eE_L \text{ and } F_- = eE_L$$

Putting $F_+ = eE_L$ and $F_- = eE_L$ into Eqs (9.51) and (9.52), we have

$$eE_L = m_+ \omega_0^2 x_1 \text{ and } eE_L = m_- \omega_0^2 x_2$$

or $\quad x_1 = \dfrac{eE_L}{m_+ \omega_0^2} \text{ and } x_2 = \dfrac{eE_L}{m_- \omega_0^2} \tag{9.53}$

Putting Eq. (9.53) into Eq. (9.48), the magnitude of the induced dipole moment will be obtained as

$$\mu_i = \frac{e^2}{\omega_0^2} \left(\frac{1}{m_+} + \frac{1}{m_-} \right) E_L \tag{9.54}$$

The ionic polarizability $\alpha_i = \dfrac{\mu_i}{E_L}$ will be given by

$$\alpha_i = \frac{e^2}{\omega_0^2}\left(\frac{1}{m_+}+\frac{1}{m_-}\right)$$

(9.55)

9.5.3 Dipolar (orientation) polarization

Dipolar polarization is also called orientation polarization. In many dielectrics such as polymers, water, some organic liquids and solids, the positive charge centre and the negative charge centre do not coincide. Hence, in these types of materials, there exist permanent dipole moments. They are called polar dielectrics. In the absence of an external electric field, these dipole moments are randomly oriented due to thermal energy as shown in Fig. 9.11(a) and under the influence of an external field, the permanent dipole moments try to align themselves nearly parallel to the external field as shown in Fig. 9.11(b).

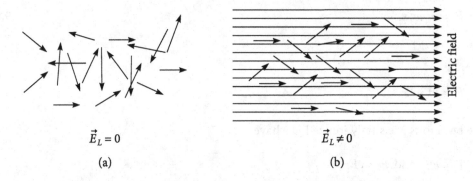

$\vec{E}_L = 0$

(a)

$\vec{E}_L \neq 0$

(b)

Electric field

Figure 9.11 Dipolar or orientation polarization. (a) The random orientation of dipoles due to thermal energy in the absence of an external electric field. (b) The dipoles try to orient themselves parallel to the external electric field

Theory of dipolar polarization

Consider the dielectric as an assembly of N number of permanent molecular dipoles per unit volume, each having dipole moment $\vec{\mu}_p$. In the absence of an external electric field since molecular dipoles are randomly oriented, the vector sum of all the dipole moments will be zero resulting in zero polarization. When an electric field is applied, dipoles try to orient themselves parallel to the electric field. The thermal energy tries to randomize the dipoles while the external electric field tries to make them parallel to the field. Thus, dipolar polarization must be some inverse function of temperature.

The magnitude of orientation polarization due to permanent molecular dipoles P_0 is given by

$$P_0 = \int_0^\pi \mu_p \cos\theta dN_\theta \qquad (9.56)$$

where

$\mu_p \cos\theta$ = component of dipole moment in the direction of the external electric field. The Fig. 9.12 may be consulted.

dN_θ = number of permanent molecular dipoles per unit volume that are oriented in the direction in between θ and $\theta + d\theta$ with respect to the external electric field.

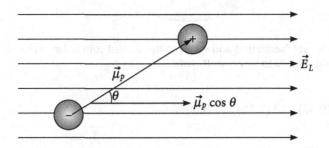

Figure 9.12 | A molecular dipole under the influence of an external electric field. $\mu_p \cos\theta$ is the component of dipole moment in the direction of the external electric field. The magnitude of the energy of interaction between the dipole having dipole moment μ_p and the field E_L is $-E_L\mu_p\cos\theta$

The number of permanent molecular dipoles per unit volume that which are oriented in the direction in between θ and $\theta + d\theta$ with respect to the external electric field dN_θ is directly proportional to the area of the shaded region in the imaginary sphere of radius r of Fig. 9.13, i.e.,

$$dN_\theta \propto 2\pi r \sin\theta \times r d\theta \qquad (9.57)$$

According to the Maxwell–Boltzmann statistics, the number of dipoles dN_θ having energy $-E_L\mu_p\cos\theta$ is proportional to the Maxwell–Boltzmann distribution function

$$e^{\frac{E_L\mu_p\cos\theta}{kT}}, \text{ i.e., } dN_\theta \propto e^{\frac{E_L\mu_p\cos\theta}{kT}} \qquad (9.58)$$

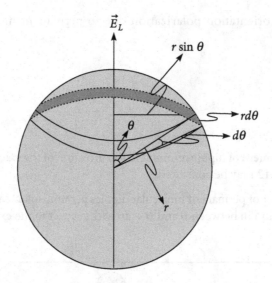

Figure 9.13 | The length (perimeter) and width of the shaded strip is $2\pi r \sin\theta$ and $rd\theta$ respectively giving the area to be $2\pi r \sin\theta \times rd\theta$

Combining Eqs (9.57) and (9.58), we have

$$dN_\theta = Ce^{\frac{E_L\mu_p\cos\theta}{kT}} 2\pi r^2 \sin\theta d\theta \tag{9.59}$$

or

$$N = \int_0^\pi dN_\theta = C2\pi r^2 \int_0^\pi e^{\frac{E_L\mu_p\cos\theta}{kT}} \sin\theta d\theta$$

or

$$C = \frac{N}{2\pi r^2 \int_0^\pi e^{\frac{E_L\mu_p\cos\theta}{kT}} \sin\theta d\theta}$$

Putting the value of C from this equation into Eq. (9.59), we have

$$dN_\theta = \frac{N2\pi r^2 e^{\frac{E_L\mu_p\cos\theta}{kT}} \sin\theta d\theta}{2\pi r^2 \int_0^\pi e^{\frac{E_L\mu_p\cos\theta}{kT}} \sin\theta d\theta} \tag{9.60}$$

Putting this equation into Eq. (9.56), the magnitude of orientation polarization P_0 is given by

$$P_0 = \frac{N\mu_p \int\limits_{0}^{\pi} e^{\frac{E_L\mu_p \cos\theta}{kT}} \cos\theta \sin\theta d\theta}{\int\limits_{0}^{\pi} e^{\frac{E_L\mu_p \cos\theta}{kT}} \sin\theta d\theta}$$

(9.61)

By putting $a = \dfrac{E_L\mu_p}{kT}$ and $x = \cos\theta$, this the above equation becomes

$$P_0 = \frac{N\mu_p \int\limits_{-1}^{1} e^{ax} x\, dx}{\int\limits_{-1}^{1} e^{ax} dx} = N\mu_p \frac{\int\limits_{-1}^{1} e^{ax} x\, dx}{\int\limits_{-1}^{1} e^{ax} dx}$$

$$= N\mu_p \frac{d}{da} \ell n \int\limits_{-1}^{1} e^{ax} dx = N\mu_p \frac{d}{da} \ell n \left(\frac{e^{ax}}{a}\Big|_{-1}^{1}\right)$$

$$= N\mu_p \frac{d}{da}\left(\ell n\left(e^a - e^{-a}\right) - n\mu_p \ell n a\right)$$

$$= N\mu_p \frac{d}{da} \ell n\left(e^a - e^{-a}\right) - N\mu_p \frac{d}{da} \ell n a$$

$$= N\mu_p \frac{e^a + e^{-a}}{e^a - e^{-a}} - N\mu_p \frac{1}{a} = N\mu_p \left(\frac{e^a + e^{-a}}{e^a - e^{-a}} - \frac{1}{a}\right)$$

or $\quad P_0 = N\mu_p \left(\coth a - \dfrac{1}{a}\right)$

(9.62)

The Langevin function $L(a)$ is defined as

$$L(a) = \coth a - \frac{1}{a}$$

(9.63)

The Langevin function $L(a)$ is plotted in Fig. 9.14.

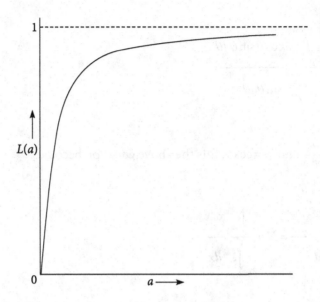

Figure 9.14 | Plotting of the Langevin function $L(a) = \coth a - \dfrac{1}{a}$. For small a values, it is almost a straight line

Thus, in terms of Langevin function $L(a)$, the magnitude of orientation polarization P_0 due to permanent dipoles is given by

$$P_0 = N\mu_p L(a) \tag{9.64}$$

Now, even at room temperature [a low temperature], $a = \dfrac{E_L \mu_p}{kT}$ is much less than 1, i.e., of the order of 10^{-13}. Therefore, in general, $a \ll 1$ then

$$L(a) \approx \frac{a}{3} \quad \text{since} \quad L(a) = \coth a - \frac{1}{a} = \frac{1}{a} + \frac{a}{3} - \frac{a^3}{45} + \ldots - \frac{1}{a}$$

Therefore, the magnitude of orientation polarization P_0 due to permanent dipoles will become

$$P_0 = \frac{N\mu_p^2 E_L}{3kT} \tag{9.65}$$

As we have discussed earlier, this the above equation shows that

$$P_0 \propto \frac{1}{T}$$

Orientation polarization due to permanent dipoles decreases with increase of temperature. At high temperature, the thermal energy is so large that orientation polarization is negligible. The variations of orientation polarization with inverse of absolute temperature and local field are shown in Fig. 9.15(a) and 9.15(b) respectively.

From Eq. (9.65), orientation polarizability α_0 can be written as

$$\alpha_0 = \frac{\mu_p^2}{3kT} \tag{9.66}$$

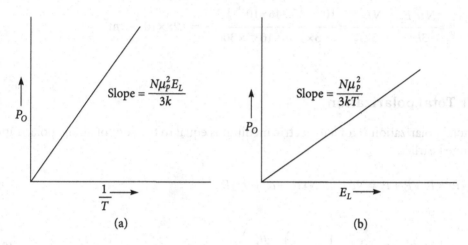

(a)   (b)

Figure 9.15 (a) Variation of dipolar polarization with inverse of absolute temperature. (b) Variation of dipolar polarization with local electric field that depends upon the externally applied field

In the derivation of Eqs (9.65) or (9.66), we have neglected induced dipole moments; however, in reality, they contribute to dipolar polarization. The polarizability contributed by the induced dipole moments to dipolar polarizability is called deformation polarizability α_d. In general, then, the total molecular polarizability α is given by

$$\alpha = \alpha_d + \frac{\mu_p^2}{3kT} \tag{9.67}$$

This equation is known as the Langevin–Debye equation and has been of great importance in interpreting molecular structures.

Example 9.7

Determine the orientation polarization of HCl vapour at room temperature when it is subjected to an electric field of 10^6 V/m. Given that the concentration of HCl molecule = 10^{27} molecules/m³, permanent dipole moment of HCl molecule = 1.04 Debye [1Debye = 3.33 × 10^{-30} Cm].

Solution

The data given are

$E = 10^6$ V/m, $N = 10^{27}$ molecules/m^3,

$\mu_p = 1.04$ Debye $= 3.463 \times 10^{-30}$ Cm, and $T = 300$ K

The orientation polarization is given by

$$P_0 = \frac{N\mu_p^2 E_L}{3kT} = \frac{N\mu_p^2 E}{3kT} = \frac{10^{27} \times \left(3.346 \times 10^{-30}\right)^2 10^6}{3 \times 1.38 \times 10^{-23} \times 300} = 9.28 \times 10^{-7} \, \text{C/m}^2$$

9.5.4 Total polarization

The total polarization P_T of a dielectric medium is equal to the sum of all the polarizations explained earlier.

$$P_T = P_e + P_i + P_0 = N\alpha E_L = N\left(\alpha_e + \alpha_i + \alpha_0\right)E_L \tag{9.68}$$

or $\quad P_T = N\left(4\pi\varepsilon_0 R^3 + \dfrac{e^2}{\omega_0^2}\left(\dfrac{1}{m_+} + \dfrac{1}{m_-}\right) + \dfrac{\mu_p^2}{3kT}\right)E_L \tag{9.69}$

Electronic polarization exists in all dielectrics. In ceramic and polymer dielectrics, all the three types of polarization explained earlier exist and Eq. (9.69) holds good. If the dielectric has no ionic bonding and there is no permanent dipole moment, polarization is only due to electronic polarization, i.e.,

$$P_T = 4\pi\varepsilon_0 R^3 N E_L \tag{9.70}$$

Equation (9.69) shows that only polar dielectrics are temperature dependent.

Putting Eq. (9.14) into Eq. (9.68), we can have

$$\varepsilon_0\left(\varepsilon_r - 1\right) = \frac{N\mu_p^2}{3k}\frac{1}{T} + N\left(\alpha_e + \alpha_i\right) \tag{9.71}$$

This above equation is of the form $y = mx + c$ with $y = \varepsilon_0(\varepsilon_r - 1)$, $x = 1/T$, $m = \dfrac{N\mu_p^2}{3k}$, and $c = N(\alpha_e + \alpha_i)$. If Eq. (9.71) is plotted, it will give a straight line as shown in the Fig. 9.16.

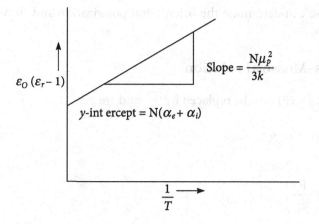

Figure 9.16 | Plotting of equation $\varepsilon_0 \left(\varepsilon_r - 1 \right) = \dfrac{N\mu_p^2}{3k} \dfrac{1}{T} + N\left(\alpha_e + \alpha_i \right)$ with inverse of absolute temperature

From the slope of the straight line, the dipole moment of a molecule of gas can be experimentally obtained. From the y-intercept, the contribution to the dielectric constant due to ionic and electronic polarization can be determined.

Again Eq. (9.69) can be written as

$$P_T = \frac{N\mu_p^2 E_L}{3k} \frac{1}{T} + N\alpha_e E_L + N\alpha_i E_L \qquad (9.72)$$

This equation is of the form $y = mx + c$ with $y = P_T$, $x = \dfrac{1}{T}$, $m = \dfrac{N\mu_p^2 E_L}{3k}$, and $c = N(\alpha_e + \alpha_i)$ E_L. If Eq. (9.72) is plotted, it will give a straight line as shown in Fig. 9.17.

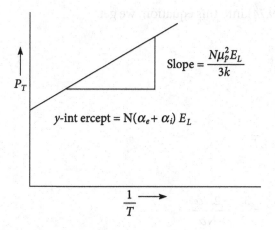

Figure 9.17 | Plotting of equation $P_T = \dfrac{N\mu_p^2 E_L}{3k} \dfrac{1}{T} + N\alpha_e E_L + N\alpha_i E_L$.

From this plot, we can determine the orientation polarization and sum of electronic and ionic polarization.

9.5.5 Clausius–Mossotti relation

Now the $\bar{E}$ in Eq. (9.18) is to be replaced by $\bar{E}_L$ and we get

$$\vec{P} = N\alpha\bar{E}_L \quad (\alpha = \alpha_e + \alpha_i + \alpha_0) \tag{9.73}$$

or $$\vec{P} = N\alpha\left(\vec{E} + \frac{\vec{P}}{3\varepsilon_0}\right)$$

or $$\vec{P} = \left(\frac{N\alpha}{1 - \dfrac{n\alpha}{3\varepsilon_0}}\right)\vec{E} \tag{9.74}$$

Equation (9.74) is the corrected form of Eq. (9.18). The term $1 - \dfrac{N\alpha}{3\varepsilon_0}$ in the denominator of Eq. (9.74) is less than unity and hence, contributes largely to the increase of polarization. Thus, polarization increases when we take into account $\bar{E}_L$, the local field or microscopic field.

Now we know from Eq. (9.11) that

$$\bar{D} = \vec{P} + \varepsilon_0\bar{E}$$

Putting $\vec{P}$ from Eq. (9.74) into this equation, we get

$$\bar{D} = \frac{N\alpha\bar{E}}{1 - \dfrac{N\alpha}{3\varepsilon_0}} + \varepsilon_0\vec{E}$$

or $$\varepsilon\vec{E} = \frac{3N\alpha\varepsilon_0\vec{E}}{3\varepsilon_0 - N\alpha} + \varepsilon_0\vec{E}$$

or $$\frac{\varepsilon}{\varepsilon_0} = \frac{3N\alpha}{3\varepsilon_0 - N\alpha} + 1 = \frac{3\varepsilon_0 + 2N\alpha}{3\varepsilon_0 - N\alpha}$$

$$\text{or} \qquad \varepsilon_r = \frac{1 + \dfrac{2N\alpha}{3\varepsilon_0}}{1 - \dfrac{N\alpha}{3\varepsilon_0}} \qquad\qquad (9.75)$$

This equation is the generalization of Eq. (9.19) taking into account the effect of the local field. In cases in which atomic or molecular concentration is small as in the case of gases, Eq. (9.75) reduces to Eq. (9.19).

From Eq. (9.75), we have

$$\varepsilon_r - 1 = \frac{3N\alpha}{3\varepsilon_0 - N\alpha}$$

and $\quad \varepsilon_r + 2 = \dfrac{9\varepsilon_0}{3\varepsilon_0 - N\alpha}$

Taking the ratio of these two equations, we get

$$\frac{\varepsilon_r - 1}{\varepsilon_r + 2} = \frac{N\alpha}{3\varepsilon_0} = \frac{\displaystyle\sum_j N_j \alpha_j}{3\varepsilon_0} \qquad\qquad (9.76)$$

This equation is called the Clausius–Mossotti relation. It relates the macroscopic parameter dielectric constant to the microscopic parameter polarizability. The Clausius–Mossotti relation holds good only for those solids which have cubic symmetry.

In the Eq. (9.76), N is the number of atoms or molecules per unit volume. If m is the mass of each atom/molecule, then the density ρ will be

$$\rho = mN$$

$$\text{or} \qquad N = \frac{\rho}{m}$$

If M is the molar mass and N_A is Avogadro's number, the mass of each atom/molecule m will be given by

$$m = \frac{M}{N_A}$$

Thus, we have

$$N = \frac{\rho N_A}{M} \tag{9.77}$$

Putting Eq. (9.77) into Eq. (9.76), we have

$$\frac{\varepsilon_r - 1}{\varepsilon_r + 2} = \frac{\rho N_A \alpha}{3 M \varepsilon_0}$$

or $$\alpha = \left(\frac{\varepsilon_r - 1}{\varepsilon_r + 2} \right) \frac{3 \varepsilon_0 M}{\rho N_A} \tag{9.78}$$

This equation shows that polarizability α can be determined from the values of ε_r, M and ρ.

We can also express polarizability α in terms of the refractive index of the dielectric medium n. From Maxwell's electromagnetic theory, we know that the speed of light 'v' in any medium having permittivity ε and magnetic permeability μ is given by

$$v = \frac{1}{\sqrt{\varepsilon \mu}}$$

Naturally, the speed of light in vacuum c will be given by

$$c = \frac{1}{\sqrt{\varepsilon_0 \mu_0}}$$

Combining these two equations, we have

$$\frac{c}{v} = \frac{\sqrt{\varepsilon \mu}}{\sqrt{\varepsilon_0 \mu_0}} = \sqrt{\frac{\varepsilon}{\varepsilon_0} \frac{\mu}{\mu_0}} = \sqrt{\varepsilon_r \mu_r}$$

The refractive index of a medium n is given by c/v. Thus, we have

$$n = \frac{c}{v} = \sqrt{\varepsilon_{re} \mu_r} \tag{9.79}$$

For a dielectric medium, since $\mu_r \approx 1$, this equation reduces to

$$n = \sqrt{\varepsilon_{re}} \tag{9.80}$$

It is important to remember that the dielectric constant ε_{re} in the equation $\varepsilon_{re} = n^2$ is only due to electronic polarization. Electronic polarization is predominant and ionic and dipolar polarization are negligibly small at optical frequencies in the range of electromagnetic radiation. From Eq. (9.80), we conclude that the contribution of electronic polarization to the dielectric constant of the matter at optical frequencies is numerically equal to the square root of its refractive index. For this reason, we changed the symbol from ε_r to ε_{re} in the Eq. (9.80). The measurement of refractive index n at optical frequency shows that the dielectric constant ε_r arises almost entirely from the electronic polarizability α_e. The equation is only applicable for very low loss material at optical frequencies.

In ionic polarization, there is a relative change in the mean positions of the atomic nuclei within the molecules. This is generally a small effect and is observed at radio frequencies, but not at optical. Therefore, the contribution of ionic polarization to the dielectric constant at optical frequency is negligibly small. The same is the fate with dipolar polarization in the range of optical frequencies.

In summary, the contribution of dipolar polarizability α_0 and ionic polarizability α_i to dielectric constants are negligibly small at high frequencies because of the inertia of the molecules and ions. Thus, dielectric constant at optical frequencies arises almost entirely from electronic polarizability. Therefore, at optical frequencies, we can neglect α_0 and α_i in the expression $\alpha = \alpha_e + \alpha_i + \alpha_0$. Under these conditions from the Clausius–Mossotti relation, we get

$$\frac{\varepsilon_{re} - 1}{\varepsilon_{re} + 2} = \frac{N\alpha_e}{3\varepsilon_0} = \frac{\sum_j N_j \alpha_j}{3\varepsilon_0} \quad \text{at optical frequencies.} \tag{9.81}$$

At optical frequencies, $\varepsilon_{re} = n^2$ and putting this into the aforementioned equation, we have

$$\frac{n^2 - 1}{n^2 + 2} = \frac{N\alpha_e}{3\varepsilon_0} = \frac{\sum_j N_j \alpha_{ej}}{3\varepsilon_0} \tag{9.82}$$

This equation is a relation between electronic polarizability and static refractive index of the medium. Equation (9.82) is called the Lorentz equation.

Example 9.8

The dielectric constant of water is 81 and its refractive index is 1.33. What is the percentage contribution of ionic polarizability and dipolar polarizability to the dielectric constant?

Solution

The percentage of contribution of ionic polarizability α_i and dipolar polarizability α_0 is

$$\frac{\alpha_i + \alpha_0}{\alpha_e + \alpha_i + \alpha_0} \times 100 = \left(1 - \frac{\alpha_e}{\alpha_e + \alpha_i + \alpha_0}\right) \times 100 \tag{A}$$

We know that [Clausius–Mossotti relation]

$$\frac{\varepsilon_r - 1}{\varepsilon_r + 2} = \frac{N(\alpha_e + \alpha_i + \alpha_0)}{3\varepsilon_0}$$

and its refractive index counter part is $\dfrac{n^2 - 1}{n^2 + 2} = \dfrac{N\alpha_e}{3\varepsilon_0}$. Combining these two equations, we have

$$\frac{n^2 - 1}{n^2 + 2} \times \frac{\varepsilon_r + 2}{\varepsilon_r - 1} = \frac{N\alpha_e}{3\varepsilon_0} \times \frac{3\varepsilon_0}{N(\alpha_e + \alpha_i + \alpha_0)} = \frac{\alpha_e}{\alpha_e + \alpha_i + \alpha_0} \tag{B}$$

Putting (B) into (A), we get

$$\frac{\alpha_i + \alpha_0}{\alpha_e + \alpha_i + \alpha_0} \times 100 = \left(1 - \frac{n^2 - 1}{n^2 + 2} \times \frac{\varepsilon_r + 2}{\varepsilon_r - 1}\right) \times 100$$

$$= \left(1 - \frac{(1.33)^2 - 1}{(1.33)^2 + 2} \times \frac{81 + 2}{81 - 1}\right) \times 100 = 79\%$$

Example 9.9

The polarizability of ammonia molecule is found by measuring the dielectric constant approximately as 2.42×10^{-39} C^2m/N at 309 K and 1.74×10^{-39} C^2m/N at 448 K. Calculate the orientation polarizability of ammonia molecule at each temperature.

Solution

The data given are

$$\alpha_1 = 2.42 \times 10^{-39}\, C^2\text{m/N} \quad \text{at} \quad T_1 = 309\text{K}$$

$$\alpha_2 = 1.74 \times 10^{-39}\, C^2\text{m/N} \quad \text{at} \quad T_2 = 448\text{K}$$

The total molecular polarizability α of a polar gas is $\alpha = \alpha_e + \alpha_i + \alpha_0$ with $\alpha = \dfrac{\mu^2}{3kT}$, i.e.,

$$\alpha - (\alpha_e + \alpha_i) = \frac{\mu^2}{3kT}$$

or $\quad \alpha_1 - (\alpha_e + \alpha_i) = \dfrac{\mu^2}{3kT_1}$ and $\alpha_2 - (\alpha_e + \alpha_i) = \dfrac{\mu^2}{3kT_2}$

or $\dfrac{\alpha_1 - (\alpha_e + \alpha_i)}{\alpha_2 - (\alpha_e + \alpha_i)} = \dfrac{T_2}{T_1}$

or $\alpha_e + \alpha_i = \dfrac{\alpha_2 T_2 - \alpha_1 T_1}{T_2 - T_1}$

But $\alpha_0 = \alpha - (\alpha_e + \alpha_i)$

or $\alpha_0 = \alpha - \dfrac{\alpha_2 T_2 - \alpha_1 T_1}{T_2 - T_1}$

or $\alpha_{01} = \alpha_1 - \dfrac{\alpha_2 T_2 - \alpha_1 T_1}{T_2 - T_1}$ and $\alpha_{02} = \alpha_2 - \dfrac{\alpha_2 T_2 - \alpha_1 T_1}{T_2 - T_1}$

Putting the earlier mentioned data into these two equations, we have

α_0 at 309K = 2.192×10^{-39} C^{2}m/N

α_0 at 448K = 1.512×10^{-39} C^{2}m/N

9.6 Effect of Temperature on Dielectrics

Equation (9.65) shows that the polarization of polar dielectrics are temperature dependent. For example, let us take polar dielectric nitrobenzene whose variation of dielectric constant with temperature is shown in Fig. 9.18.

Figure 9.18 shows that as we decrease the temperature towards the melting point, the dielectric constant of nitrobenzene increases as expected from Eq. (9.71). At the melting point, when nitrobenzene becomes solid, the permanent dipole moments do not have much degree of freedom and are arrested. Hence, at or below the melting point, orientation polarization ceases to contribute to the total polarization. In this region, electronic and ionic polarization contribute to the dielectric constant (refer to Eq. (9.71)) as a result of which dielectric constant decreases abruptly to a lower value at melting point. Below the melting point, to the solid nitrobenzene, only electronic and ionic polarizations contributes to its dielectric behaviour. As electronic and ionic polarizations are temperature independent, the dielectric constant of solid nitrobenzene is independent of temperature below the melting point.

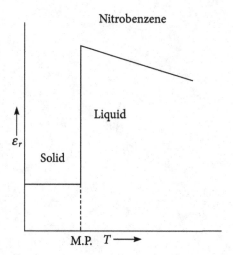

Figure 9.18 | Variation of dielectric constant with temperature for of a polar dielectric nitrobenzene

Next consider the variation of dielectric constant with temperature of polar polymers as shown in Fig. 9.19.

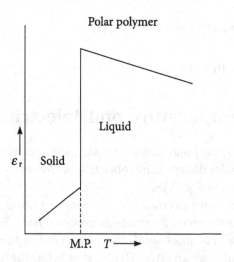

Figure 9.19 | Variation of dielectric constant with temperature of a polar polymer

In polar polymers the dielectric constant increases slightly with increasing temperature in the solid state as shown in Fig. 9.19. As the temperature increases due to thermal energy, the dipoles becomes less inhibited and so get just mobile enough to get oriented in the field direction contributing to orientation polarization or dielectric constant. At the melting point, dielectric constant of polar polymers increases suddenly and then decreases with increase of temperature.

9.7 Effect of Frequency on Dielectrics

In previous sections, dielectric polarization by static or dc electric field ($\omega = 0$) has been discussed. Now let us analyze dielectric polarization under the action of an ac field ($\omega \neq 0$).

When an ac field is applied to dielectrics, displacements of charges, ions and the dipole moments vary in direction and magnitude periodically. This displacement depends strongly on frequency of the applied ac field and so are the electronic, ionic and orientation polarization. Since movement of electrons or electronic polarization is a very fast process, it can appear from minimum to very high frequency ac fields, even up to UV range. The displacement of positive and negative charges in ionic solids is a moderate process. Hence, ionic polarization exists up to moderate frequency of ac fields and vanishes at very high frequency. Dipolar polarization, which involves arrested displacement of dipoles, is a very slow process. It is the slowest process among the three types of charge displacement. Therefore, orientation polarization manifests only under low frequency ac fields. Finally, we can conclude that dielectric constant depends upon the frequency of the applied field for frequencies more than 10^6 hz. The polarization can follow the ac field for frequencies less than 10^6 hz without any significant lag and the imaginary part of the dielectric constant will be zero. Hence, in this case, permittivity is independent of the frequency and has the same magnitude as in a static field.

The role of frequency of applied ac fields in different polarizations is given in the following table.

Polarizability	Low frequency	Medium frequency	High frequency
Electronic	✓	✓	✓
Ionic	✓	✓	✗
Orientation	✓	✗	✗

In the following sections, we will discuss the effects of ac field on electronic, ionic, and orientation polarizabilities.

9.7.1 Electronic polarizability

In Section 9.5.1, we have discussed electronic polarization under static electric field ($\omega = 0$) and under ideal conditions. The displacement of electron cloud was without radiation contradicting the fact that accelerated charges emit radiation. When an electric field is applied, the electron cloud gets acceleration, losing energy in the form of electromagnetic radiation. Moreover, it loses energy due to collision. Thus, its motion is damped. Since there is a force of attraction between the positive nucleus and the negative electron cloud, displacement of electron cloud obeys Hook's law for small displacements. Considering all these factors, the equation motion of an electron under the action of an external electric field $E = E_0 \cos \omega t$ is given as

$$m_e \frac{d^2 x}{dt^2} = -kx - b\frac{dx}{dt} + \left(-eE_0 \cos \omega t\right) \tag{9.83}$$

The details of this equation have already been discussed. In Eq. (9.83),

$$m_e \frac{d^2x}{dt^2} = \text{Total force (Newton's second law)},$$

$$-kx = \text{Restoring force (Hooke's law)}$$

$$-b\frac{dx}{dt} = \text{Damping force}$$

$$-eE_0 \cos \omega t = \text{Electric field force on the electron}$$

From Eq. (9.83), we have

$$\frac{d^2x}{dt^2} + \frac{k}{m_e}x + \frac{b}{m_e}\frac{dx}{dt} = -\frac{eE_0}{m_e}\cos \omega t$$

or $$\frac{d^2x}{dt^2} + 2\gamma\frac{dx}{dt} + \omega_0^2 x = F_0 \cos \omega t \qquad (9.84)$$

Here we have the following substitution

$$\frac{b}{2m_e} = \gamma = \text{damping coefficient},$$

$$\sqrt{\frac{k}{m_e}} = \omega_0 = \text{natural angular frequency}$$

and $$-\frac{eE_0}{m_e} = F_0$$

Equation (9.84) is very similar to Eq. (1.36, Part-I). Following the same methods as described there, the steady state complex solution of Eq. (9.84) is

$$x(t) = \frac{F_0 e^{i\omega t}}{\omega_0^2 - \omega^2 + 2i\gamma\omega} \qquad (9.85)$$

The complex induced dipole moment as a function of time is

$$\mu_e(t) = -ex(t) = \frac{\dfrac{e^2}{m_e}E_0 e^{i\omega t}}{\omega_0^2 - \omega^2 + 2i\gamma\omega} = \frac{\dfrac{e^2}{m_e}}{\omega_0^2 - \omega^2 + 2i\gamma\omega}E_0 e^{i\omega t}$$

Thus, complex polarizability is obtained as

$$\alpha_e = \frac{\dfrac{e^2}{m_e}}{\omega_0^2 - \omega^2 + 2i\gamma\omega} \tag{9.86}$$

or

$$\alpha_e = \frac{e^2}{m_e}\left(\frac{\omega_0^2 - \omega^2 - 2i\gamma\omega}{\left(\omega_0^2 - \omega^2\right)^2 + 4\gamma^2\omega^2}\right)$$

or

$$\alpha_e = \frac{e^2}{m_e}\left(\frac{\omega_0^2 - \omega^2}{\left(\omega_0^2 - \omega^2\right)^2 + 4\gamma^2\omega^2}\right) - i\frac{e^2}{m_e}\left(\frac{2\gamma\omega}{\left(\omega_0^2 - \omega^2\right)^2 + 4\gamma^2\omega^2}\right) \tag{9.87}$$

This equation is in the form of

$$\alpha_e = \alpha'_e - i\alpha''_e$$

with

$$\alpha'_e = \frac{e^2}{m_e}\left(\frac{\omega_0^2 - \omega^2}{\left(\omega_0^2 - \omega^2\right)^2 + 4\gamma^2\omega^2}\right) \tag{9.88}$$

and

$$\alpha''_e = \frac{e^2}{m_e}\left(\frac{2\gamma\omega}{\left(\omega_0^2 - \omega^2\right)^2 + 4\gamma^2\omega^2}\right) \tag{9.89}$$

Here α'_e and α''_e are the real and imaginary parts of electronic polarizability respectively.

Dependence of α_e' on the frequency ω of applied electric field

i.   Under the action of a dc field, $\omega = 0$ gives $\alpha_e' = \dfrac{e^2}{m_e \omega_0^2}$

ii. $\omega > \omega_0$ gives $\alpha_e' = $ negative

iii. $\omega = \omega_0$ gives $\alpha_e' = $ zero

iv. $\omega < \omega_0$ gives $\alpha_e' = $ positive

v. $\omega \ll \omega_0$ gives $\alpha_e' = \dfrac{e^2}{m_e \omega_0^2}$ (neglecting $4\gamma^2\omega^2$ and ω^2). This is the same as case(i). It proves that at low frequency, the applied field contributes nothing to electronic polarizability.

vi.   When $\omega_0 \approx \omega$, we may have $\omega_0 - \omega = \Delta\omega$ and $\omega_0 - \omega = 2\omega_0$. Hence, Eq. (9.88) gives

$$\alpha_e' \approx \frac{e^2}{2m_e\omega_0}\left(\frac{\Delta\omega}{\Delta\omega^2 + 4\gamma^2}\right) \tag{9.90}$$

On applying the principle of maxima and minima of functions on this above equation, we find that for $\omega = \omega_0 - 2\gamma$, α_e' is maximum $\left(= \dfrac{e^2}{8m_e\omega_0\gamma}\right)$ and for $\omega = \omega_0 + 2\gamma$, it is minimum

$\left(= -\dfrac{e^2}{8m_e\omega_0\gamma}\right)$.

Thus, the difference between the positions of maximum and minimum is 4γ. Therefore, damping coefficient is a measure of the separation between the maximum and minimum in the dispersion curve. The variation of α_e' with ω is shown in Fig. 9.20.

Dependence of α_e'' on the frequency ω of the applied electric field

i.   Under the action of a dc field, $\omega = 0$ gives $\alpha_e'' = 0$. Thus, for a static field, polarizability is not complex; its imaginary part is absent.

ii.   Under the action of a high frequency field, i.e., when $\omega \to \infty$, $\alpha_e'' = 0$

iii. For

$$\omega = \omega_0, \alpha_e'' = \frac{e^2}{2m_e\omega_0\gamma},$$

iv. $\omega \ll \omega_0$ gives $\alpha_e'' = 0$ (neglecting $4\gamma^2\omega^2$ and ω^2).

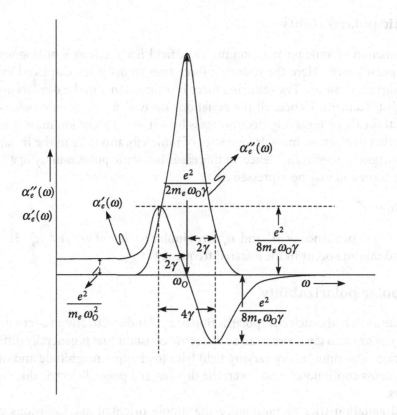

Figure 9.20 | The variation of α_e' and α_e'' with ω known as the dispersion curve is shown schematically

v.   For $\omega \approx \omega_0$, we may have $\omega_0 - \omega = \Delta\omega$ and $\omega_0 - \omega = 2\omega_0$. Hence, Eq. (9.89) gives

$$\alpha_e'' \approx \frac{\gamma e^2}{m_e \omega_0 \left(\Delta\omega^2 + 4\gamma^2 \right)} \tag{9.91}$$

On applying the principle of maxima of functions on this equation, we find that for $\Delta\omega = 0$, i.e., for $\omega = \omega_0$, α_e'' is maximum $\left(= \dfrac{e^2}{2m_e \omega_0 \gamma} \right)$.

All these above equations are valid for bound electric charges. The expression for dielectric constant and susceptibility can be obtained by putting these expressions into Eqs (9.19) and (9.20) and thus, their variations with frequency of the applied ac field can be inferred.

9.7.2 Ionic polarizability

The phenomenon of ionic polarization under ac field $E = E_0 \cos \omega t$ is analogous to that of electronic polarization. Here the restoring force tries to drive the displaced ions back to their equilibrium position. The restoring force is proportional to the displacement of ions for small displacement. Hence, all the equations derived for electronic polarization will be valid in this case by replacing electron mass by ion mass. Since ion mass is many times more than that of electron mass, frequencies of ionic vibrations lie in the infrared part of the electromagnetic spectrum. Hence, in this case also, ionic polarizability of a molecule is complex in form and may be expressed as

$$\alpha_i = \alpha_i' - i\alpha_i'' \tag{9.92}$$

The frequency dependence of α_i' and α_i'' are similar to that of α_e' and α_e''. However, the maxima and minima occur in the infrared frequency range.

9.7.3 Dipolar polarizability

Now we shall analyze the dielectric polarization of a polar dielectric by an alternating electric field $E = E_0 \cos \omega t$. This gives rise to an ac dielectric constant that is generally different from the static case. The sinusoidally varying field tries to changes magnitude and direction of dipole moments continuously; however, the dipoles in a polar dielectric do not have free movements.

Thermal agitation tries to randomize the dipole orientations. Collisions and lattice vibrations also help the randomization of the dipole orientations. Moreover, dipoles are not independent of mutual interactions. As a result, dipoles cannot respond instantaneously to the changes in the applied field. There exist some minimum reorientation time. Above all, even if all these factors are absent, if the applied electric field changes at high frequencies, dipoles cannot follow the field and remain randomly oriented with zero net polarization!

At low frequencies, the dipoles follow the field and the non-zero polarization vector changes accordingly. We need to calculate polarizability as a function of frequency so that the dielectric constant can be determined by using the Clausius–Mossotti relation.

The dipolar relaxation equation is given as

$$\frac{d\mu_e(t)}{dt} = \frac{\mu_e(t) - \alpha_d(0)E}{\tau} \tag{9.93}$$

In this equation, τ is the relaxation time, which is defined as the average time during which 63% of the dipoles are randomized. It is independent of time and is a function of material properties and temperature. Putting the applied field $E = E_0 \cos \omega t$ in the above Eq. (9.93), we get

$$\frac{d\mu_e(t)}{dt} = \frac{\mu_e(t) - \alpha_d(0)E_0 \cos \omega t}{\tau} = \frac{\mu_e(t) - \alpha_d(0)E_0 e^{i\omega t}}{\tau}$$

or $\qquad \dfrac{d\mu_e(t)}{dt}=\dfrac{\mu_e(t)}{\tau}+\dfrac{\alpha_d(0)}{\tau}E_0e^{i\omega t}$ $\qquad\qquad\qquad\qquad\qquad$ (9.94)

Solving this differential equation, the instantaneous dipole moment $\mu_e(t)$ is obtained as

$$\mu_e(t)=\dfrac{\alpha_d(0)}{1+i\omega\tau}E_0e^{i\omega t}=\dfrac{\alpha_d(0)}{1+i\omega\tau}E \qquad\qquad\qquad (9.95)$$

Comparing the above equation with Eq. (9.16), orientation polarizability under alternating field $\alpha_d(\omega)$ is obtained as

$$\alpha_d(\omega)=\dfrac{\alpha_d(0)}{1+i\omega\tau} \qquad\qquad\qquad\qquad\qquad (9.96)$$

This equation shows that at low frequencies, $(\omega\tau<<1)$ $\alpha_d(\omega)\approx\alpha_d(0)$ and dipole moments are in phase with the applied electric field. At high frequencies, there is a phase difference between the applied electric field and dipole moments and hence, polarization cannot follow the electric field. Equation (9.96) shows that since polarizability is complex under an alternating field, the dielectric constant according to Eq. (9.19) will be complex under an alternating field. Hence, we can express complex dielectric function $\varepsilon_r(\omega)$ in the following form

$$\varepsilon_r(\omega)=\varepsilon_r'(\omega)-i\varepsilon_r''(\omega) \qquad\qquad\qquad\qquad (9.97)$$

According to (9.96), orientation polarizability $\alpha_d(\omega)$ is maximum for $\omega=0$, i.e., for dc fields. Hence, according to Eq. (9.19), the maximum value of $\varepsilon_r(\omega)$ is $\varepsilon_r'(0)$ for dc fields. According to Eq. (9.96), at very high frequencies (i.e., $\omega\to\infty$), $\alpha_d(\omega)\to 0$, and as a result of which maximum value of $\varepsilon_r(\omega)$, i.e., $\varepsilon_r'(0)$ decreases to 1.

At high frequencies $(\omega\to\infty)$, dipoles cannot follow the field and remain randomly oriented with zero net polarization contributing nothing to the dielectric constant. Thus, in this case, the imaginary part $\varepsilon_r''(\omega)$ reduces to zero. Again at low frequencies, according to Eq. (9.96), $\varepsilon_r''(\omega)$ is absent. Therefore, in both these cases, as $\omega\to\infty$ and $\omega\to 0$, $\varepsilon_r''(\omega)$ decreases to zero value and peaks for $\omega\tau=1$. The variation of real and imaginary parts of the complex dielectric constant is shown in Fig. 9.21.

With each cycle of reversal, the dipoles attempt to reorient with the alternating field against random collisions. The imaginary part $\varepsilon_r''(\omega)$ measures the energy lost in the dielectric medium during the reorientation process. The absorption of electrical energy by a dielectric material subjected to an alternating electric field is termed dielectric loss. A low dielectric loss is desired at the frequency of utilization for most electronic devices. This energy is lost as heat. If we want to store a charge in a capacitor, dielectric loss is not good;

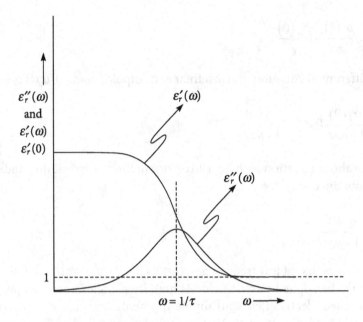

Figure 9.21 Variation of $\varepsilon_r'\left(\omega\right)$ and $\varepsilon_r''\left(\omega\right)$ with ω

However, it is good in microwave ovens to prepare food items. Dielectric losses are measured by a parameter known loss tangent. The loss tangent or dissipation factor is defined by

$$\tan\delta = \frac{\varepsilon_r''\left(\omega\right)}{\varepsilon_r'\left(\omega\right)} \tag{9.98}$$

For $\varepsilon_r'\left(\omega\right) \gg \varepsilon_r''\left(\omega\right)$, Eq. (9.98) boils down to

$$\delta = \frac{\varepsilon_r''\left(\omega\right)}{\varepsilon_r'\left(\omega\right)} \tag{9.99}$$

This equation shows that loss tangent is a function frequency of the applied field. Though it peaks at around $\omega = (1/\tau)$, it depends upon the material properties. Typically, for liquid and solid media, $\omega = (1/\tau)$ is in the microwave frequencies range.

9.8 Dielectric Breakdown

Ideally, dielectrics are perfect insulators. They are used as dielectrics in capacitors to increase their capacitance and as insulators in electric and electronic circuits. When a dielectric is subjected to potential difference, it does not conduct electricity; however, as the applied potential difference increases to a certain maximum value, it suddenly starts to conduct

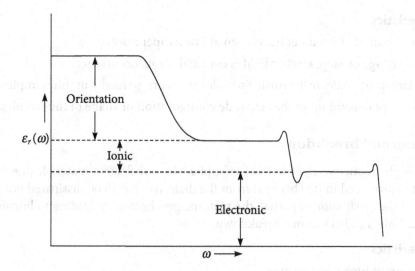

Figure 9.22 | Variation of dielectric constant $\varepsilon_r(\omega)$ with frequency ω of an applied alternating electric field. Electronic, ionic and orientation polarization contributions to the dielectric constant have been shown

electricity. This phenomenon in which a dielectric material conducts electricity under a certain high potential difference is called dielectric breakdown and the corresponding high potential difference is called breakdown voltage. The minimum electric field at which a dielectric conducts electricity is called dielectric strength and is measured in volt/meter. The following three types of mechanism play a role in the dielectric breakdown phenomenon.

i. Avalanche breakdown/intrinsic breakdown.

ii. Thermal breakdown.

iii. Electrochemical breakdown.

iv. Defect breakdown.

9.8.1 Avalanche breakdown

Dielectrics are perfect insulators. They have a completely filled valence band and a completely empty conduction band. The band gap between the conduction band and the valence band is more than 5 eV. When a dielectric is subjected to high electric field, the electron in the valence band gains sufficient energy to be excited to the conduction band above the valence band. These electrons are accelerated highly by the high electric field and by collisions, they excite more electrons to the conduction band. Thus, more and more electrons are released to the conduction band resulting in an avalanche of conduction electrons. Ultimately, the dielectric becomes highly conducting and conducts large currents.

Characteristics

i.   It can occur at any temperature, even at low temperatures.

ii.   High voltage or large electric field is essential for its occurrence.

iii. Avalanche breakdown/intrinsic breakdown occurs generally in thin samples.

iv.   It does not depend upon the electrode configuration or shape of the samples.

9.8.2 Thermal breakdown

When a high frequency electric field is applied to a dielectric material, due to energy loss, heat is produced in it. This heat from the dielectric has to be dissipated out from the material. If the dissipation is partial, the material gets heated up leading to burning of the dielectric. This is called thermal breakdown.

Characteristics

i.   It occurs at high temperatures.

ii.   The dielectric strength depends upon the shape and size of the dielectric material.

iii. Breakdown time is of the order of milliseconds.

9.8.3 Electrochemical breakdown

When the temperature of a dielectric increases, the mobility of the ions inside the dielectric increases, resulting in an electrochemical reaction. When the mobility of ions increases, leakage current will increase, thereby, decreasing the insulation resistance. This will result in a dielectric breakdown called electrochemical breakdown.

Characteristics

i.   Electrochemical breakdown is determined by leakage currents, density of ions, temperatures and permanent dipoles in the material.

ii.   To avoid electrochemical breakdown, the dielectric should not be operated at high temperatures.

iii. To avoid electrochemical breakdown, the dielectric should be 100% pure.

9.8.4 Defect breakdown

This type of breakdown occurs in dielectric materials that have defects like cracks and pores. These cracks and pores will be filled with gases or air. At high electric fields, the local field at the pores and cracks will be so high that gas discharges will occur causing breakdown of the dielectric.

Characteristics

i.   It occurs at high voltage.

ii.   It cannot occur in dielectrics that have having no cracks and pores.

9.9 Ferroelectric Materials

In polar dielectrics, in the absence of an external electric field, the permanent dipoles are randomly oriented resulting in zero net polarization. In the presence of an external field, the permanent electric dipoles in polar dielectrics are oriented nearly parallel to the applied field [see Fig. 9.11] resulting in net polarization given by [see Eq. 9.65]

$$P_0 = \frac{N\mu_p^2 E_L}{3kT}.$$

This equation shows that $P_0 \propto E_L$, i.e., polarization varies linearly with the local field which in turn depends upon the externally applied field. Polarization becomes zero the moment the electric field is removed since the plot of P_0 versus E_L is a straight line passing through the origin [see Fig. 9.15(b)]. In other words, orientations of permanent dipoles are randomized when the external field is removed.

There are certain polar dielectrics in which orientations of permanent dipoles are not randomized but remain oriented in the field direction even when the external field is removed. These types of polar dielectrics are called ferroelectrics or ferroelectric materials. Typical examples of ferroelectric materials are $BaTiO_3$ (barium titanate) and $KNbO_3$ (potassium niobate). In ferroelectrics, polarization is not a linear function of the applied field. Ferroelectrics exhibit hysteresis curve (shown in Fig. 9.23) similar to that shown by of ferromagnetic materials and hence, the name ferroelectrics.

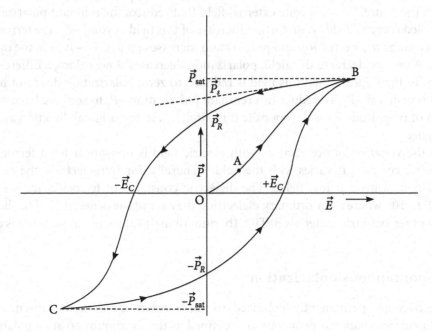

Figure 9.23 | Hysteresis effect in ferroelectrics

Ferroelectrics are characterized by ferroelectric hysteresis, spontaneous polarization, irreversibility of polarization and ferroelectric transition temperature.

9.9.1 Ferroelectric hysteresis

When a ferroelectric is subjected to an external field, permanent dipoles orient themselves in the direction of the field. Most permanent dipoles remain oriented even after the electric field is removed. This is due to the presence of strong microscopic electric fields within the material in small regions. These small microscopic regions within which all the dipoles orient in the same direction are called ferroelectric domains. Ferroelectric materials therefore may be imagined to consist of a large number of ferroelectric domains, each domain having a specific polarization direction. When an external field is applied, all the domains orient in the field direction. Thus, the domains in the direction of the external field grow in size.

The variation of polarization with external field called ferroelectric hysteresis is depicted in Fig. 9.23. Ferroelectric hysteresis depends upon temperature. Let us assume that, at a certain temperature, the ferroelectric is completely depolarized at the beginning. As we go on increasing the external field, polarization of the ferroelectric increases non-linearly from zero and attains a maximum value $\vec{P}_{sat}$ at B called the saturation polarization. When we decrease the external field, polarization decreases slowly along a different path as shown in the figure. When the external field is reduced to zero, polarization does not become zero but remains $\vec{P}_R$, called the remnant polarization. To reduce this remnant polarization $\vec{P}_R$ to zero, we have to apply an external field of magnitude E_C in the opposite direction as shown in the figure. This opposite external field that reduces the remnant polarization to zero is called coercive field. With further increase of this field beyond $-E_C$, the ferroelectric domains change their direction and polarization increases up to C $(= -\vec{P}_{sat})$ in the opposite manner. When we decrease this field, polarization decreases slowly along a different path as shown in the figure. When this field is reduced to zero, polarization does not become zero but remains at $-P_R$. To reduce this remnant polarization $-P_R$ to zero, we have to apply this field of magnitude E_C in the opposite direction, i.e., in the original direction as shown in the figure.

Since the variation of polarization with electric field is non-linear for a ferroelectric, its dielectric constant ε_r varies with the field. Generally, for ferroelectrics, the dielectric constant is measured at low fields. The dielectric constants of ferroelectrics are of the order 10^4 to 10^5, whereas for ordinary dielectrics, they are of the order of 10. The dielectric constant of ferroelectric materials $BaTiO_3$ (barium titanate) at room temperature is as high as 5000.

9.9.2 Spontaneous polarization

Ferroelectrics are permanently polarized to some extent even in the absence of an external field. Spontaneous polarization is defined as the maximum possible polarization in the absence of an external electric field. The maximum possible value of the remnant polarization is the spontaneous polarization. This has been shown as $\vec{P}_S$ in Fig. 9.23. Spontaneous polarization in ferroelectric materials is a function of temperature as shown in Fig. 9.24. By absorbing thermal energy, dipoles get disoriented, resulting in the decrease of spontaneous polarization. The absorption of thermal energy is more at higher temperature.

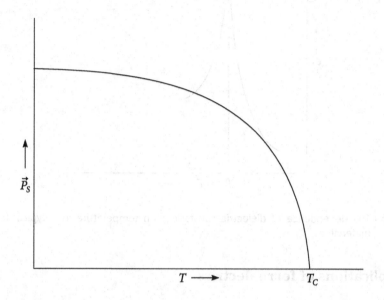

Figure 9.24 Variation of spontaneous polarization $\vec{P}_s$ of a ferroelectric with temperature. The temperature T_C above which spontaneous polarization vanishes is called the ferroelectric Curie temperature or ferroelectric transition temperature

The ferroelectric Curie temperature or ferroelectric transition temperature T_C of a ferroelectric is defined as the temperature above which spontaneous polarization vanishes. Above the ferroelectric Curie temperature, the dielectric constant of a ferroelectric varies with temperature according to the relation

$$\varepsilon_r \approx \frac{C}{T - T_C} \qquad (9.100)$$

where

$\quad C = $ constant.

$\quad T_C = $ ferroelectric transition temperature.

The plot in Fig. 9.25 shows the dependence of dielectric constant ε_r on temperature in a ferroelectric material.

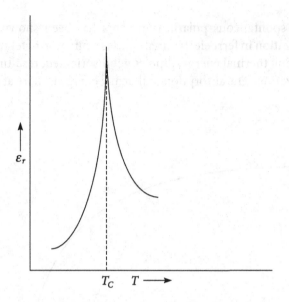

Figure 9.25 | The dependence of dielectric constant ε_r on temperature in a typical ferroelectric material

9.9.3 Applications of ferroelectrics

Ferroelectric materials have several practical applications due to their special characteristics. These materials can be used in variable capacitors, gyroscopes, acoustics sensors and actuators, microgenerators and memory devices.

	Characteristics	Applications
1	Large dielectric constants	Capacitors of large capacitance
2	Large piezoelectric constants	Sonar detectors, strain sensors, actuators
3	Bi-stable orientational states	Binary memories
4	Pyroelectricity	Infrared detectors, infrared imaging
5	Large non-linear polarization	Optical memory display, optical wave guide

The slowness of switching speed coupled with the requirement of high electric field for the material at present makes ferroelectric materials less attractive for use as memory devices as compared with ferrimagnetic materials.

9.10 Piezoelectrics

A unit cell is said to have a centre of symmetry at O if when $\bar{r}$ is a position vector with respect to O of any charge, then $-\bar{r}$ is a position vector for another same type of charge (including sign). Any point on any charge may be a centre of symmetry under this condition. Figure 9.26(a) explains the centre of symmetry. Crystals having centre of symmetry are called centrosymmetric crystals; otherwise, they are called non-centrosymmetric crystals. This implies that non-centrosymmetric crystals are polar and centrosymmetric crystals are non-polar without finite polarization or dipole moment. There are 11 classes of centrosymmetric crystals and 21 classes of non-centrosymmetric crystals of which 20 are polar.

In general, all materials undergo a small change in dimension when subjected to external stimuli like force, electric field, mechanical stress, change in temperature and so on. The word "piezo" in Greek means pressure and therefore piezoelectric materials or simply piezoelectrics mean materials that can be polarized by pressure or mechanical stress. The phenomenon of development of positive and negative charges by mechanical stress on two opposite faces of a crystal formed by polar molecules with a non-centrosymmetric structure is called the piezoelectric effect. This means that a mechanical stress applied to the crystal specimen will create an overall polarization and hence, a voltage across it. The reverse of the stress direction will cause the reverse of the polarity of the polarization and hence, the voltage. The piezoelectric effect is convertible. This means that an applied electric field will create a mechanical strain, expansion, or contraction, depending on the direction of the field.

An applied electric field always causes mechanical distortion in the geometric shape of the material, because matter consists of charged nuclei surrounded by a compensating electron cloud. The polarization induced by the applied field will cause changes in charge distribution and hence mechanical distortion. This phenomenon is called electrostriction.

Under the action of mechanical stress, atoms in centrosymmetric crystals are slightly displaced symmetrically. Since ionic displacements are symmetrical about the centre of symmetry, centres of charge of negative and positive charges in the unit cell remain coincident; the charge distribution inside the dielectric is not altered appreciably. Therefore, no polarization occurs in centrosymmetric crystals under the action of mechanical stress.

The dipole moments in crystals formed by polar molecules with a non-centrosymmetric structure may just mutually cancel each other in the material under the unstrained condition. Since in non-centrosymmetric crystals, ionic displacements by stress about the centre of symmetry are non-symmetrical, the centres of charge of negative and positive ions in the unit cell are different. Thus, unidirectional dipoles along the direction of stress are produced and the dielectric as a whole is polarized. The net result is the development of positive and negative charges on two opposite faces of the dielectrics giving rise to an electric field. In this case, the relation between the electric field and the mechanical strain is linear in the first approximation. The direction of this polarization depends on the direction of the applied stress. In the following, we will try to explain the mechanism of the piezoelectric effect in an idealized simplet way.

9.10.1 Simple molecular model of the piezoelectric effect

Based on the concepts explained earlier, an extremely simplified molecular model has been depicted in Fig. 9.26.

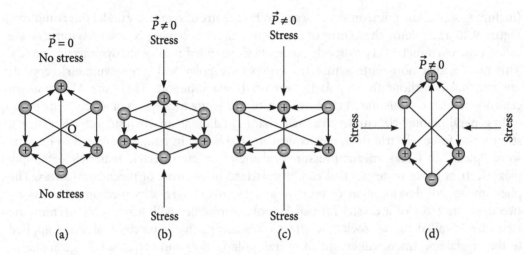

Figure 9.26 Polarization by mechanical stress. Arrow marks connecting the ions are polarization vectors. (a) In the absence of stress, the centres of charge for positive and negative ions coincide at O. (b) & (c) Under an applied stress along the vertical direction, the centres of charge of positive and negative ions are different, which gives rise to a net dipole moment along the vertical direction, (d) When the stress is along the horizontal direction, there may not be a net dipole moment in that direction but along the vertical direction. This figure also illustrates that a hexagonal unit cell has no centre of symmetry; the opposite direction of O contains opposite ions violating the condition for centre of symmetry

In Fig. 9.26(a), we have illustrated the cancellation of dipole moments without any external stress. Figure 9.26(b, c, d) explains how net dipole moment appears under the action of external stress. As per vector addition theorem, the net dipole moment appears in the vertical direction except in Fig. 9.26(a), where it is zero.

Few examples of piezoelectrics are quartz (crystalline form of SiO_2), barium titanate ($BaTiO_3$), berlinite ($AlPO_4$), gallium orthophosphate ($GaPO_4$) and lead zirconate titanate (PZT).

9.10.2 Applications of piezoelectric effect

The phenomenon of developments of positive and negative charges by mechanical stress on two opposite faces of certain crystals is called piezoelectric effect. This means that a mechanical stress applied to the crystal specimen will create an overall polarization and hence, a voltage across it. In converse piezoelectric effect, application of an electrical field creates mechanical stress in the crystal. Thus, by utilizing the piezoelectric effects, interconversion of electrical and mechanical energy is possible. The best known applications of piezo crystals are discussed briefly here.

i. *High voltage* Some substances like quartz can generate potential differences of thousands of volts through direct piezoelectric effect.

ii. *Gas lighter* The most common application of piezoelectrics to generate a potential is the dielectric gas lighter. Pressing the button of the lighter causes a spring-loaded hammer to hit a piezoelectric crystal, producing a sufficiently high voltage that electric current flows across a small spark gap, thus igniting the gas. The same technique is used commercially in vehicles for ignition.

iii.   *Ultrasonic generators*   Ultrasonic waves can be produced by feeding high frequency electrical signals to a piezoelectric crystal in the proper way.

iv.   *Piezoelectric presser sensor*   Detection of pressure variations in the form of sound is the most common presser sensor application, for example, piezoelectric microphones. Sound waves bend the piezoelectric material, creating a changing voltage.

v.   *Ultrasound imaging*   Piezoelectric sensors are used with high frequency sound in ultrasonic transducers for medical imaging. For many sensing techniques, the sensor can act as both a sensor and an actuator. Ultrasonic transducers, for example, can inject ultrasound waves into the body, receive the returned wave, and convert it to an electrical signal to form the image.

vi.   *Sonar sensors*   Piezoelectric elements are also used in the detection and generation of sonar waves. Applications include power monitoring in high power applications such as medical treatment, sonochemistry and industrial processing.

vii.   *Chemical and biological sensors*   Piezoelectric microbalances are used as very sensitive chemical and biological sensors. Piezoelectrics are also used as strain gauges.

viii. *Music instruments*   Piezoelectric transducers are used in electronic drum pads to detect the impact of the drummer's sticks.

ix.   *Automotive application*   Automotive engine management systems use a piezoelectric transducer to detect detonation by sampling the vibrations of the engine block. Ultrasonic piezosensors are used in the detection of acoustic emissions in acoustic emission testing.

x.   *Piezoresistive silicon devices*   The piezoresistive effect of semiconductors has been used for sensor devices employing all kinds of semiconductor materials such as germanium, polycrystalline silicon, amorphous silicon, and single crystal silicon. Since silicon is today the material of choice for integrated digital and analog circuits, the use of piezoresistive silicon devices has been of great interest. It enables the easy integration of stress sensors with bipolar and CMOS circuits.

xi.   *Piezoresistors*   Piezoresistors are resistors made from a piezoresistive material and are usually used for measurement of mechanical stress. They are the simplest form of the piezoresistive device.

xii.   *Piezoresistive effect*   It is the changing electrical resistance of a material due to applied mechanical stress. The piezoresistive effect differs from the piezoelectric effect. In contrast to the piezoelectric effect, the piezoresistive effect only causes a change in resistance; it does not produce an electric potential.

Besides the applications mentioned here, there are also other applications of the piezoelectric effect in crystal oscillators, crystal speakers, ultrasonic therapy, electronic clocks, frequency controllers in radio transmitters, record player pick ups, actuators and so on.

9.11 Pyroelectrics

The word "pyro" in Greek means fire and therefore, pyroelectric materials or pyroelectrics are materials which produce electricity by fire, i.e., heat. The development of opposite electrical charges on different parts of a crystal that is subjected to temperature change is called pyroelectric effect. Pyroelectric effect is rigorously defined as the phenomenon of temperature dependence of spontaneous polarization in certain anisotropic solids. Out of 20 non-centrosymmetric crystal classes, 10 classes contain a unique polar axis (i.e., the direction of the spontaneous polarization) under the unstrained condition. This means that such crystals are already spontaneously polarized in a certain temperature range. Thus, they exhibit both piezoelectric and pyroelectric effects.

If a pyroelectric specimen is heated non-uniformly, then the temperature gradient will also develop a mechanical stress. Therefore, in this case, the specimen may have mixed piezoelectric and pyroelectric effects. However, under a uniform heating condition, the resultant piezoelectric effect vanishes and only pyroelectric effect remains. When the temperature increases uniformly, the pyroelectric crystal expands and the relative distances of ions increase, increasing the dipole moments or polarization. This will develop free electric charges at the surfaces because the spontaneous polarization inside the crystal is changed by the heat.

Pyroelectricity is the electricity, i.e., electric current produced in an external closed circuit due to pyroelectric effects. It results mainly from the temperature dependence of the spontaneous polarization of the polar materials. A change in temperature will cause a corresponding change in polarization and hence, a change in the compensating charges on the metallic electrodes deposited on its surfaces. This change will produce a current called pyroelectricity in an external circuit as explained here below by taking an idealized example.

A thin, parallel-sided sample of pyroelectrics, such as a tourmaline crystal or a ceramic disk of barium titanate is taken. It is cut in such a manner that its crystallographic symmetry axis is perpendicular to the flat surfaces. The unit cells of pyroelectric materials have a dipole moment. The spontaneous polarization vector shown in the figure produces a layer of bound charge on each flat surface of the sample. Nearby free charges such as electrons or ions will be attracted to the sample as shown in Fig. 9.27(a).

Now the conducting electrodes are then attached to the surfaces and connected through an ammeter having a low internal resistance. If the temperature of the sample is constant, then so is spontaneous polarization and no current flows through the circuit. An increase in temperature causes the spontaneous polarization to decrease and so does is the quantity of bound charges. The re-distribution of free charges to compensate for the change in bound charge results in a current flow. This is the pyroelectric current in the circuit. If the temperature of the sample is increased, instead of decreasing, the current's sign would be reversed. Note that the pyroelectric effect is only observable during the period in which the temperature changes i.e. $\frac{dT}{dt} \neq 0$. In an open circuit, the free charges would simply remain on the electrodes and a voltage can be measured.

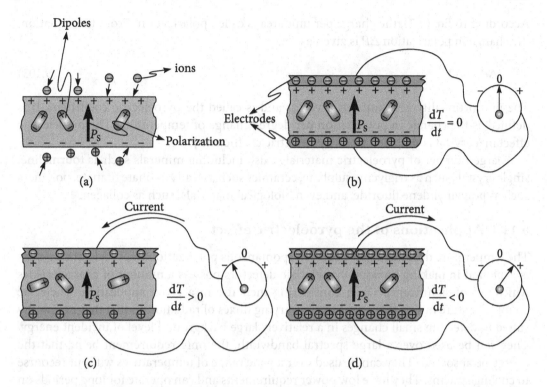

Figure 9.27 | Change of temperature produces pyroelectricity. (a) Thin sheet of spontaneously polarized tourmaline crystal, (b) No current exists, (c) Current when temperature increases, (d) Current when temperature decreases

If the pyroelectric specimen is not connected to an external circuit and is kept perfectly insulated from its surroundings, the surface charges on the electrodes released due to a change in temperature will flow inside the material through the intrinsic electrical conductivity of the material.

Pyroelectric effect is the ability of the pyroeletric to develop induced charges on the crystal surface when it is heated or cooled. The effect is produced due to the development of positive and negative charges on opposite ends of a crystal's polar axis as a result of temperature change. The amount of charge developed ΔQ on the surface of the specimen is proportional to the temperature change ΔT and surface area S. Thus, we have

$$\Delta Q = \gamma S \Delta T \qquad (9.101)$$

The surface charge density in this case is given as

$$\frac{\Delta Q}{S} = \gamma \Delta T \qquad (9.102)$$

According to Eq. (9.7), the charge per unit area is called polarization. From this equation, the change in polarization ΔP is given as

$$\Delta P = \gamma \Delta T \tag{9.103}$$

The proportionality constant γ is a vector and is called the pyroelectric coefficient. It is defined as the change in polarization per degree change of temperature. The pyroelectric effect in a crystal is specified by its pyroelectric coefficient γ.

A large number of pyroelectric materials exist, including minerals such as tourmaline, single crystals such as triglycine sulphate, ceramics such as lead zirconate titanate, polymers such as polyvinylidene fluoride, and even biological materials, such as collagen.

9.11.1 Applications of the pyroelectric effect

The temperature dependence property of spontaneous polarization in pyroelectric crystals is exploited in making sensors. Pyroelectric detectors possess a number of characteristics which are of significance when considering their use in a given application. Their "**ac coupled**" nature makes them sensitive to varying fluxes of radiation so that they are ideally suited to detecting small changes in a relatively large background level of incident energy. They can be used over a large spectral bandwidth, the only requirement being that the energy be absorbed. They can be used over a wide range of temperatures without recourse to cooling systems. They have low power requirements and can operate for long periods on battery power, just enough to drive the field effect transistor that amplifies the signal. Most importantly, they are generally low-cost devices.

i.   *Pyroelectric thermometer*   A pyroelectric thermometer can determine the change of temperature by measuring the voltage induced by the separation of the charges.

ii. *Intruder alarms*   This is an ideal application for pyroelectric detectors. In the absence of an intruder, the interior of an unoccupied building present a fairly constant thermal scene. An intruder moving into the area surveyed by the detector provides a varying flux of IR radiation, which can be detected and used to trigger an alarm.

iii. *Pollution monitoring and gas analysis*   The concentrations of gases in the atmosphere can be measured from the strength of particular lines in their absorption spectra. The analysis systems generally employ a modulated broadband source of IR illuminating two pyroelectric detectors equipped with filters at the chosen wavelengths. The concentration of the pollutants can be measured by this system.

iv. *Fire alarms*   To detect fires, it is normal to operate at shorter wavelengths than for intruder alarms. Systems usually sense the flicker frequency of the flames at 5-40 Hz to avoid false alarms.

v.   *Pyroelectric thermal imaging*   Thermal imaging utilizes the different powers radiated in the far-infrared by objects in the scene that are at different temperatures. Such techniques can be used for "multi-colour" IR imaging by which absolute temperatures in the scene can be measured remotely by their radiances in the different bands. This has application in industrial process control, in the military field and so on.

9.12 Dielectrics as Electrical Insulators

There are a variety of dielectrics available nowadays. Ideally, dielectrics are perfect insulators. In choosing dielectrics as insulating materials in electrical machines and devices, at least two properties, namely, breakdown voltage and their working temperature have to be taken into account. The voltage at which a device is working must be less than the breakdown voltage of the dielectrics. If the device containing the dielectric is working under a high frequency electric field, large heat will be produced due to dielectric loss in the material damaging the dielectric. The dielectric constants of a few dielectrics are also frequency dependent. Therefore, it is necessary to chose proper dielectrics as insulators for specific applications.

Insulating materials are classified into seven temperature classes according to their temperature limits.

	Classifications	Operating temperature limit.	Materials
1	Class Y	90°C	Un-impregnated paper, cotton, silk, vulcanized natural rubber, aniline, urea, thermoplastics
2.	Class A	105°C	Cotton, silk, paper impregnated with oil or varnish, laminated materials with cellulose filler phenolic resins.
3	Class E	120°C	Polyurethane, epoxy resins, varnishes, cellulose triacetate, phenol formaldehyde, polyethylene teraphthalate.
4	Class B	130°C	Mica, glass, asbestos fibers, and fabrics bonded and impregnated with shellac bitumen, epoxy, phenol, melamine formaldehyde.
5	Class F	155°C	Same as Class B, but with resins which are approved for class F operations
6	Class H	180°C	Composite inorganic materials impregnated with silicone, rubber.
7	Class C	180°C	Mica, asbestos, ceramics, glass, quartz, teflon polytetrafluoro ethylene

Questions

9.1   What do you mean by dielectric materials/dielectrics?

9.2   What is a dielectric material?

9.3   What do you mean by polar and non-polar molecules?

9.4   What is an electric dipole?

9.5   What is dielectric polarization?

9.6   What do you mean by induced charges?

9.7   What are the parameters of a dielectric?

9.8   What do you mean by dielectric constant?

9.9   What is a dipole moment?

9.10   What do you mean by dielectric polarization?

9.11   What do you mean by polarizability?

9.12   What do you mean by deformation polarizability?

9.13   Show that the unit of polarizability is Fm^2.

9.14   Explain why polarizability α is a microscopic parameter.

9.15   Prove that polarization is equal to surface charge density of induced charges.

9.16   What is Gauss' law for polarization vector $\vec{P}$?

9.17   Why is there is a negative sign in the equation $\oint_S \vec{P}.\vec{ds} = -q_i$?

9.18   Why cannot free space be electrically polarized?

9.19   What is the main difference between electric field intensity and electric displacement?

9.20   Why is electric susceptibility direction dependent in some materials?

9.21   Derive $\varepsilon_r = 1 + \dfrac{N\alpha}{\varepsilon_0}$

9.22   Prove that $\varepsilon_r = 1 + \dfrac{N\alpha}{\varepsilon_0}$ from $\varepsilon_r = \dfrac{1 + \dfrac{2N\alpha}{3\varepsilon_0}}{1 - \dfrac{N\alpha}{3\varepsilon_0}}$, i.e., Eq. (9.37) is the generalization of Eq. (9.19).

9.23   Express the dielectric constant of a dielectric in terms of density and molar mass of the dielectric.

9.24   Show that the dielectric constant increases linearly with density.

9.25   Show that the dielectric constant increases linearly with polarizability.

9.26   Why is the electric field inside a dielectric not same as that of an applied electric field?

9.27   What is Lorentz field?

9.28   What is depolarizing field?

9.29   Why do equations $\varepsilon_r = 1 + \dfrac{N\alpha}{\varepsilon_0}$ and $\varepsilon_r = 1 + \dfrac{\rho N_A}{M\varepsilon_0}\alpha$ not hold good in case of solids and liquids?

9.30   Derive a relation between applied field E_0 and macroscopic field E.

9.31   Distinguish between microscopic field E_L and macroscopic field E

9.32   Derive the Clausius–Mossotti relation.

9.33   Explain how polarization increases when we take into account the local field or microscopic field, instead of the macroscopic field.

9.34   Derive a relation between dielectric constant, a macroscopic parameter and polarizability, a microscopic parameter.

9.35   What are the different types of dielectric polarization?

9.36   What is the origin of electronic polarization?

9.37 Show that electronic polarizability is proportional to cube of the atomic radius.

9.38 Prove that the dielectric constant of diamond is $\varepsilon_r = 1 + 4\pi NR^3$, where R is the radius of carbon atom and N is the number of carbon atoms per unit volume.

9.39 What is the origin of ionic polarization?

9.40 What is ionic polarization? What is ionic polarizability?

9.41 Derive an expression for ionic polarizability?

9.42 What do you mean by polar dielectrics?

9.43 What is the origin of dipolar polarization?

9.44 Explain how loss tangent is a necessary evil.

9.45 Show mathematically that orientation polarization of polar materials is not complex under the action of a low frequency electric field.

9.46 Show that dielectric constant of a polar material is not a function of the applied ac field at high frequency.

9.47 Explain why the dielectric constant is unaffected by ionic and orientation polarizabilities under very high frequency electric field.

9.48 Explain the role of frequency of the applied electric field on the value of dielectric constant of a material.

9.49 Show mathematically that electronic polarizability is not complex under the action of a low frequency electric field.

9.50 Describe the behaviour of dielectric materials under applied ac fields at around the natural frequency of electrons.

9.51 Explain mathematically how dipolar polarization is an inverse function of temperature.

9.52 Explain the temperature variation of dielectric constant of a polar dielectric.

9.53 Explain the temperature variation of dielectric constant of a polar polymer.

9.54 Distinguish between dielectric constant and dielectric strength.

9.55 What are the mechanisms of dielectric breakdown phenomenon?

9.56 Explain how avalanche breakdown of dielectrics occurs?

9.57 What are the characteristics of avalanche breakdown of dielectrics?

9.58 Explain how thermal breakdown of dielectrics occurs?

9.59 What are the characteristics of thermal breakdown of dielectrics?

9.60 Explain how electrochemical breakdown of dielectrics occurs?

9.61 What are the characteristics of electrochemical breakdown of dielectrics?

9.62 Explain how defects breakdown of dielectrics occurs?

9.63 What are the characteristics of defects breakdown of dielectrics?

9.64 What are the factors at least which must be taken into account to use dielectrics as electrical insulators?

9.65 Why are ferroelectric materials called so?

9.66 What are the characteristics of ferroelectric materials?

9.67   What is the origin of ferroelectric hysteresis?

9.68   Explain the ferroelectric hysteresis curve.

9.69   What do you mean by saturation polarization in ferroelectric materials?

9.70   What do you mean by remnant polarization in ferroelectric materials?

9.71   What do you mean by coercive electric field in ferroelectric materials?

9.72   Mention a few applications of ferroelectrics.

9.73   Show the variation of ferroelectric dielectric constant with temperature.

9.74   Show the variation of spontaneous polarization of a ferroelectric with temperature.

9.75   What do you mean by piezoelectric materials/piezoelectrics?

9.76   What is piezoelectric effect?

9.77   What is the origin of piezoelectric effect?

9.78   What is the origin of electrostriction effect or inverse piezoelectric effect?

9.79   Mention a few applications of piezoelectric effect

9.80   What is pyroelectric effect?

9.81   What do you mean by pyroelectrics or pyroelectric materials?

9.82   Explain the explanation for origin of pyroelectric effect

9.83   Mention a few applications of the pyroelectric effect.

Problems

9.1   Determine the percentage of ionic polarizability in a sodium chloride crystal having refractive index 1.5 and static dielectric constant 5.6.   [Ans   51%]

9.2   The polarizability of ammonia molecule is found by measuring the dielectric constant approximately as 2.5×10^{-39} C^2 m/N at 300 K and 2.00×10^{-39} C^2m/N at 400 K. Calculate the orientation polarizability of ammonia molecule at each temperature.
            [Ans $\alpha_0 = 2 \times 10^{-39}$ C^2m/N at 300 K $\alpha_0 = 1.5 \times 10^{-39}$ C^2m/N at 400 K]

9.3   The atomic radius of Se atom is 0.12 nm. Calculate the electronic polarizability of Se atom.   [Ans 1.9226×10^{-40} Fm^2]

9.4   The electronic polarizability of helium is 1.8×10^{-39} C^2m/N. Calculate the atomic radius of the helium atom and permittivity if there are 2.7×10^{25} atoms/m^3.
            [Ans 5.46×10^{-11} m, 1.000057776]

9.5   Calculate the shift of electron clouds with respect to the nucleus in argon atom when an electric field of 10^5 V/m is applied. The polarizability of argon is 1.8×10^{-40} C^2m/N .
            [Ans 6.25×10^{-18} m]

9.6   The lattice parameter of NaCl (simple cubic) is 5.64×10^{-10} m containing four atoms per unit cell. The electronic polarizability of Na^+ is 1.6×10^{-41} Fm^2 and that of Cl^- is 3.41×10^{-40} Fm^2. Calculate the dielectric constant of NaCl.   [Ans   2.26]

9.7   Calculate the polarization of helium gas from the following data: Applied electric field = 6×10^5 V/m, polarizability = 1.8×10^{-41} C^2m/N, concentration atoms = 2.6×10^{25} atoms/m^3. Also calculate the dipole moments of helium atoms and the separation between positive and negative charges.

[Ans. 2.808×10^{-10} C/m^2, 1.08×10^{-35} Cm, 3.375×10^{-17} m]

9.8   When an electric field of 300 Volt/cm is applied to water H_2O, calculate the dipole moment of one water molecule. Data given are: density = 1.0 gm/cc, static dielectric constant = 80.   [Ans 6034×10^{-34} Cm]

9.9   When an electric field 300 V/cm is applied to water, it is polarized to an extent of 2.124×10^{-5} C/m^2. Calculate the local field acting on the water molecule.

9.10   When an electric field 300Volt/cm is applied to water, it is polarized to an extent of 2.124×10^{-5} C/m^2. Calculate the polarizability of water if there are 3.35×10^{28} molecules/m^3.   [Ans 4.04×10^{-40} Cm^2/V]

Multiple Choice Questions

1.   The function of a dielectric is to obstruct the flow of electric current while the function of an insulator is to store electrical energy.
      (i)   True   (ii) False

2.   In polar and non-polar molecules, total net charge is
      (i)   Not zero   (ii) Zero
      (iii) Equal to charge of the nucleus   (iv) Equal to charge of electrons.

3.   In polarized dielectrics, the total net charge is
      (i)   Zero   (ii) Not zero
      (iii) Negative   (iv) Positive

4.   Which of the following is not a synonym of dielectric constant?
      (i)   Dielectric strength   (ii) Specific inductive capacity
      (iii) Relative permittivity   (iv) Dielectric coefficient

5.   The units of dielectric polarization is
      (i)   No units   (ii) A/m^2
      (iii) C/m^2 (iv) Am^2

6.   The unit of permanent dipole moment is
      (i)   Newton   (ii) Debye
      (iii) Curie   (iv) None of the above.

7.   The unit of polarizability is
      (i)   Cm^2 (ii) Cm
      (iii) Fm^2 (iv) F/m^2

8. Which of the following is not a unit of polarizability?

    (i) Fm^2 (ii) C^2m/N

    (iii) Cm (iv) Cm^2/V

9. Direction of electric displacement vector and electric field intensity is always same in all materials.

    (i)   True   (ii)   False

10. The relation between microscopic field E_L and macroscopic field E in free space is

    (i) $E_L = E$ (ii) $E_L > E$

    (iii) $E_L < E$ (iv) $E_L >> E$

11. What is the unit of $\dfrac{\vec{P}}{\varepsilon_O}$?

    (i)   N/C   (ii)   C/N

    (iii) $C^2/(Nm^2)$ (iv)   None of the above

12. When a dielectric material is subjected to an electric field E_0, the macroscopic electric field inside the dielectric is

    (i)   Equal to E_0 (ii)   Less than E_0

    (iii) More than E_0 (iv)   Equal to zero

13. Clausius–Mossotti relation holds good for all solids.

    (i)   True   (ii)   False

14. The electronic polarizability of a monoatomic gas is proportional to

    (i) R (ii) R^2

    (iii) R^3 (iv) R^4

    (v) R = atomic radius

15. Which types of polarization exist in all types of dielectrics?

    (i)   Ionic polarization   (ii)   Dipolar polarization

    (iii) Electronic polarization   (iv)   None.

16. In water, what type of polarization can exist predominantly?

    (i)   Ionic polarization   (ii)   Dipolar polarization

    (iii) Electronic polarization   (iv)   All of the above

17. Which types of polarization is a direct function of atomic radius?

    (i)   Ionic polarization   (ii)   Dipolar polarization

    (iii) Electronic polarization   (iv)   None of the above

18. Which types of polarization depends on temperature inversely?

    (i)   Ionic polarization   (ii)   Dipolar polarization

    (iii) Electronic polarization   (iv)   None of the above

19. In which types of dielectrics do all types of polarization exist?

    (i) Ceramics and polymers   (ii) Polar dielectrics

    (iii) Alkali halides   (iv) Organic liquids

20. Dielectrics have completely filled valence band and completely empty conduction band.

    (i) True   (ii) False

21. The variation of ferroelectric polarization with electric field is non-linear.

    (i) True   (ii) False

22. Which of the following materials exhibit spontaneous polarization?

    (i) Ferroelectrics   (ii) Paraelectrics

    (iii) Piezoelectrics   (iv) Pyroelectrics

23. All dielectric crystals which lack a centre of symmetry are

    (i) Ferroelectrics   (ii) Paraelectrics

    (iii) Piezoelectrics   (iv) pyroelectrics

24. Piezoelectrics must be centrosymmetric crystals.

    (i) True   (ii) False

25. Piezoelectric effect can be used to measure

    (i) Force   (ii) Strain

    (iii) Acceleration   (iv) All of the above

Answers

1 (ii)	2 (ii)	3 (i)	4 (i)	5 (iii)	6 (ii)	7 (iii)	8 (i & iv)
9 (ii)	10 (i)	11 (i)	12 (ii)	13 (ii)	14 (iii)	15 (iii)	16 (ii)
17 (iii)	18 (ii)	19 (ii)	20 (i)	21 (i)	22 (iii)	23 (i)	24 (ii)
25 (ii)							

10 Electronic Theory of Solids

10.1 Introduction

Many physical properties of solids can be understood with help of the free electron model. According to this model, the valance electrons of the constituent atoms become conduction electrons and move freely throughout the solid. The aggregates of free electrons constitute an electron gas or electron cloud. The interpretation of properties of solids by the free electron model was developed long before the advent of quantum physics, a versatile tool of physical science. The classical theory has several conspicuous successes. However, as it is natural, the simplicity of the theory puts limitation on its success. Many properties of solids like specific heats, magnetic susceptibility of solids, superconductivity and the like cannot be explained by the classical theory of free electron model. With the advent of quantum physics, almost all the properties of solids could be explained in minute detail. Here the electron gas is subjected to Pauli's exclusion principle and Fermi–Dirac statistics instead of Maxwell–Boltzmann statistics as in classical theory. This new formulation constitutes the quantum theory of free electrons.

The free electron model of metals gives us a good insight into the heat capacity, thermal and electrical conductivity, susceptibility and electrodynamics of metals. However, the model fails to explain the distinction between metals, semi-metals (materials with a very small overlap between the bottom of the conduction band and the top of the valence band), semiconductors and insulators; the occurrence of positive values of Hall coefficient; the relation between conduction electrons in the metal and the valence electrons of free atoms and many transport properties of solids. The band theory of solids, a simple theory, then was developed to account for all these facts. We shall apply these theories to explain a few properties of solids in detail as far as the scope of this book permits.

10.2 Free Electron Theory of Metals

Metals are generally characterized by high electrical conductivity and thermal conductivity. Many of the special properties of metals are attributed to valence electrons. Valence electrons are free to move throughout the solid and constitute the electron gas or electron cloud or simply Fermi gas. In the free electron theory, the fundamental postulate is that potential field due to ion cores is uniform throughout the solid. The potential energy value is negative since there is a force of attraction between the positive ion cores and electrons. For mathematical simplicity, the reference potential energy level may be taken zero.

10.2.1 Classical free electron theory of metals

The following are the important postulates of the classical free electron theory of metals or the Drude–Lorentz theory of metals.

i.   Electrons are free to move within the solids.

ii.   Free electrons are responsible for electrical conduction.

iii. The electrostatic interaction between electrons and lattice ions is completely neglected.

iv.   The interactions among electrons are neglected.

v.   All the free electrons inside the metal have the same potential energy at all points.

Classically, in the absence of an external electric field, electrons behave just like ideal gas molecules. They have random motions and the average velocity $<v>$ of an electron, called drift velocity, comes out to be zero and the net electric current is zero. When an electric field is applied to free electrons, they drift colliding with each other, in the direction opposite to the direction of the applied electric field. The average time lapse between two collisions is called relaxation time τ and the average displacement between two collisions is called mean free path λ. Therefore, we can define average velocity or drift velocity $<v>$ as

$$<v> = \frac{\lambda}{\tau} \qquad (10.1)$$

Let E be the electric field intensity applied to a conductor containing N number of free electrons per unit volume. Let $-e$ and m be the charge and mass of an electron respectively. During the motion of the electrons, they suffer collisions with each other and these collisions introduce a frictional force $\frac{m<v>}{\tau}$ to the electrons during their motion under the influence of the applied electric field. Besides the frictional force, a natural $m\frac{d<v>}{dt}$ force acts on the moving electron. Therefore, the total force acting on a moving electron in a metal is $m\frac{d<v>}{dt} + \frac{m<v>}{\tau}$. The electric force exerted on an electron is $-eE$. Now applying Newton's law of motion to the moving electron, the equation of motion of the moving electron in the metal is obtained as

$$m\frac{d<v>}{dt}+\frac{m<v>}{\tau}=-eE \qquad (10.2)$$

or $\quad \tau\frac{d<v>}{dt}=-\left(<v>+\frac{eE}{m}\tau\right)$

or $\quad \dfrac{d<v>}{\dfrac{eE}{m}\tau+<v>}=-\dfrac{dt}{\tau}$

Integrating both sides of this equation, we get

$$\ell n\left(\frac{eE}{m}\tau+<v>\right)=-\frac{t}{\tau}+\ell n\,C$$

or $\quad \dfrac{eE\tau}{m}+<v>=Ce^{-\frac{t}{\tau}} \qquad (10.3)$

At the time of applying the electric field ($t = 0$), the average velocity of an electron is zero; i.e.,

At $t = 0$, $<v> = 0$

Applying these boundary condition to Eq. (10.3), we get

$$C=\frac{eE\tau}{m}$$

Putting the value of C into Eq. (10.3), we get

$$<v>=\frac{eE\tau}{m}\left(e^{-\frac{t}{\tau}}-1\right) \qquad (10.4)$$

Let the number of electrons per unit volume be N. In dt time, electrons travel a distance of $<v>\,dt$. This volume contains $A<v>\,dt \times N$ number of free electrons and all the $A<v>\,dt \times N$ number of electrons cross the cross-sectional area A in time dt. In other words, $A<v>\,dt \times N \times (-e)$ amount of charges cross the cross-sectional area A in time dt.

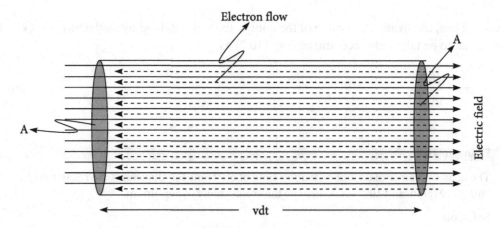

Figure 10.1 Calculation of electric current *I* in terms of drift velocity <*v*>. The electric field is applied from left to right and electrons move from right to left. Electrons travel a distance of *vdt* in *dt* time, i.e., the electrons which are at the left end will reach the right end in time *dt* if its length is *vdt*

The amount of charges crossing the cross-sectional area *A* in unit time, i.e., electric current is $A < v > N(-e)$. Thus, we have

$$I = -ANe < v >$$

or　Current density $J = -Ne < v >$　(10.5)

Putting the value of <*v*> from Eq. (10.4) into Eq. (10.5), we have

$$J = \frac{Ne^2 E\tau}{m}\left(1 - e^{-\frac{t}{\tau}}\right)$$　(10.6)

Again according to Maxwell's electromagnetic theory, $J = \sigma E$. Hence, this equation becomes

$$\sigma E = \frac{Ne^2 E\tau}{m}\left(1 - e^{-\frac{t}{\tau}}\right)$$

or　$$\sigma = \frac{Ne^2 \tau}{m}\left(1 - e^{-\frac{t}{\tau}}\right)$$　(10.7)

The relaxation time τ is of the order of 10^{-14} s. When an electric field is applied to the conductor, conductivity σ of the material increases and attains steady state value within

10^{-14} s. Thus, the steady state value of the conductivity is obtained by neglecting $e^{-\frac{t}{\tau}}$ ($e^{-\frac{t}{\tau}}$ is too small to be taken into account) in Eq. (10.7) as

$$\sigma = \frac{Ne^2\tau}{m} \qquad\qquad (10.8)$$

Example 10.1

The density, atomic mass and relaxation time of electrons for aluminum are respectively 2.7 gm/cm^3, 27amu, and 10^{-14} sec. Calculate the conductivity of aluminum.

Solution

Data given are atomic mass = 27 amu, relaxation time = $\tau = 10^{-14}$ s and density = 2.7 gm/cm^3 = 2.7×10^3 kg/m^3.

The density of aluminum is 2.7×10^3 kg/m^3, i.e., 2.7×10^3 kg mass is contained in unit volume.

or $\dfrac{2.7\times10^3}{1.66054\times10^{-27}}$ amu $= 1.62\times10^{30}$ amu mass is contained in unit volume.

The atomic mass of aluminum atom is 27 amu, i.e., mass of one aluminum atom is 27 amu. Hence, 27 amu corresponds to 1 aluminum atom. 1.62×10^{30} amu will correspond to

$\dfrac{1.62\times10^{30}}{27} = 6\times10^{28}$ aluminum atoms.

Thus, the number of aluminum atoms per unit volume $N = 6 \times 10^{28}$ atoms/m^3.
Since each atoms contributes three electrons to the electron gas, the number of electrons per unit volume of the electron gas will be

$$N = 18\times10^{28}\ \frac{\text{electrons}}{\text{m}^3}$$

[Alternatively, $N = \dfrac{\rho N_A}{M_{At.}} \times$ number of free electrons per atom

$$= 18\times10^{22}\ \frac{\text{electrons}}{\text{cm}^3} = 18\times10^{28}\ \frac{\text{electrons}}{\text{m}^3}\ ]$$

The conductivity of a material is given by

$$\sigma = \frac{Ne^2\tau}{m} = 18\times10^{28} \times \frac{(1.6\times10^{-19})^2}{9.11\times10^{-31}} \times 10^{-14}\ \Omega\text{m} = 5.06\times10^7\ \Omega\text{m}$$

Resistivity

The resistivity of a material ρ is defined as the resistance offered by a body made of the material having unit length and unit cross-sectional area. The reciprocal of resistivity is defined as the conductivity σ, i.e., $\sigma = 1/\rho$. Hence, So the resistivity of a material ρ is obtained as

$$\rho = \frac{1}{\sigma} \tag{10.9}$$

Putting the value of σ from Eq. (10.7) into this equation, we get

$$\rho = \frac{m}{Ne^2\tau} \tag{10.10}$$

The variation of resistivity with temperature is explained as follows.

Replacing τ in this equation by $\tau = \dfrac{\lambda}{<v>}$, we get

$$\rho = \frac{m<v>^2}{Ne^2\lambda<v>}$$

We know that $\dfrac{1}{2}m<v>^2 = \dfrac{3}{2}kT$ or $m<v>^2 = 3kT$

Therefore, $$\rho = \frac{3kT}{Ne^2\lambda<v>} \tag{10.11}$$

Thus, for pure metals

$$\rho \propto T$$

Electron mobility μ_e

The average steady state value of the electron velocity as obtained from Eq. (10.4) is

$$<v> = \frac{eE\tau}{m} \tag{10.12}$$

The electron mobility μ_e is defined as the average velocity of the electron per unit electric field. Therefore, from Eq. (10.12), we have

$$\frac{<v>}{E} = \frac{e\tau}{m}$$

or $\mu_e = \dfrac{e\tau}{m}$ (10.13)

The expression for relaxation time as obtained from Eq. (10.8) is given by

$$\tau = \dfrac{m\sigma}{Ne^2}$$ (10.14)

Putting Eq. (10.14), into (10.13) we get

$$\mu_e = \dfrac{e}{m}\dfrac{m\sigma}{Ne^2}$$ (10.15)

or $\sigma = Ne\mu_e$ (10.16)

This equation gives the electrical conductivity in terms of electron mobility.

Thermal conductivity K

The thermal conductivity K determines the rate of heat flow through a material and is defined as the negative ratio of the amount of heat flow per unit time to temperature gradient. Mathematically, it is defined as

$$K = -\dfrac{Q}{\dfrac{dT}{dx}}$$ (10.17)

where

$Q =$ amount of heat flow per unit time.

$\dfrac{dT}{dx}$ = temperature gradient.

The negative sign indicates that heat flows from a higher temperature to a lower temperature.

Let there be N number of electrons per unit volume with average velocity <v>. Every electron in the electron gas has random motion in three-dimensional space defined by the X, Y, Z rectangular coordinate system. Therefore, out of N number of electrons, on the average, $N/3$ number of electrons move along the X-axis; out of $N/3$ number of electrons moving along the X-axis, $N/6$ number of electrons move along the positive X-axis. The number of electrons entering into the cross-section at A per second towards C will be $N/6$ <v>. We know that the average kinetic energy of an electron at a region having temperature T is $3/2\ kT$.

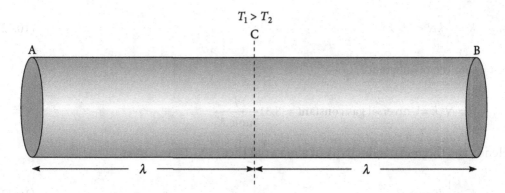

Figure 10.2 | Derivation of expression for thermal conductivity

Consider a cross-sectional area C at a distance equal to the mean free path λ of the electrons between the two ends A and B. The distance between the two ends A and B will be 2λ. The temperature of the end A is T_1 and the temperature of the end B is T_2. The amount of energy entering the cross-section of the rod at A towards C per second Q_1 will be

$$Q_1 = \frac{3}{2}kT_1\frac{N}{6}<v>=\frac{1}{4}kT_1N<v> \tag{10.18}$$

Similarly, the amount of energy entering the cross-section of the rod at B towards C per second Q_2 will be

$$Q_2 = \frac{1}{4}kT_2N<v> \tag{10.19}$$

The net amount of energy passing the cross-sectional area per unit time will be obtained by combining Eqs (10.18) and (10.19) as

$$Q = Q_1 - Q_2 = \frac{1}{4}kT_1N<v>-\frac{1}{4}kT_2N<v>=\frac{1}{4}kN<v>(T_1-T_2) \tag{10.20}$$

The distance between two ends A and B is 2λ and they are at the temperatures T_1 and T_2 respectively. The average temperature gradient along the conductor will be

$$\frac{T_1-T_2}{2\lambda} \tag{10.21}$$

The thermal conductivity of a material K as defined earlier is

$$K = \frac{Q}{\dfrac{T_1-T_2}{2\lambda}} = \frac{kN<v>(T_1-T_2)}{4\dfrac{T_1-T_2}{2\lambda}}$$

or $K = \frac{1}{2}kN <v> \lambda$ (10.22)

For one mole of substance,

$$R = N_A k = \text{Universal gas constant} = 8.31 \frac{J}{\text{mole } K}$$

Hence, the thermal conductivity of a material [Eq. (10.22)] becomes

$$K = \frac{1}{2}R <v> \lambda$$ (10.23)

According to the classical free electron theory, the electronic specific heat of a metal is given by

$$C_{Ve} = \frac{3}{2}R$$ (10.24)

Hence, the thermal conductivity of a material becomes

$$K = \frac{1}{3}C_{Ve} <v> \lambda$$ (10.25)

Wiedemann–Franz law

The Wiedemann–Franz law states that at low temperatures, the ratio of thermal conductivity K to electrical conductivity σ of a metal is directly proportional to the absolute temperature. Mathematically,

$$\frac{K}{\sigma} \propto T$$

or $\frac{K}{\sigma T} = L$ (10.26)

The proportionality constant L is known as the Lorentz number.

From the expressions for thermal conductivity K (Eq. 10.23) and electrical conductivity σ (Eq. 10.11), we have

$$\frac{K}{\sigma} = \frac{(1/2)kN <v> \lambda}{\dfrac{Ne^2 <v> \lambda}{3kT}} = \frac{3}{2}\frac{k^2 T}{e^2}$$

or $\quad \dfrac{K}{\sigma T} = \dfrac{3}{2}\dfrac{k^2}{e^2} = L$ (10.27)

$L = 3/2\, k^2/e^2$ is known as the Lorentz number. Putting the values of k and e into this equation, the Lorentz number is obtained as

$$L = \frac{3 \times (1.38 \times 10^{-23})^2}{2 \times (1.6 \times 10^{-19})^2}\frac{W\Omega}{K^2} = 1.22 \times 10^{-8}\frac{W\Omega}{K^2}$$

The experimental value of the Lorentz number is $2.44 \times 10^{-8}\ W\Omega/K^2$. Thus, the experimental value differs from the theoretical value. This discrepancy shows the failure of the classical free electron theory of metals.

Example 10.2

A copper wire of length 0.5 m and diameter 0.3 mm has a resistance 0.12 Ω at 300 K. Calculate the Lorentz number according to the classical free electron theory of metals if the thermal conductivity of copper at 300 K is 390 W/mK.

Solution

The data given are $\ell = 0.5\,\mathrm{m}$, $d = 0.3$ mm $= 3 \times 10^{-4}$ m, $R = 0.12$ Ω, $K = 390$ W/mK, and $T = 300$ K.

The electrical conductivity of a conductor is given by

$$\sigma = \frac{\ell}{AR} = \frac{\ell}{\dfrac{\pi d^2}{4}R} = \frac{4 \times 0.5}{\pi\left(0.3 \times 10^{-3}\right)^2 \times 0.12}\Omega^{-1}\ \mathrm{m}^{-1} = 589 \times 10^{7}\,\Omega^{-1}\ \mathrm{m}^{-1}$$

The Lorentz number L is given by

$$L = \frac{K}{\sigma T} = \frac{390\ \mathrm{Wm^{-1}\,K^{-1}}}{5.98 \times 10^{7}\,\Omega^{-1}\mathrm{m}^{-1} \times 300\,\mathrm{K}} = 2.17 \times 10^{-8}\ W\Omega K^{-2}$$

Ohm's law

Ohm's law states that at constant temperature, the potential difference between the two ends of a conductor is directly proportional to the amount of current flowing through the conductor. Mathematically,

$$V \propto I$$

or $\quad V = RI$ (10.28)

The proportionality constant R is called the resistance of the conductor. As we know from Eq. (10.8),

$$\sigma = \frac{Ne^2}{m}\tau$$

or $$\sigma E = \frac{Ne^2\tau}{m}E$$

or $$J = \frac{Ne^2\tau}{m}E$$

or $$\frac{I}{A} = \frac{Ne^2\tau}{m}\frac{V}{\ell}, \quad \ell \text{ being the length of the conductor}$$

or $$V = \frac{m}{Ne^2\tau}\frac{I\ell}{A} = \frac{I\ell}{\sigma A} = \frac{\rho\ell}{A}I = RI \quad \left(\because \rho = \frac{1}{\sigma}\right)$$

$V = RI$ Ohm's law.

This equation shows that Ohm's law can be derived by using the classical free electron theory of metals.

10.2.2 Advantages and disadvantages

Following are the advantages and disadvantages of classical free electron theory.

Advantages

The classical free electron theory was used to derive expressions for
i.   Electrical conductivity.
ii.   Thermal conductivity.
iii. Weidemann–Franz law.
iv.   Ohm's law.
v.   Current density.

Disadvantages

The classical free electron theory was not completely successful in explaining
i.   Specific heat of metals.

ii.   Superconducting properties of materials.

iii. Photoelectric effect, Compton effect, black body radiation and so on.

iv.   Temperature dependence of electrical conductivity of metals.

v.   Thermal conductivity.

10.2.3 Quantum theory of free electrons

Since electrons are quantum particles, it is most appropriate to apply the concepts of quantum theory rather than classical theory to explain the behaviour of electrons in metals. In addition to the postulates of classical free electron theory of metals, in the quantum theory of free electrons or Sommerfeld's theory of free electrons, the following assumptions are assumed to be valid.

i.   Metal is treated as a three-dimensional potential box [$V = 0$ at $0 \leq x \leq a, 0 \leq y \leq b$ and $0 \leq z \leq c$, i.e., inside the box and $V = \infty$ at $x > a, y > b$ and $z > c$, [i.e., outside the box]. Inside the potential box, the conduction electrons can move freely obeying Pauli's exclusion principle.

ii.   The interactions among the electrons are neglected.

The motion of the electrons is described by Schrödinger's time independent equation

$$\nabla^2 \psi + \frac{2m}{\hbar^2}(E - V)\psi = 0 \tag{10.29}$$

Since the electrons are assumed to be free inside the potential box, Schrödinger's equation becomes

$$\nabla^2 \psi + \frac{2mE}{\hbar^2}\psi = 0 \tag{10.30}$$

This equation can be solved by the method of separation of variables as detailed here.
Transforming Eq. (10.30) into Cartesian coordinates, we have

$$\frac{\partial^2 \psi}{\partial x^2} + \frac{\partial^2 \psi}{\partial y^2} + \frac{\partial^2 \psi}{\partial z^2} + \frac{2mE}{\hbar^2}\psi = 0 \tag{10.31}$$

This is a partial differential equation in three independent variables x, y, and z. To solve this equation, it is necessary to obtain three independent differential equations corresponding to three variables x, y, and z. Now let

$$\psi(x, y, z) = X(x)Y(y)Z(z) \tag{10.32}$$

Substituting $\psi(x, y, z) = X(x)Y(y)Z(z)$ into Eq. (10.31), we can have

$$YZ\frac{\partial^2 X}{\partial x^2} + ZX\frac{\partial^2 Y}{\partial y^2} + XY\frac{\partial^2 Z}{\partial z^2} + \frac{2mE}{\hbar^2}XYZ = 0$$

or $$\frac{1}{X}\frac{\partial^2 X}{\partial x^2}+\frac{1}{Y}\frac{\partial^2 Y}{\partial y^2}+\frac{1}{Z}\frac{\partial^2 Z}{\partial z^2}+\frac{2mE}{\hbar^2}=0 \qquad (10.33)$$

In this equation, the first term is a function of x [i.e., independent of y and z], the second term is a function of y [i.e., independent of x and z] and the third term is a function of z [i.e., independent of x and y] only. From Eq. (10.33), we have

$$\frac{1}{X}\frac{\partial^2 X}{\partial x^2}=-\frac{1}{Y}\frac{\partial^2 Y}{\partial y^2}-\frac{1}{Z}\frac{\partial^2 Z}{\partial z^2}-\frac{2mE}{\hbar^2}=-\alpha_1^2 \qquad (10.34)$$

(Since the value of $\dfrac{\partial^2 X}{\partial x^2}$ does not depend on the values of $\dfrac{\partial^2 Y}{\partial y^2}$ and $\dfrac{\partial^2 Z}{\partial z^2}$)

or $$\frac{1}{X}\frac{\partial^2 X}{\partial x^2}+\alpha_1^2=0 \qquad (10.35)$$

Similarly,

$$\frac{1}{Y}\frac{\partial^2 Y}{\partial y^2}=-\frac{1}{X}\frac{\partial^2 X}{\partial x^2}-\frac{1}{Z}\frac{\partial^2 Z}{\partial z^2}-\frac{2mE}{\hbar^2}=-\alpha_2^2 \qquad (10.36)$$

or $$\frac{1}{Y}\frac{\partial^2 Y}{\partial y^2}+\alpha_2^2=0 \qquad (10.37)$$

and

$$\frac{1}{Z}\frac{\partial^2 Z}{\partial z^2}=-\frac{1}{X}\frac{\partial^2 X}{\partial x^2}-\frac{1}{Y}\frac{\partial^2 Y}{\partial y^2}-\frac{2mE}{\hbar^2}=-\alpha_3^2 \qquad (10.38)$$

or $$\frac{1}{Z}\frac{\partial^2 Z}{\partial z^2}+\alpha_3^2=0 \qquad (10.39)$$

Addition of Eqs (10.34), (10.36) and (10.38) gives

$$\alpha_1^2+\alpha_2^2+\alpha_3^2=\frac{2mE}{\hbar^2} \qquad (10.40)$$

Solutions of Eqs (10.35), (10.37), and (10.39) give respectively

$$X=A_1\cos\alpha_1 x+B_1\sin\alpha_1 x$$

$$Y = A_2 \cos\alpha_2 y + B_2 \sin\alpha_2 y$$

$$Z = A_3 \cos\alpha_3 z + B_3 \sin\alpha_3 z$$

Thus, substituting these values of X, Y, and Z into Eq. (10.32), we get

$$\psi = (A_1 \cos\alpha_1 x + B_1 \sin\alpha_1 x) \times (A_2 \cos\alpha_2 y + B_2 \sin\alpha_2 y) \times (A_3 \cos\alpha_3 z + B_3 \sin\alpha_3 z) \quad (10.41)$$

In this equation, A_1, A_2, A_3, B_1, B_2, and B_3 are constants whose values can be obtained by applying the boundary conditions, i.e., properties of the wave function ψ to Eq. (10.41). The boundary conditions appropriate for the wave function ψ in this case are

$$\psi = 0 \text{ for } a \le x \le 0, \; b \le y \le 0 \text{ and } c \le z \le 0 \qquad (10.42)$$

When separated, Eq. (10.42) becomes

i. $\psi = 0$ for $x = 0$, $y = 0$ and $z = 0$ $\qquad\qquad (10.43)$

and

ii. $\psi = 0$ for $x = a$, $y = b$ and $z = c$ $\qquad\qquad (10.44)$

where a, b and c are the length, breadth and height of the three-dimensional potential box. Now putting the boundary condition, Eq. (10.43) into Eq. (10.41), we set

$$A_1 = A_2 = A_3 = 0$$

Thus, Eq. (10.41) becomes

$$\psi = B_1 \sin\alpha_1 x \times B_2 \sin\alpha_2 y \times B_3 \sin\alpha_3 z \qquad (10.45)$$

Out of the three constants, B_1, B_2, and B_3, no one can be zero, i.e., $B_1 \ne 0$, $B_2 \ne 0$, and $B_3 \ne 0$. If any one of B_1, B_2, and B_3 is zero, the wave function ψ becomes zero implying the absence of electrons inside the potential box, contrary to our assumptions. Again putting the boundary condition, Eq. (10.44) into Eq. (10.45), we set

$$B_1 \sin\alpha_1 a = 0$$

or $\sin\alpha_1 a = 0$ since $B_1 \ne 0$

or $\alpha_1 a = n_x \pi$

or $\alpha_1 = \dfrac{n_x \pi}{a}$ with $n_x = 1, 2, 3, 4, \ldots$ (10.46)

Similarly,

$\alpha_2 = \dfrac{n_y \pi}{a}$ with $n_y = 1, 2, 3, 4, \ldots$ (10.47)

and

$\alpha_3 = \dfrac{n_z \pi}{a}$ with $n_z = 1, 2, 3, 4, \ldots$ (10.48)

Here n_x, n_y, and n_z are called quantum numbers. Putting the values of α_1, α_2, and α_3 from Eqs (10.46)–(10.48) into Eq. (10.45), we get

$$\psi_{n_x n_y n_z} = B_1 B_2 B_3 \sin \frac{n_x \pi}{a} x \sin \frac{n_y \pi}{b} y \sin \frac{n_z \pi}{c} z$$

or $$\psi_{n_x n_y n_z} = N \sin \frac{n_x \pi}{a} x \sin \frac{n_y \pi}{b} y \sin \frac{n_z \pi}{c} z$$ (10.49)

where $B_1 B_2 B_3 = N$ is the normalization constant.

The value of N can be determined by applying the normalization condition, i.e.,

$$\int_0^a \int_0^b \int_0^c \psi_{n_x n_y n_z} \psi^*_{n_x n_y n_z} \, dx\,dy\,dz = 1$$

or $$N^2 \int_0^a \int_0^b \int_0^c \sin^2 \frac{n_x \pi x}{a} \sin^2 \frac{n_y \pi y}{b} \sin^2 \frac{n_z \pi z}{c} \, dx\,dy\,dz = 1$$

or $$N^2 \int_0^a \sin^2 \frac{n_x \pi x}{a} dx \int_0^b \sin^2 \frac{n_y \pi y}{b} dy \int_0^c \sin^2 \frac{n_z \pi z}{c} dz = 1$$

or $$N^2 \frac{a}{2} \frac{b}{2} \frac{c}{2} = 1$$

or $$N = \frac{2\sqrt{2}}{\sqrt{abc}}$$ (10.50)

Thus, putting $N = \dfrac{2\sqrt{2}}{\sqrt{abc}}$ into Eq. (10.49), the wave function is normalized and becomes

$$\psi_{n_x n_y n_z} = \frac{2\sqrt{2}}{\sqrt{abc}} \sin\frac{n_x \pi}{a} x \sin\frac{n_y \pi}{b} y \sin\frac{n_z \pi}{c} z \qquad (10.51)$$

Now by putting the values of α_1, α_2 and α_3 from Eqs (10.46)–(10.48) into (10.40), we can have

$$\frac{n_x^2 \pi^2}{a^2} + \frac{n_y^2 \pi^2}{b^2} + \frac{n_z^2 \pi^2}{c^2} = \frac{2mE_{n_x n_y n_z}}{\hbar^2} \qquad (10.52)$$

or $\quad E_{n_x n_y n_z} = \dfrac{h^2}{8m}\left(\dfrac{n_x^2}{a^2} + \dfrac{n_y^2}{b^2} + \dfrac{n_z^2}{c^2}\right) \qquad (10.53)$

Figure 10.3 shows the energy levels, degree of degeneracy and quantum numbers of an electron enclosed in a cubical potential box $[a = b = c = L]$.

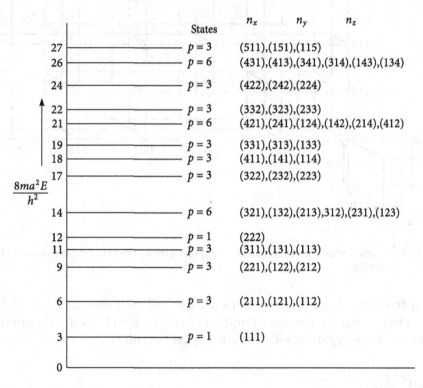

Figure 10.3 | Energy levels, degree of degeneracy and quantum numbers of an electron enclosed in a cubical potential box. $E_0 = (h^2/8mL^2)$ and $E = E_0\left(n_x^2 + n_y^2 + n_z^2\right)$

The number of allowed energy states lying between E and $E + dE$ can be evaluated by using Eqs (10.53) in the following way. Though the precise solution of this problem is rigorous, we find the solution when n_x, n_y and n_z are large. Imagine a three-dimensional lattice as shown in Fig. 10.4 of which unit cells have sides $1/a$, $1/b$ and $1/c$ respectively. The energy is obtained from

$$\frac{8mE}{h^2} = \left(\frac{n_x^2}{a^2} + \frac{n_y^2}{b^2} + \frac{n_z^2}{c^2} \right)$$

(10.54)

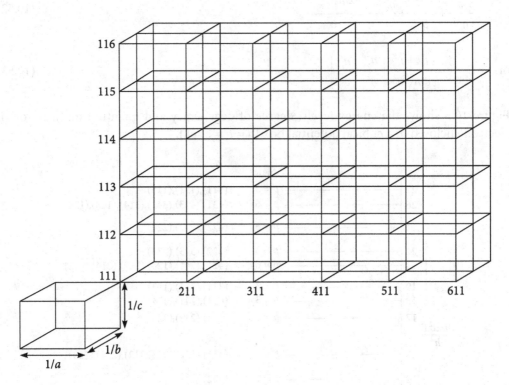

Figure 10.4 | A geometrical representation of energy levels for a particle in a cuboid of sides $1/a$, $1/b$, and $1/c$

It is clear from this above equation that each unit cell gives one energy state. Now the problem is to find out the number of unit cells between E and $E + dE$. Equation (10.54) shows that all the energy states will lie inside a sphere of radius

$$r = \sqrt{\frac{8mE}{h^2}}.$$

(10.55)

Hence, the number of energy eigenvalues lying in the energy interval dE is equal to the number of unit cells in the first quadrants of the spherical shell lying inside a thickness dr.

The elemental volume of this first quadrant is

$$\frac{1}{8}4\pi r^2 dr = \frac{1}{2}\pi\frac{8m}{h^2}E\sqrt{\frac{8m}{h^2}}\frac{1}{2}E^{-\frac{1}{2}}dE = 2\pi\frac{m}{h^3}\sqrt{8m}E^{\frac{1}{2}}dE = 2\pi\left(\frac{2m}{h^2}\right)^{\frac{3}{2}}E^{\frac{1}{2}}dE$$

We have chosen the first quadrant only because n_x, n_y and n_z have been taken to be positive numbers. The volume of the unit cell is

$$\frac{1}{a}\times\frac{1}{b}\times\frac{1}{c} = \frac{1}{abc}$$

Now $1/abc$ volume corresponds to one unit cell, i.e., one energy state and so unit volume corresponds to abc energy states. Therefore,

$$2\pi\left(\frac{2m}{h^2}\right)^{\frac{3}{2}}E^{\frac{1}{2}}dE$$

volume will correspond to

$$2\pi\left(\frac{2m}{h^2}\right)^{\frac{3}{2}}E^{\frac{1}{2}}dE\times abc$$

energy states. Hence, the number of energy states between E and $E+dE$ will be

$$2\pi abc\left(\frac{2m}{h^2}\right)^{\frac{3}{2}}E^{\frac{1}{2}}dE = 2\pi V\left(\frac{2m}{h^2}\right)^{\frac{3}{2}}E^{\frac{1}{2}}dE; \qquad (10.56)$$

(See Eq. (10.68))

Again from Eq. (10.52), we can have

$$E = \frac{\hbar^2}{2m}\left(\frac{\pi^2}{a^2}n_x^2 + \frac{\pi^2}{b^2}n_y^2 + \frac{\pi^2}{c^2}n_z^2\right) \qquad (10.57)$$

or $\quad E = \frac{\hbar^2}{2m}\left(k_x^2 + k_y^2 + k_z^2\right) = \frac{\hbar^2 k^2}{2m} \qquad (10.58)$

where $k^2 = k_x^2 + k_y^2 + k_z^2$ with $k_x = \frac{\pi}{a}n_x$, $k_y = \frac{\pi}{a}n_y$, and $k_z = \frac{\pi}{a}n_z$. Here k_x, k_y, and k_z are the x, y and z components of the propagation vector $\vec{k}$.

From Eq. (10.58), we can also have

$$E = \frac{1}{2m}\left(\hbar^2 k_x^2 + \hbar^2 k_y^2 + \hbar^2 k_z^2\right)$$

or $\quad E = \frac{1}{2m}\left(p_x^2 + p_y^2 + p_z^2\right) = \frac{p^2}{2m}$ (10.59)

where $p^2 = p_x^2 + p_y^2 + p_z^2$ with $p_x = \hbar k_x$, $p_y = \hbar k_y$, and $p_z = \hbar k_z$. Here $p_x, p_y,$ and p_z are the $x, y,$ and z components of the momentum vector $\vec{P}$.

According to classical mechanics, the energy of electrons at 0 K is zero. That means no electrons have any energy at 0 K. The situation is completely different in quantum mechanics. The lowest possible value, the quantum numbers n_x, n_y and n_z can have is 1, i.e., $n_x = 1, n_y = 1$ and $n_z = 1$. Out of the quantum numbers n_x, n_y, and n_z, none can have zero value, i.e., $n_x \neq 0, n_y \neq 0,$ and $n_z \neq 0$. If any one of n_x, n_y and n_z has zero value, the wave function ψ becomes zero implying the absence of electrons inside the potential box, contrary to our assumptions. Now putting the lowest possible values of n_x, n_y and n_z, i.e., $n_x = 1, n_y = 1$ and $n_z = 1$ into Eq. (10.57), the lowest possible energy level E_0 obtained is

$$E_O = \frac{\hbar^2 \pi^2}{2m}\left(\frac{1}{a^2} + \frac{1}{b^2} + \frac{1}{c^2}\right)$$

For a cubical potential box, $a = b = c = L$. Therefore, for a cubical potential box, this equation becomes

$$E_O = \frac{3\pi^2 \hbar^2}{2mL^2}$$ (10.60)

Thus, the electrons occupy energy level from minimum energy level

$$E_O = \frac{3\pi^2 \hbar^2}{2mL^2}$$

to Fermi level at 0 K. According to Pauli's exclusion principle, each and every energy level is occupied by a maximum of two electrons, one with spin 'up' and the other one with spin 'down'.

Degeneracy of energy states

Each combination of different quantum numbers n_x, n_y and n_z give rise to different energy states. Let us represent the energy states defined by the quantum numbers n_x, n_y and n_z by

$(n_x n_y n_z)$. The energy of the energy states defined by the equal values n_x, n_y and n_z is unique; no other energy states can have this energy. For example, the energy of the energy state defined by (1 1 1) is $\dfrac{3\pi^2\hbar^2}{2mL^2}$ and is unique. Similarly, the energy states defined by (2 2 2) and (3 3 3) have energies

$$\frac{12\pi^2\hbar^2}{2mL^2} = \frac{6\pi^2\hbar^2}{mL^2} \text{ and } \frac{27\pi^2\hbar^2}{2mL^2}$$

respectively. The energy states defined by (2 2 2), (3 3 3) and so on, have energies that are unique; no other energy states can have this energy. These energy states are called non-degenerate energy states.

On the other hand, there is a possibility that more than one energy state can have the same energy. The number of quantum states with the same energy is called degree of degeneracy. For example, the energy of the energy states defined by (1 2 3), (1 3 2), (2 3 1), (2 1 3) (3 2 1), and (3 1 2) have the same energy of

$$\frac{7\pi^2\hbar^2}{mL^2}.$$

Since these six energy states have the same energy value, these energy states are said to be degenerate energy states. States having the same energy but different values of quantum numbers are called degenerate states. The energy states (2 2 1), (2 1 2), (1 2 2) and (3 1 1), (1 3 1), (1 1 3) are two more examples of degenerate energy states.

Example 10.3

What should be the energy of the quantum state which has 14 for the sum of the squares of the quantum numbers n_x, n_y and n_z? What is the degeneracy of this state?

Solution

According to the question, $n_x^2 + n_y^2 + n_z^2 = 14$. It implies that $n_x = 1, n_y = 2$ and $n_z = 3$.
   The energy of the quantum state is given by

$$E = \frac{\pi^2\hbar^2}{2mL^2}\left(n_x^2 + n_y^2 + n_z^2\right) = \frac{\pi^2\hbar^2}{2mL^2}\times 14 = \frac{7\pi^2\hbar^2}{mL^2}$$

(1 2 3), (1 3 2), (2 3 1), (2 1 3) (3 2 1), and (3 1 2) quantum states have the same energy of $\dfrac{7\pi^2\hbar^2}{mL^2}$. Therefore, the degeneracy of the quantum state is 6.

10.3   Statistical Distribution Functions

In general, the statistical distribution function is defined by

$$F(\varepsilon) = \frac{1}{e^{\alpha} e^{\frac{\varepsilon}{kT}} + \beta} \tag{10.61}$$

There are three different statistics commonly found in physics; they are

i.   Maxwell–Boltzmann statistics is applicable to identical distinguishable particles such as atoms or molecules of a gaseous system. For Maxwell–Boltzmann statistics, $\beta = 0$ in Eq. (10.61) and the Maxwell–Boltzmann distribution function is given by

$$F_{MB}(\varepsilon) = \frac{1}{e^{\alpha} e^{\frac{\varepsilon}{kT}}} = A e^{-\frac{\varepsilon}{kT}} \tag{10.62}$$

ii.   Bose–Einstein statistics is applicable to identical indistinguishable particles that do not obey Pauli's exclusion principle like photons. These particles are called bosons. For Fermi–Dirac statistics, $\beta = 1$ in Eq. (10.61) and the Bose–Einstein distribution function is given by

$$F_{BE}(\varepsilon) = \frac{1}{e^{\alpha} e^{\frac{\varepsilon}{kT}} - 1} \tag{10.63}$$

iii. Fermi–Dirac statistics is applicable to identical indistinguishable particles that obey Pauli's exclusion principle. These particles are called fermions (electrons). For Fermi–Dirac statistics, $\beta = 1$ in Eq. (10.61) and the Fermi–Dirac distribution function is given by

$$F_{FD}(\varepsilon) = \frac{1}{e^{\alpha} e^{\frac{\varepsilon}{kT}} + 1} \tag{10.64}$$

10.3.1 Fermi–Dirac distribution function $F(\varepsilon)$

The distribution function $F(\varepsilon)$ is defined as the average number of particles in each state of energy ε. It can also be defined as the probability of occupancy of an available energy state or level having energy ε by an electron. The Fermi–Dirac distribution function $F(\varepsilon)$ applicable to electron gas (in general, fermions) obeying quantum restrictions is given by

$$F(\varepsilon) = \frac{1}{e^{\alpha} e^{\frac{\varepsilon}{kT}} + 1} \tag{10.65}$$

The quantity α depends upon the properties of the particular statistical system and may be a function of the temperature. The +1 term in the denominator of Eq. (10.65) is due to the uncertainty principle. Whatever may be the values of α, ε, and T, $F(\varepsilon)$ cannot exceed 1. Fermi energy ε_F is defined as the energy of the level that has 50% probability of occupation by an electron at any temperature. Thus, we have

$$\varepsilon = \varepsilon_F \text{ at } P(E)=50\% = \frac{1}{2}$$

Equation (10.65) becomes

$$F(\varepsilon) = \frac{1}{e^{\alpha} e^{\frac{\varepsilon_F}{kT}} + 1} = \frac{1}{2}$$

Solving this equation for α, we have

$$\alpha = -\frac{\varepsilon_F}{kT}$$

Putting the value of α back into the equation, the Fermi–Dirac distribution function in terms of Fermi energy ε_F becomes

$$F(\varepsilon) = \frac{1}{e^{\frac{\varepsilon_F}{kT}} e^{\frac{\varepsilon}{kT}} + 1} = \frac{1}{e^{\frac{\varepsilon - \varepsilon_F}{kT}} + 1} \tag{10.66}$$

Example 10.4

Evaluate the Fermi–Dirac distribution function for an energy kT above Fermi energy.

Solution

The datum given is $\varepsilon - \varepsilon_F = kT$
Hence, we have

$$F(\varepsilon) = \frac{1}{e^{\frac{\varepsilon - \varepsilon_F}{kT}} + 1} = \frac{1}{e^{\frac{kT}{kT}} + 1} = 0.269$$

Evaluation of the Fermi–Dirac distribution function $F(\varepsilon)$ at 0K

Suppose the electron gas is at 0K, i.e., –273.15°C. For all the energy levels below the Fermi level, $\varepsilon < \varepsilon_F$, i.e., $\varepsilon - \varepsilon_F$ is negative. Hence, the Fermi–Dirac distribution function becomes

$$F(\varepsilon) = \frac{1}{e^{\frac{-(\varepsilon_F - \varepsilon)}{0}} + 1} = 1$$

This calculation shows that all the energy levels below the Fermi levels at absolute zero are completely filled since the probability of occupancy is 1.

Let us calculate the probability of occupancy of any energy level above the Fermi level at absolute zero. In this case, for all the energy levels above the Fermi level, $\varepsilon < \varepsilon_F$, i.e., $\varepsilon - \varepsilon_F$ is positive. Hence, the Fermi–Dirac distribution function becomes

$$F(\varepsilon) = \frac{1}{e^{\frac{\varepsilon - \varepsilon_F}{kT}} + 1} = 0$$

This calculation shows that all the energy levels above the Fermi levels at absolute zero is completely empty since the probability of occupancy is zero. Thus, at absolute zero, all the energy states up to ε_F are completely filled and all the energy states above ε_F are completely empty.

Characteristics of fermi level

i.   The Fermi level is any energy level having the probability that it is exactly half filled with electrons.

ii.   Levels of lower energy than the Fermi level are completely filled with electrons at 0 K.

iii.   Levels of higher energy than the Fermi level are completely empty at 0 K.

iv.   When materials with different individual Fermi levels are placed in contact, some electrons flow from the material with the higher Fermi level into the other material. This transfer of electrons raises the lower Fermi level and lowers the higher Fermi level. When the transfer is complete, the Fermi levels of the two materials are equal.

Explanation

i.   The function $F(\varepsilon)$ represents the probability of occupation of an available energy level by an electron with energy ε. ε_F is the Fermi energy or Fermi level. Fermi energy is defined as the energy of the highest filled level at 0K. For an ordinary range of temperatures (0 K to 10000 K), the Fermi level is nearly constant for metals. At 0K, the system is in the ground state and all the levels below the Fermi level are filled or occupied by the electrons; all the levels above the Fermi level are not occupied by the electrons. Therefore, at 0K, the probability of occupation of an energy level below the Fermi level by an electron is 1 and the probability of occupation of an energy level above the Fermi level by an electron is 0.

ii.   At temperatures above 0 K, some electrons just below the Fermi level, getting thermal energy, move to the higher energy level above the Fermi level. Therefore, at temperatures above 0K, the probability of occupation of an energy level just below the Fermi level by an electron is less than 1 and the probability of occupation of an energy level just above the Fermi level by an electron is more than 0. At temperatures much above 0K, more electrons below the Fermi level, getting thermal energy, move to the higher energy level above the Fermi level.

iii.   At $\varepsilon = \varepsilon_F$, $F(\varepsilon) = 0.5$ since ε_F is independent of the ordinary range of temperatures for metals. Therefore, we can define the Fermi level in metals ε_F as the energy level for which the probability of occupancy of the said energy level by an electron is 0.5 at any temperatures.

iv.   Graphically, the Fermi–Dirac distribution function $F(\varepsilon)$ is represented in Fig. 10.5 incorporating all the discussions made here.

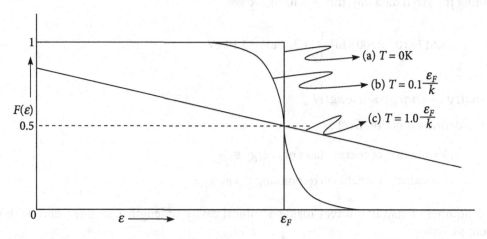

Figure 10.5 | The Fermi–Dirac distribution function $F(\varepsilon)$ of electron gas at three temperatures. Plot (a) shows that at 0 K, all the energy states up to $\varepsilon = \varepsilon_F$ are completely filled and above ε_F, they are completely empty. Plot (b) shows that at low temperatures of the order of $T = 0.1(\varepsilon_F/k)$, some electrons just below the Fermi energy level ε_F acquire energy and jump to energy levels just above ε_F. Plot (c) shows that at very high temperatures of the order of $T = 1.0(\varepsilon_F/k)$, some electrons from any state below the Fermi energy level ε_F may move into higher states above ε_F.

Example 10.5

Find the energy level in sodium for which probability of occupation is 0.75. Fermi energy of sodium is 3.13 eV at 300 K.

Solution

The data given are $F(\varepsilon) = 0.75$, $\varepsilon_F = 3.13$ and $T = 300$ K

The probability of occupation of the energy level $\varepsilon \, \varepsilon_F$ at any temperature is given by

$$F(\varepsilon) = \frac{1}{e^{\frac{\varepsilon - \varepsilon_F}{kT}} + 1}$$

Putting the given data into this equation, we have

$$0.75 = \frac{3}{4} = \frac{1}{e^{\frac{\varepsilon - 3.13}{kT}} + 1}$$

or $e^{\frac{\varepsilon - 3.13}{kT}} = \frac{4}{3} - 1 = \frac{1}{3}$

or $\varepsilon = kT\ell \, \text{n}\left(\frac{1}{3}\right) + 3.13 \text{ eV}$

Putting the given data into this equation, we get

$$\varepsilon = 8.617 \times 10^{-5} \times 300 \times \ln\left(\frac{1}{3}\right) + 3.13 eV = 3.10 \text{ eV}$$

Density of energy statesg(ε)

Let us define a function $G(\varepsilon)$ as

$G(\varepsilon)$ = number of energy states of energy ε

= statistical weight corresponding to energy ε

The number of standing waves inside a cubical cavity of length L is given here without proof as

$$G(j)dj = \pi j^2 dj \text{ , where } j = \frac{2L}{\lambda} \tag{10.67}$$

We know from quantum mechanics [problems of three-dimensional potential well] that each standing wave corresponds to an energy state. Thus, the number of energy states between ε and $\varepsilon + d\varepsilon$, $G(\varepsilon)d\varepsilon$ can be evaluated in the following way.

$$j = \frac{2L}{\lambda} = \frac{2Lp}{h} \quad \because p = \frac{h}{\lambda} \text{ de Broglie relation}$$

However, for non-relativistic velocity [The drift velocity of electron in metal is purely non-relativistic!]

$$j = \frac{2L\sqrt{2m\varepsilon}}{h} \quad (\because \; p = \sqrt{2m\varepsilon}\,)$$

or $j^2 = \frac{8L^2 m\varepsilon}{h^2}$ and $dj = \frac{L\sqrt{2}m^{\frac{1}{2}}\varepsilon^{-\frac{1}{2}}}{h}d\varepsilon$

Thus, putting these two equations into the RHS of Eq. (10.67), we have

$$G(\varepsilon)d\varepsilon = \pi \frac{8L^2 m\varepsilon}{h^2} \frac{L\sqrt{2}m^{\frac{1}{2}}\varepsilon^{-\frac{1}{2}}}{h}d\varepsilon$$

$$= \frac{8\sqrt{2}V\pi m^{\frac{3}{2}}\varepsilon^{\frac{1}{2}}}{h^3}d\varepsilon \qquad (10.68)$$

or $\quad \dfrac{G(\varepsilon)d\varepsilon}{V} = \dfrac{8\sqrt{2}\pi m^{\frac{3}{2}}\varepsilon^{\frac{1}{2}}}{h^3}d\varepsilon \qquad (10.69)$

$G(\varepsilon)d\varepsilon$ is the number of energy states between ε and $\varepsilon + d\varepsilon$. Therefore, $G(\varepsilon)d\varepsilon/V$ represents the number of energy states between ε and $\varepsilon + d\varepsilon$ per unit volume, i.e., density of energy states $g(\varepsilon)d\varepsilon$ between and $\varepsilon + d\varepsilon$. Hence, we define mathematically, the density of energy states as

$$g(\varepsilon) = \frac{G(\varepsilon)}{V}$$

Therefore, from Eq. (10.69), we have

$$g(\varepsilon)d\varepsilon = \frac{8\sqrt{2}\pi m^{\frac{3}{2}}\varepsilon^{\frac{1}{2}}}{h^3}d\varepsilon \qquad (10.70)$$

or $\quad g(\varepsilon)d\varepsilon = C\varepsilon^{\frac{1}{2}}d\varepsilon \qquad (10.71)$

where

$$C = \frac{8\sqrt{2}\pi m^{\frac{3}{2}}}{h^3} = 6.8 \times 10^{27}\frac{1}{m^3 eV^{-3/2}} \qquad (10.72)$$

This expression is the number of energy states per unit volume between energy levels ε and $\varepsilon + d\varepsilon$.

Since $F(\varepsilon)$ is the average number of particles per energy state of energy ε, $F(\varepsilon)g(\varepsilon)$ will be the number of particles per unit volume $n(\varepsilon)$ having energy ε, i.e.,

$$n(\varepsilon) = F(\varepsilon)g(\varepsilon)$$

= number of particles per unit volume having energy ε.

The number of particles per unit volume between the energy levels ε and $\varepsilon + d\varepsilon$ will be given by

$$n(\varepsilon)d\varepsilon = F(\varepsilon)g(\varepsilon)d\varepsilon$$

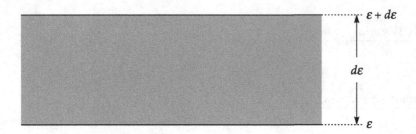

Figure 10.6 | Illustration of energy levels ε_F and $\varepsilon + d\varepsilon$

The number of particles per unit volume N having energy from 0 to ∞ will be given by

$$N = \int_0^\infty n(\varepsilon)d\varepsilon = \int_0^\infty F(\varepsilon)g(\varepsilon)d\varepsilon \qquad (10.73)$$

Here N is the number of particles per unit volume and is called the density of free electrons.

The number of particles per unit volume between the energy levels ε and $\varepsilon + d\varepsilon$ is $n(\varepsilon)d\varepsilon$ $= F(\varepsilon)g(\varepsilon)d\varepsilon$, where the energy of each particle is ε. Therefore, the energy of $n(\varepsilon)d\varepsilon = F(\varepsilon)$ $g(\varepsilon)d\varepsilon$ number of particles will be given by

$$\varepsilon n(\varepsilon)d\varepsilon = \varepsilon F(\varepsilon)g(\varepsilon)d\varepsilon \quad \because \quad \varepsilon = \text{energy of each particle.}$$

Hence, the total energy per unit volume E (energy density) of all the particles will be given by

$$E = \int_0^\infty \varepsilon n(\varepsilon)d\varepsilon = \int_0^\infty \varepsilon F(\varepsilon)g(\varepsilon)d\varepsilon \qquad (10.74)$$

Evaluation of N, ε_F and E at 0K

At 0K, the highest occupied level is the Fermi level ε_F. Within ε_F, $F(\varepsilon) = 1$ and above ε_F, $F(\varepsilon) = 0$. Therefore, at 0K, the number of particles per unit volume N from this equation is given by

$$N = \int_0^\infty F(\varepsilon)g(\varepsilon)d\varepsilon = \int_0^{\varepsilon_F} F(\varepsilon)g(\varepsilon)d\varepsilon + \int_{\varepsilon_F}^\infty F(\varepsilon)g(\varepsilon)d\varepsilon$$

At 0 K, $F(\varepsilon) = 0$ for $\varepsilon_F < \varepsilon < \infty$. Therefor, the number of particles per unit volume at 0 K N_0 will be given by

$$N_0 = \int_0^{\varepsilon_F} F(\varepsilon)g(\varepsilon)d\varepsilon + \int_{\varepsilon_F}^\infty 0 g(\varepsilon)d\varepsilon = \int_0^{\varepsilon_F} 1 \times g(\varepsilon)d\varepsilon$$

or $\quad N_0 = \int\limits_0^{\varepsilon_F} g(\varepsilon)d\varepsilon$ $\qquad$ (10.75)

Putting the values of $g(\varepsilon)$ from Eq. (10.71) into Eq. (10.75), we get

$$N_0 = C \int\limits_0^{\varepsilon_F} \varepsilon^{\frac{1}{2}} d\varepsilon$$

or $\quad N_0 = \dfrac{2C}{3} \varepsilon_{F0}^{\frac{3}{2}}$ $\qquad$ (10.76)

or $\quad \varepsilon_{F0} = \left(\dfrac{3N_0}{2C}\right)^{\frac{2}{3}}$ at 0 K $\qquad$ (10.77)

Putting the value of C from Eq. (10.72) into Eq. (10.77), we get

$$\varepsilon_{F0} = \dfrac{h^2}{8m}\left(\dfrac{3N_0}{\pi}\right)^{\frac{2}{3}} \qquad (10.78)$$

The energy density E_0 in between energy levels 0 and ε_F at 0 K is obtained from Eq. (10.74) as

$$E_0 = \int\limits_0^{\varepsilon_F} \varepsilon \times 1 \times g(\varepsilon)d\varepsilon + \int\limits_{\varepsilon_F}^{\infty} 0 g(\varepsilon)\varepsilon d\varepsilon = \int\limits_0^{\varepsilon_F} \varepsilon \times g(\varepsilon)d\varepsilon \text{ at 0 K}$$

or $\quad E_0 = \int\limits_0^{\varepsilon_F} \varepsilon g(\varepsilon)d\varepsilon$ at 0 K

Putting the values of $g(\varepsilon)$ from Eq. (10.71) into this equation, we get

$$E_0 = C \int\limits_0^{\varepsilon_F} \varepsilon \varepsilon^{\frac{1}{2}} d\varepsilon = C \int\limits_0^{\varepsilon_F} \varepsilon^{\frac{3}{2}} d\varepsilon$$

or $\quad E_0 = \dfrac{2C}{5} \varepsilon_{F0}^{\frac{5}{2}}$ is the energy per unit volume at 0 K.

The number of particles per unit volume at 0 K as obtained from Eq. (10.77) is

$$N_0 = \frac{2C}{3}\varepsilon_{F0}^{\frac{3}{2}}$$

(10.79)

Therefore, the average energy $<E>$ of an electron at 0 K will be

$$<E> = \frac{E_0}{N_0} = \frac{\dfrac{2C}{5}\varepsilon_{F0}^{\frac{5}{2}}}{\dfrac{2C}{3}\varepsilon_{F0}^{\frac{3}{2}}}$$

or $<E> = \dfrac{3}{5}\varepsilon_{F0}$

(10.80)

Example 10.6

Find the Fermi energy in copper at 0K on the assumption that each copper atom contributes one electron to the electron gas. The density and atomic mass of copper are 8.94×10^3 kg/m^3 and 63.5 amu respectively.

Solution

The density of copper is 8.94×10^3 kg/m^3, i.e.,

8.94×10^3 kg is contained in unit volume.

or $\dfrac{8.94 \times 10^3}{1.66054 \times 10^{-27}}$ amu $= 5.38 \times 10^{30}$ amu is contained in unit volume.

The atomic mass of copper atom is 63.5 amu, i.e., mass of one copper atom is 63.5 amu. Hence, 63.5 amu corresponds to 1 copper atom

5.38×10^{30} amu will correspond to $\dfrac{8.38 \times 10^{00}}{63.5} = 8.47 \times 10^{28}$ copper atoms.

Thus, the number of copper atoms per unit volume

$$N_0 = 8.47 \times 10^{28} \frac{\text{atoms}}{\text{m}^3}.$$

Since each atoms contributes one electron to the electron gas, the number of electrons per unit volume of the electron gas will be

$$N_0 = 8.47 \times 10^{28} \frac{\text{electrons}}{\text{m}^3}$$

[Alternatively $N_0 = \dfrac{\rho N_A}{M_{At.}} \times$ number of free electrons per atom $= 8.47 \times 10^{28} \dfrac{\text{electrons}}{\text{m}^3}$]

Putting this value of N_0 into the equation $\varepsilon_{F0} = \dfrac{h^2}{8m} \left(\dfrac{3N_0}{\pi} \right)^{\frac{2}{3}}$, the Fermi energy at 0 K is obtained as

$$\varepsilon_{F0} = \dfrac{h^2}{8m} \left(\dfrac{3N_0}{\pi} \right)^{\frac{2}{3}} = \dfrac{\left(6.626 \times 10^{-34} \right)^2}{8 \times 9.11 \times 10^{-31}} \times \left(\dfrac{3 \times 8.47 \times 10^{28}}{22/7} \right)^{\frac{2}{3}} J = 7.04 \text{ eV}$$

Thus, at 0K, there would be electrons with energies up to 7.04 eV in copper. Hence, the electron gas in copper is degenerated.

Example 10.7

Calculate the Fermi energy in aluminum at 0K on the assumption that each aluminum atom contributes three electrons to the electron gas. The density and atomic mass of aluminum are 2.7 g/cm³ and 27 amu respectively.

Solution

The data given are density of aluminum is $\rho = 2.7 \times 10^3$ kg/m³
Atomic mass of aluminum $M_A = 27$ amu

$$N_0 = \dfrac{\rho N_A}{M_{At.}} \times \text{number of free electrons per atom} = 18 \times 10^{28} \dfrac{\text{electrons}}{\text{m}^3}]$$

Putting this value of N_0 into the equation $\varepsilon_{F0} = \dfrac{h^2}{8m} \left(\dfrac{3N_0}{\pi} \right)^{\frac{2}{3}}$, the Fermi energy at 0 K is obtained as

$$\varepsilon_{F0} = \dfrac{h^2}{8m} \left(\dfrac{3N_0}{\pi} \right)^{\frac{2}{3}} = \dfrac{\left(6.626 \times 10^{-34} \right)^2}{8 \times 9.11 \times 10^{-31}} \times \left(\dfrac{3 \times 18 \times 10^{28}}{22/7} \right)^{\frac{2}{3}} \text{Joule} = 11.64 \text{ eV}$$

Example 10.8

The Fermi energy of silver at 0 K is 5.51 eV. What is the average energy of free electrons in silver at 0 K?

Solution

The datum given is $\varepsilon_{F0} = 5.51$ eV

The average energy of free electrons in electron gas at 0 K is given by $<E> = 3/5\ \varepsilon_{F0}$. Thus the average energy of free electrons in silver at 0 K will be

$$<E> = \frac{3}{5} \times 5.51\ eV = 3.31\ eV$$

Evaluation of N, ε_F and E at any temperature T K

From Eq. (10.73), we have

$$N = \int_0^\infty F(\varepsilon)g(\varepsilon)d\varepsilon$$

Putting the value of $g(\varepsilon)$ from Eq. (10.71) into this equation, we get

$$N = C \int_0^\infty F(\varepsilon)\varepsilon^{\frac{1}{2}}d\varepsilon$$

Integrating this by parts, we get

$$N = CF(\varepsilon) \int_0^\infty \varepsilon^{\frac{1}{2}}d\varepsilon - C \int_0^\infty \left[\frac{dF(\varepsilon)}{d\varepsilon} \int_0^\infty \varepsilon^{\frac{1}{2}}d\varepsilon \right]d\varepsilon$$

$$= C\left(F(\varepsilon)\frac{2}{3}\varepsilon^{\frac{3}{2}} \right)\Big|_0^\infty - C \int_0^\infty \left[\frac{dF(\varepsilon)}{d\varepsilon}\frac{2}{3}\varepsilon^{\frac{3}{2}} \right]d\varepsilon$$

$$= 0 - C\frac{2}{3} \int_0^\infty \left[\frac{dF(\varepsilon)}{d\varepsilon}\varepsilon^{\frac{3}{2}} \right]d\varepsilon$$

or $$N = -C\frac{2}{3} \int_0^\infty \left[\frac{dF(\varepsilon)}{d\varepsilon}\varepsilon^{\frac{3}{2}} \right]d\varepsilon$$

Solving this expression asymptotically, we get

$$N = C\left[\frac{2}{3}\varepsilon_F^{\frac{3}{2}} + \frac{\pi^2}{12}\frac{k^2T^2}{\varepsilon_F^{\frac{1}{2}}} + ... \right] \tag{10.81}$$

Equation (10.81) gives the number of particles per unit volume N having energy from 0 to

∞

From Eq. (10.81), we can also have

$$\frac{3N}{2C} = \frac{3}{2}\left[\frac{2}{3}\varepsilon_F^{\frac{3}{2}} + \frac{\pi^2}{12}\frac{k^2T^2}{\varepsilon_F^{\frac{1}{2}}} + ...\right] \approx \varepsilon_F^{\frac{3}{2}}\left[1 + \frac{3}{2}\frac{\pi^2}{12}\frac{k^2T^2}{\varepsilon_F^2}\right]$$

or

$$\left(\frac{3N}{2C}\right)^{\frac{2}{3}} = \varepsilon_F\left[1 + \frac{\pi^2}{12}\frac{3}{2}\frac{k^2T^2}{\varepsilon_F^2}\right]^{\frac{2}{3}}$$

or

$$\varepsilon_{F0} = \varepsilon_F\left[1 + \frac{\pi^2}{12}\frac{3}{2}\frac{k^2T^2}{\varepsilon_F^2}\right]^{\frac{2}{3}}$$

or

$$\varepsilon_F = \varepsilon_{F0}\left[1 + \frac{\pi^2}{12}\frac{3}{2}\frac{k^2T^2}{\varepsilon_F^2}\right]^{-\frac{2}{3}}$$

Expanding this equation by the binomial theorem up to the second term, we have

$$\varepsilon_F = \varepsilon_{F0}\left[1 - \frac{2}{3}\frac{\pi^2}{12}\frac{3}{2}\frac{k^2T^2}{\varepsilon_F^2}\right] = \varepsilon_{F0}\left[1 - \frac{\pi^2}{12}\frac{k^2T^2}{\varepsilon_F^2}\right]$$

This equation shows that the difference between ε_F and ε_{F0} is very small. Hence, we can replace ε_F by ε_{F0} in the RHS of the equation and obtain

$$\varepsilon_F = \varepsilon_{F0}\left[1 - \frac{\pi^2}{12}\frac{k^2T^2}{\varepsilon_{F0}^2}\right] \tag{10.82}$$

Eq. (10.82) gives the Fermi energy at any temperature T.

Putting the values of $g(\varepsilon)$ from Eq. (10.71) into Eq. (10.74), we get

$$E = C\int_0^\infty \varepsilon F(\varepsilon)\varepsilon^{\frac{1}{2}}d\varepsilon = C\int_0^\infty F(\varepsilon)\varepsilon^{\frac{3}{2}}d\varepsilon$$

$$= CF(\varepsilon)\int_0^\infty \varepsilon^{\frac{3}{2}}d\varepsilon - C\int_0^\infty\left[\frac{dF(\varepsilon)}{d\varepsilon}\int_0^\infty \varepsilon^{\frac{3}{2}}d\varepsilon\right]d\varepsilon$$

$$= C\left(F(\varepsilon)\frac{2}{5}\varepsilon^{\frac{5}{2}}\Big|_0^{\infty} - C\int_0^{\infty}\left[\frac{dF(\varepsilon)}{d\varepsilon}\frac{2}{5}\varepsilon^{\frac{5}{2}}\right]d\varepsilon\right)$$

$$= 0 - C\frac{2}{5}\int_0^{\infty}\left[\frac{dF(\varepsilon)}{d\varepsilon}\varepsilon^{\frac{5}{2}}\right]d\varepsilon$$

or $\quad E = -C\dfrac{2}{5}\displaystyle\int_0^{\infty}\varepsilon^{\frac{5}{2}}\dfrac{dF(\varepsilon)}{d\varepsilon}d\varepsilon$

Solving this expression asymptotically, we get

$$E = C\left[\frac{2}{5}\varepsilon_F^{\frac{5}{2}} + \frac{\pi^2}{4}k^2T^2\varepsilon_F^{\frac{1}{2}}\right] \qquad (10.83)$$

The average energy per electron at any temperature will be given by

$$<E> = \frac{E}{N} = \frac{C}{N}\left[\frac{2}{5}\varepsilon_F^{\frac{5}{2}} + \frac{\pi^2}{4}k^2T^2\varepsilon_F^{\frac{1}{2}}\right]$$

However, according to Eq. (10.82), putting the value of ε_F into the this equation, we get

$$<E> = \frac{C}{N}\left[\frac{2}{5}\varepsilon_{F0}^{\frac{5}{2}}\left(1 - \frac{\pi^2}{12}\frac{k^2T^2}{\varepsilon_{F0}^2}\right)^{\frac{5}{2}} + \frac{\pi^2}{4}k^2T^2\varepsilon_{F0}^{\frac{1}{2}}\left(1 - \frac{\pi^2}{12}\frac{k^2T^2}{\varepsilon_{F0}^2}\right)^{\frac{1}{2}}\right]$$

From Eq. (10.77), we know that $\dfrac{N}{C} = \dfrac{2}{3}\varepsilon_{F0}^{\frac{3}{2}}$. Putting this value of N/C into this above equation, we get

$$<E> = \frac{3\varepsilon_{F0}^{-\frac{3}{2}}}{2}\left[\frac{2}{5}\varepsilon_{F0}^{\frac{5}{2}}\left(1 - \frac{\pi^2}{12}\frac{k^2T^2}{\varepsilon_{F0}^2}\right)^{\frac{5}{2}} + \frac{\pi^2}{4}k^2T^2\varepsilon_{F0}^{\frac{1}{2}}\left(1 - \frac{\pi^2}{12}\frac{k^2T^2}{\varepsilon_{F0}^2}\right)^{\frac{1}{2}}\right]$$

$$= \frac{3\varepsilon_{F0}^{-\frac{3}{2}}}{2}\left[\frac{2}{5}\varepsilon_{F0}^{\frac{5}{2}}\left(1 - \frac{\pi^2}{12}\frac{5}{2}\frac{k^2T^2}{\varepsilon_{F0}^2}\right) + \frac{\pi^2}{4}k^2T^2\varepsilon_{F0}^{\frac{1}{2}}\left(1 - \frac{\pi^2}{12}\frac{1}{2}\frac{k^2T^2}{\varepsilon_{F0}^2}\right)\right]$$

$$= \frac{3\varepsilon_{F0}^{-\frac{3}{2}}}{2}\left[\frac{2}{5}\varepsilon_{F0}^{\frac{5}{2}} - \frac{\pi^2}{12}k^2T^2\varepsilon_{F0}^{\frac{1}{2}} + \frac{\pi^2}{4}k^2T^2\varepsilon_{F0}^{\frac{1}{2}} - \frac{\pi^4}{96}k^4T^4\varepsilon_{F0}^{-\frac{3}{2}}\right]$$

$$\approx \frac{3\varepsilon_{F0}^{-\frac{3}{2}}}{2}\left[\frac{2}{5}\varepsilon_{F0}^{\frac{5}{2}} - \frac{\pi^2}{12}k^2T^2\varepsilon_{F0}^{\frac{1}{2}} + \frac{\pi^2}{4}k^2T^2\varepsilon_{F0}^{\frac{1}{2}}\right]$$

$$= \left[\frac{3}{5}\varepsilon_{F0} + \frac{\pi^2}{4}k^2T^2\varepsilon_{F0}^{-1}\right]$$

or $\quad <E> = \frac{3}{5}\varepsilon_{F0}\left[1 + \frac{5\pi^2}{12}\frac{k^2T^2}{\varepsilon_{F0}^2}\right]$ $\qquad\qquad$ (10.84)

Eq. (10.84) gives the average energy of one electron at any temperature T.

10.3.2 Electronic specific heat

The specific heat at constant volume per unit mass C_V is defined as the derivative of energy with respect to temperature, i.e.,

$$C_V = \frac{dE}{dT} \qquad\qquad (10.85)$$

The average energy of a free electron at temperature T according to Eq. (10.84) is given by the equation

$$<E> = \frac{3}{5}\varepsilon_{F0}\left[1 + \frac{5\pi^2}{12}\frac{k^2T^2}{\varepsilon_{F0}^2}\right]$$

If there is N number of electrons per unit volume, the energy per unit volume will be

$$<E> = \frac{3N}{5}\varepsilon_{F0}\left[1 + \frac{5\pi^2}{12}\frac{k^2T^2}{\varepsilon_{F0}^2}\right]$$

Therefore, the electronic specific heat per unit mass at constant volume C_{VE} is obtained by differentiating this expression with respect to temperature as

$$C_{VE} = \frac{d}{dT}\left[\frac{3N}{5}\varepsilon_{F0} + \frac{\pi^2}{4}\frac{Nk^2T^2}{\varepsilon_{F0}}\right] = \frac{\pi^2Nk^2T}{2\varepsilon_{F0}} \qquad\qquad (10.86)$$

Expression (10.86) shows that the electronic specific heat is proportional to the absolute temperature. It has been found that the theory agrees very well with the experimental conclusions.

If N in this expression becomes the Avogadro's number, the expression gives the electronic molar specific heat. Thus, the electronic molar specific heat at constant volume C_{VEM} is given as

$$C_{VEM} = \frac{\pi^2 N_A k^2 T}{2\varepsilon_{F0}} = \frac{\pi^2 RkT}{2\varepsilon_{F0}} \tag{10.87}$$

According to the Dulong–Petit law, the molar specific heat is given by

$$C_V = 3R \tag{10.88}$$

At room temperature (300 K), $kT = 0.0259$ eV and ε_{F0} is a few electron volts, say 10 eV. Hence, at room temperature

$$\frac{kT}{\varepsilon_{F0}} = \frac{0.0259 \text{ eV}}{\approx 10 \text{ eV}} = 0.00259$$

$$C_{VEM} = \frac{\pi^2 R}{2} \times 0.00259 = 0.0128R \tag{10.89}$$

The electronic contribution to molar heat capacity is of the order of 0.0128 R which is negligible compared with the room temperature lattice contribution 3 R [Dulong–Petit law].

10.3.3 Thermal conductivity

Thermal conductivity according to the classical free electron theory is given by

$$K = \frac{1}{3} C_{VE} <v> \lambda$$

Putting $C_{VE} = \dfrac{\pi^2 N k^2 T}{2\varepsilon_{F0}}$ into this equation, we get

$$K = \frac{1}{3}\frac{\pi^2 N k^2 T}{2\varepsilon_{F0}} <v><v>\tau = \frac{1}{3}\frac{\pi^2 N k^2 T}{2\varepsilon_{F0}} <v>^2 \tau \tag{10.90}$$

Only the electrons near the Fermi level contribute to thermal conductivity like that of electrical conductivity. Hence, the speed of electrons near the Fermi level, called Fermi speed v_F, will be obtained from

$$\frac{1}{2}mv_F^2 = \varepsilon_{F0}$$

or $\quad v_F^2 = \dfrac{2\varepsilon_{F0}}{m}$ $\hfill$ (10.91)

Putting this value of v_F^2 in place of $<v>$ into the Eq. (10.90), we get

$$K = \frac{1}{3}\frac{\pi^2 Nk^2 T}{2\varepsilon_{F0}}\frac{2\varepsilon_F}{m}\tau = \frac{\pi^2 Nk^2 T\tau}{3m}$$ $\hfill$ (10.92)

10.3.4 Wiedemann–Franz law

According to Eq. (10.8), electrical conductivity is given by

$$\sigma = \frac{Ne^2\tau}{m}$$

The appearance of $N\tau/m$ in the expressions for thermal and electrical conductivity is not surprising because both the properties depend on the actions of free electrons. Taking the ratio of thermal conductivity to electrical conductivity, we get

$$\frac{K}{\sigma} = \frac{\pi^2 Nk^2 T\tau}{3m}\frac{m}{Ne^2\tau} = \frac{\pi^2 k^2 T}{3e^2}$$

or $\quad \dfrac{K}{\sigma T} = \dfrac{\pi^2 k^2}{3e^2}$ $\hfill$ (10.93)

In this expression,

$$\frac{\pi^2 k^2}{3e^2} = 2.45\times10^{-8}\,\text{W}\Omega\text{K}^{-2}$$

is known as the Lorentz number. The experimental values of the Lorentz number of different metals at different temperatures are very close to the theoretical value 2.44×10^{-8} $\text{W}\Omega\text{K}^{-2}$.

10.4 Conductivity of Metals

Metals have a high density of conduction electrons. The aluminum atom has three valence electrons in a partially filled outer shell. In metallic aluminum, the three valence electrons per atom become conduction electrons. The number of conduction electrons is constant, depending on neither temperature nor impurities. Metals conduct electricity at all temperatures, but for most metals, the conductivity is best at low temperatures. Divalent atoms, such as magnesium or calcium, donate both valence electrons to become conduction electrons, while monovalent atoms, such as lithium or gold, donate one. As will be recalled, the number of conduction electrons alone does not determine conductivity; it depends on electron mobility as well. Silver, with only one conduction electron per atom, is a better conductor than aluminum with three, for the higher mobility of silver compensates for its fewer electrons. In metals such as sodium and aluminum, the atoms donate all their valence electrons to the conduction band. The resulting ions are small, occupying only 10–15 percent of the volume of the crystal. The conduction electrons are free to roam randomly through the remaining space.

A simple model, which often describes the properties of conduction electrons well, treats them as non-interacting either with the ions or with each other. The electrons are approximated as free particles wandering easily through the crystal. This concept was first proposed by the German scientist Arnold Johannes Wilhelm Sommerfeld. It works quite well for those metals, known as simple metals, whose conduction electrons are donated from sp states, i.e., aluminum, magnesium, calcium, zinc and lead. They are called simple because they are aptly described by the simple theory of Sommerfeld.

Transition metals are found in three rows of the periodic table; (i) the first row consists of scandium through nickel, (ii) the second row is yttrium through palladium, and (iii)the third row is lanthanum and hafnium through platinum. Within these rows, as the atomic number increases, the electrons fill the d-states in the outer shell of the atom. Transition metals in crystal forms have interesting properties. The d electrons are more tightly bound to the ion centre than are the sp electrons. While the sp valence electrons become conduction electrons and move freely through the crystal, the d electrons are bound to the ion. Neighbouring ions may covalently bond d electrons. In most cases, these d states are only partially filled. Electrons in these d states can conduct as well as those in the sp states, but the electron motion in the d states is not well approximated by the Sommerfeld model of free particles. Instead, the electrons move from ion to ion through the shared covalent bonds of the d electrons. These metals have some conduction electrons donated from sp states and others from d-states. Therefore, some electrons move freely according to the Sommerfeld model, while others move through the bonds. Each electron moves back and forth between these two modes of conduction, resulting in an electron motion that is quite complicated. An applied voltage causes the electrons of metals to accelerate and contribute to the electric current.

The electrons scatter from defects in the crystal, and the rate of scattering determines the mobility. Ions in the crystals vibrate around their lattice site, with the amplitude vibration increasing with temperature. The vibration may cause the ion to be displaced from its crystal site, providing a defect from which an electron will scatter. The resistivity of metals increases at high temperature, owing to the increase in vibrations of the ions in the crystal and the resulting increase in scattering.

10.5 Hall Effect

Development of a transverse electric field in a solid material when it carries an electric current and is placed in a magnetic field that is perpendicular to the current is called Hall effect. More precisely, if current is flowing in a solid material along the positive $+X$-axis and a magnetic field is applied along the positive $+Z$-axis, then an electric field is developed in the solid material along the Y-axis. The Hall effect is depicted in Fig. 10.7. This phenomenon was discovered in 1879 by the U.S. physicist Edwin Herbert Hall. The electric field so developed is called the Hall field. The Hall field is perpendicular to both the current and magnetic field. In the case described here, the direction of Hall field is along the positive Y-axis if electric current is constituted by the positive charge carriers (hole/protons) and along the negative Y-axis if electric current is constituted by the negative charge carriers (electrons).

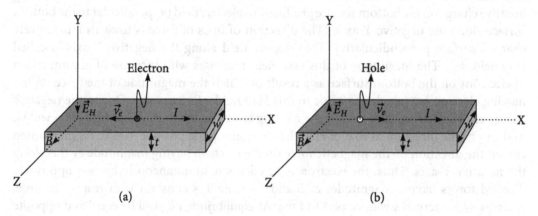

Figure 10.7 (a) Illustration of Hall effect when electric current is constituted by a flow of electrons. The direction of current and direction of applied magnetic field is shown in the figure. $\vec{v}_e$ is the velocity of electrons. The magnetic force experienced by the electron of charge $-e$ moving with velocity $\vec{v}_e$ is given by $-e(\vec{v}_e \times \vec{B}) = e(\vec{B} \times \vec{v}_e)$. Since the charge carriers are electrons, the bottom surface will be negatively charged with respect to the top surface making the direction of the electric field along the negative Y-axis. (b) Illustration of Hall effect when electric current is constituted by a flow of holes. The direction of current and direction of the applied magnetic field is shown in the figure. v_h is the velocity of holes having charge e. The magnetic force experienced by the hole of charge e moving with velocity v_h is given by $e(\vec{v}_h \times \vec{B})$. Since the charge carriers are holes, the bottom surface will be positively charged with respect to the top surface making the direction of electric field along the positive Y-axis

10.5.1 Explanation

Consider a rectangular piece of conductor carrying the current I along the positive X-axis. The current I is constituted by charge carriers, which may be positive or negative. The direction of current is the direction along which a positive charge or hole moves. Let a magnetic field of induction $\vec{B}$ be applied to the conductor along the positive Z-axis. The situation is illustrated in Fig. 10.7. The magnetic force experienced by the charge carriers e moving with velocity $\vec{v}$ is given by $e(\vec{v} \times \vec{B})$. The direction of this force $e(\vec{v} \times \vec{B})$ is found out by the rule of cross product.

Negative charge carriers

If charge carriers are electrons as in most metals and n-type semiconductors, the direction of flow of electrons, according to Fig. 10.7(a), is along the negative X-axis i.e. opposite to the current direction. The magnetic force exerted on the electron having negative charge $-e$ and moving with velocity $\vec{v_e}$ is given by $-e(\vec{v_e} \times \vec{B}) = e(\vec{B} \times \vec{v_e})$. According to the rule of cross product, the direction of $e(\vec{B} \times \vec{v_e})$ will be along the negative Y-axis and its magnitude is evB. Therefore, the electrons will be deflected in the direction of the magnetic force along the negative Y-axis towards the bottom surface of the rectangular conductor. The deflected electrons will be accumulated at the bottom surface. The accumulation of electrons having negative charge on the bottom surface produces an electric field perpendicular to the bottom surface along the negative Y-axis. [The direction of lines of force is towards a negatively charged surface perpendicularly]. This electric field along the negative Y-axis is called Hall field $\vec{E_H}$. The magnitude of this Hall field increases with increase of accumulation of electrons on the bottom surface as a result of which the magnitude of the force on the moving electron having charge e due to this Hall field $e\vec{E_H}$ increases. Due to the negative charge of the electron, the force on the moving electron due to this Hall field $e\vec{E_H}$ will be in the opposite direction as that of Hall field, i.e., along the positive Y-axis. As mentioned earlier, the direction of the magnetic force on the electron having magnitude ev_eB is along the negative Y-axis. Thus, the electron is acted upon simultaneously by two oppositely directed forces having magnitudes ev_eB and $e\vec{E_H}$. ev_eB is constant with respect to time, whereas $e\vec{E_H}$ increases with respect to time. At equilibrium, eE_H will be equal and opposite to ev_eB and further accumulation of electrons on the bottom surface of the conductor will be prevented.

Positive charge carriers

If charge carriers are protons/holes as in a few metals such as beryllium, zinc, cadmium and p-type semiconductors, the direction of flow of electrons according to Fig. 10.7(b) is along the positive X-axis, i.e., the same direction of the current. The magnetic force exerted on the hole having positive charge e and moving with velocity v_h is given by $e(v_h \times \vec{B})$. The direction of $e(v_h \times \vec{B})$ will be along the negative Y-axis and its magnitude is evB. Therefore, the holes will be deflected in the direction of the magnetic force along the negative Y-axis towards the bottom surface of the rectangular conductor. The deflected holes will be accumulated at the bottom surface. The accumulation of holes having positive charge on the bottom surface produces an electric field perpendicular to the bottom surface along the positive Y-axis. [The direction of lines of force is away from a positively charged surface perpendicularly]. This electric field along the positive Y-axis is called Hall field $\bar{E}_H$. The magnitude of this Hall field increases with increase of accumulation of holes on the bottom surface as a result of which the magnitude of the force on the moving electron having charge e due to this Hall field $e\bar{E}_H$ increases. Due to the positive charge of the holes, the force on the moving hole due to this Hall field $e\bar{E}_H$ will be in the same direction as that of the Hall field, i.e., along the positive Y-axis. As mentioned earlier, the direction of magnetic force on the hole having magnitude ev_hB is along the negative Y-axis. Thus, the hole is acted upon simultaneously by two oppositely directed forces having magnitudes ev_hB and $e\bar{E}_H$. ev_hB is constant with respect to time, whereas $e\bar{E}_H$ increases with respect to time. At equilibrium, eE_H will be equal and opposite to ev_hB and further accumulation of holes on the bottom surface of the conductor will stop.

The direction of the Hall field depends upon the nature of the charge carriers, whether positive or negative. In both the cases, charge carriers accumulate on the bottom surface and the direction of force due to Hall field $\bar{E}_H$ on the moving charge is along positive the Y-axis.

Mathematical analysis

At equilibrium, further accumulation of charges on the bottom or top surface of the conductor will stop and eE_H will be equal to evB. Thus, we have

$$eE_H = evB \tag{10.94}$$

However,

Current density $J = Nev$ $\tag{10.95}$

or $\quad ev = \dfrac{J}{N}$ $\tag{10.96}$

where

A = cross-sectional area

N = concentration of charge carriers, i.e., charge carriers per unit volume

Putting the value of ev from Eq. (10.96) into Eq. (10.94), we have

$$eE_H = \dfrac{JB}{N}$$

or $\quad E_H = \dfrac{BJ}{Ne}$

or $\quad E_H = R_H BJ$ $\tag{10.97}$

where $R_H = 1/Ne$ is called the Hall coefficient. Rigorous calculation shows that $R_H = b/Ne$ with $b = 3\pi/8$. Though the value of b does not differ much from 1, we shall take the Hall coefficient as

$$R_H = \frac{3\pi}{8}\frac{1}{Ne} \tag{10.98}$$

If charge carriers are electrons, the Hall coefficient becomes

$$R_{He} = -\frac{3\pi}{8}\frac{1}{Ne} \text{ for electrons} \tag{10.99}$$

where N is the concentration of current-carrying electrons. The Hall coefficient R_H is negative for the conductors where the charge carriers are electrons. If charge carriers are holes, the Hall coefficient becomes

$$R_{Hh} = \frac{3\pi}{8}\frac{1}{pe} \text{ for holes} \tag{10.100}$$

where p is the concentration of current-carrying holes. The Hall coefficient R_H is positive for a few metals such as beryllium, zinc, cadmium and p-type semiconductors, where charge carriers are holes.

Another interesting parameter, Hall angle, can be calculated by combining Eqs (10.97) and (10.98). We can have

$$E_H = \frac{3\pi}{8}\frac{1}{Ne}BJ = \frac{3\pi}{8}\frac{1}{Ne}BNev = \frac{3\pi}{8}Bv = \frac{3\pi}{8}\frac{BvE}{E} = \frac{3\pi}{8}B\mu E$$

or $\quad \dfrac{E_H}{E} = \dfrac{3\pi}{8}B\mu$ \hfill (10.101)

where

μ = mobility of the charge carriers

E = applied electric field = $\dfrac{\text{Applied voltage}}{\text{Length of the specimen}}$ \hfill (10.102)

The Hall angle θ_H is defined by

$$\theta_H = \tan^{-1}\frac{E_H}{E} = \tan^{-1}\frac{3\pi}{8}B\mu \tag{10.103}$$

Hall angle θ_H is a measure of the relative strength of Hall field E_H.

10.5.2 Determination of Hall coefficient R_H

As discussed earlier, the Hall field $\vec{E}_H$ is produced along the thickness of the rectangular conductor. Then due to the Hall field $\vec{E}_H$, there must be a potential difference between the top surface and the bottom surface of the rectangular conductor. The potential difference between the top surface and the bottom surface of the rectangular conductor produced due to the Hall field is called Hall voltage V_H. If the charge carriers are holes, the bottom surface will be at a higher potential than that of the top surface and opposite if the charge carriers are electrons. If t is the thickness of the rectangular conductor, then we have

$$E_H = \frac{V_H}{t}$$

or　$V_H = E_H t$

Putting the value of $\vec{E}_H$ from Eq. (10.94) into this equation, we get $E_H = \dfrac{R_H B I}{A}$

$$V_H = \frac{R_H B I t}{A}$$

or　$R_H = \dfrac{V_H A}{B I t}$

If w is the width of the rectangular conductor, then cross-sectional area $A = wt$. Putting this value of A into this equation, we get

$$R_H = \frac{V_H wt}{B I t}$$

or　$R_H = \dfrac{V_H w}{B I}$　　　　　　(10.104)

Equation (10.104) shows that Hall coefficient R_H can be determined if we can measure the Hall voltage V_H across the top and bottom surface of the rectangular conductor. B, I, and w are the experimental constants for a particular experiment. If we are calculating Hall coefficient for an n-type semiconductor, then according to Eq. (10.99), it will be negative. If we are calculating, Hall coefficient for a p-type semiconductor, then according to Eq. (10.100), it will be positive. In Eq. (10.104), the polarity of V_H will be opposite for n-type and p-type semiconductors. The sign of R_H will be different for the two types of semiconductors.

Experiment

A rectangular sample of the given material having thickness t meters and width w meters is taken and current I ampere is allowed to pass through the sample by connecting it to a battery as shown in Fig. 10.8. The sample is placed in a magnetic field of known magnetic induction in such a manner that I and the magnetic field coincides with the X-axis and Z-axis respectively. The Hall voltage is then measured by placing two probes at the centres of the top and bottom surface of the sample. If the magnetic induction B is in Wb/m^2 and Hall voltage is in volts, then the Hall coefficient R_H obtained from Eq. (10.104) will be in m^3/C.

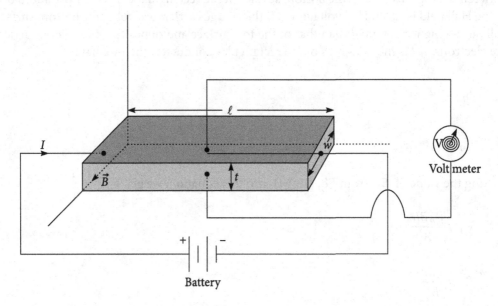

Figure 10.8 | Experimental arrangements for Hall effect. The material is taken in the form of a rectangular slab of thickness t and width w as shown in the figure

10.5.3 Applications of Hall effect

The Hall effect is commercially exploited extensively. In the following, we shall discuss a few commercial applications of the Hall effect.

i.   *Determination of semiconductor type*   According to the Eq. (10.99) and (10.100), the sign of the Hall coefficient R_H can be used to determine whether a given semiconductor is an *n*-type or a *p*-type. If the sign of the Hall coefficient is positive, then a sample material under test is *p*-type and if the sign of the Hall coefficient is negative, then the sample material under test is *n*-type.

ii. *Determination of charge carrier concentration*   By measuring the Hall coefficient R_H of a semiconductor, the charge carrier concentrations N_e and N_h of the semiconductors can be determined by using the Eqs (10.99) and (10.100).

iii. *Determination of charge carrier mobility and drift velocity*   The mobility μ of a charge carrier is defined as the velocity v of the charge carrier per unit applied electric field E. Mathematically, it is given by

$$\mu = \frac{v}{E}$$ (10.105)

From Eq. (10.95), current density J is given as

$$J = Nev$$

However, current density J in terms of electrical conductivity of the medium is given by

$$J = \sigma E$$

or $\sigma E = Nev$

or $\dfrac{E}{v} = \dfrac{Ne}{\sigma}$

Putting this value of E/V into Eq. (A), we get

$$\mu = \frac{\sigma}{Ne}$$

or $\mu = \sigma \left| R_H \right|$ (10.106)

Thus, knowing the electrical conductivity of the sample σ and measuring the Hall coefficient R_H, we can determine the mobility of charge carriers.

Drift velocity is defined as the velocity of the charge carrier in the conductor. Using the definition of charge carrier mobility μ, we have from Eq. (10.106)

$$v = \sigma \left| R_H \right| E$$

or $v = \sigma \left| R_H \right| \dfrac{V}{\ell}$

Thus, knowing the electrical conductivity of the sample σ and measuring the Hall coefficient R_H, applied voltage V and length of the sample ℓ, we can determine the drift velocity of the charge carriers.

iv. Measurement of magnetic induction: From Eq. (10.104), we can have

$$\frac{V_H}{B} = \frac{R_H I}{w} = \text{constant for a given experimental set up.}$$

This equation shows that the Hall voltage is directly proportional to the magnetic induction. Thus, Hall effect can be used as the basis for the design of a magnetic induction meter. By using Hall effect, we can measure the magnetic field in the range $10\mu T$ to $1T$.

v.   Hall effect multiplier: Equation (10.104) shows that Hall voltage is proportional to the product of two independent variables like magnetic induction and current. This property can be exploited to carry out multiplication process and to measure the power loss in a load.

vi.   Hall effect sensors: Hall effect semiconductor sensors are available commercially in the market to sense magnetic field as small as one milli Tesla. Hall voltage is produced whenever there is a transverse magnetic field to an electric current in the semiconducting materials.

vii. Hall effect electronic switches: The Hall effect concept is used in manufacturing magnetically activated electronic switches. Magnetically activated electronic switches are used as non-contacting key boards and panel switches.

viii. Measurement of power in a plane polarized electromagnetic wave: A semiconductor sample is placed in such a manner that the electric component E of the electromagnetic wave is parallel to the semiconductor. This electric field E will drive a current I to flow in the semiconductor. Since the magnetic component H of the electromagnetic wave is perpendicular to the current in the semiconductor, transverse Hall field/Hall voltage will be developed. This Hall voltage V_H is proportional to the product EH. [Hall field=

$$E_H = \frac{3\pi}{8}\frac{1}{Ne}B \times NEv = \frac{3\pi}{8}Bv = \frac{3\pi}{8}BE\mu \text{ and Hall voltage } V_H = E_H t = \frac{3\pi\mu t}{8}\frac{HE}{\mu_m}. \text{ Here}$$

μm is the magnetic permeability of the semiconductor]. The product EH is nothing but the magnitude of the Poynting vector, which is the rate of power flow in the electromagnetic wave. Thus, the Hall effect can be used to determine the rate of power flow in an electromagnetic wave.

Example 10.8

A current of 5 A is established in a rectangular slab of copper of width 2 cm and 0.5 cm thickness. A magnetic field of induction 1.5 T is applied perpendicular to both current and the plane of the slab. The concentration of free electrons in copper is $8.4 \times 10^{28}\,\text{m}^{-3}$. Calculate the Hall voltage across the slab.

Solution

The data given are $B = 1.5$ T, $I = 5$ A, $w = 2$ cm $= 0.02$ m, $t = 0.5$ cm $= 0.005$ m, and $N = 8.4 \times 10^{28}\,\text{m}^{-3}$

The Hall coefficient is

$$R_H = \frac{3\pi}{8}\frac{1}{Ne} = \frac{3\pi}{8}\frac{1}{8.4\times10^{28}\times1.6\times10^{-19}}\frac{\text{m}^3}{\text{C}} = 8.33\times10^{-11}\frac{\text{m}^3}{\text{C}}$$

The Hall voltage is given by

$$V_H = \frac{R_H BI}{w} = \frac{8.33 \times 10^{-11} \times 1.5 \times 5}{0.02} V = 3.12 \times 10^{-8} V$$

Example 10.9

A current of 50 A is established in a rectangular slab of copper of width 2 cm and 0.5 cm thickness. A magnetic field of induction 1.5 T is applied perpendicular to both current and the plane of the slab. The Hall voltage developed across the slab is measured to be 3.29 × 10⁻⁷ volt. Calculate the concentration of free electrons in copper.

Solution

The data given are B = 1.5 T, I = 50 A, w = 2 cm = 0.02 m, t = 0.5 cm = 0.005 m, and V_H = 3.29 × 10⁻⁷ V

The Hall coefficient is obtained as

$$R_H = \frac{V_H w}{BI} = \frac{3.29 \times 10^{-7} \times 0.02}{1.5 \times 50} \frac{m^3}{C} = 8.76 \times 10^{11} \frac{m^3}{C}$$

The concentration of free electrons is given by

$$N = \frac{3\pi}{8} \frac{1}{R_H e} = \frac{3\pi}{8} \times \frac{1}{8.76 \times 10^{-11} \times 1.6 \times 10^{-19}} m^{-3} = 8.40 \times 10^{28} m^{-3}$$

Questions

10.1 What is Fermi gas?

10.2 Why should the potential energy of an electron in the Fermi gas be negative?

10.3 What is mean free path of an electron in Fermi gas?

10.4 What is electrical resistivity?

10.5 What is electrical conductivity?

10.6 Why, when an electric field is applied to the electron gas in a metal, do the electrons try to move in the opposite direction to that of the electric field?

10.7 What do you mean by thermal conductivity of a material?

10.8 Define thermal conductivity

10.9 Explain how the value of the Lorentz number indicates the failure of the classical free electron theory of metals.

10.10 Show that the unit of the Lorentz number is WΩ/K².

10.11 Prove that $kg^{\frac{3}{2}} (Js)^{-3} = m^{-3} J^{\frac{3}{2}}$

10.12 State and derive Wiedemann–Franz law

10.13 What is Ohm's law?

10.14 Why is quantum theory more appropriate for electron gas?

10.15 Define Fermi energy.

10.16 What is the role of the uncertainty principle in defining Fermi–Dirac distribution function?

10.17 Prove that the probability of occupancy of an energy level below Fermi level at 0 K is 1.

10.18 Prove that the probability of occupancy of an energy level above the Fermi level at 0 K is 0.

10.19 Explain how the probability of occupancy of an energy level above the Fermi level at temperature more than 0 K is not 0.

10.20 Explain how the probability of occupancy of an energy level below Fermi level at 0 K is 1.

10.21 Define density of energy states.

10.22 Calculate the value of C in Eq. (10.94)

10.23 Show that the wavelength associated with an electron having an energy equal to Fermi energy at 0 K will be given by

$$\lambda = 2\left(\frac{\pi}{3N_O}\right)^{\frac{1}{3}}$$

10.24 What is the difference between current conduction in copper and zinc?

10.25 What is the quantum mechanical correction to the Hall coefficient?

10.26 What is Hall voltage?

10.27 What is Hall coefficient?

10.28 How is the Hall effect, affected by the sign of the charge carrier?

10.29 What are the assumptions of the classical free electron theory of metals?

10.30 How does a frictional force act on the electron during its motion in a solid?

10.31 Derive an expression for current density in terms of drift velocity.

10.32 Derive an expression for electrical conductivity in terms of relaxation time.

10.33 What is relaxation time of an electron in Fermi gas? What is its order of magnitude?

10.34 Derive an expression for electron mobility.

10.35 Derive an expression for the conductivity of metals in terms of relaxation time based on the classical free electron theory of metals.

10.36 Derive an expression for electrical conductivity in terms of electron mobility.

10.37 Derive an expression for electrical conductivity in terms of the Boltzmann constant.

10.38 Derive an expression for relaxation time in terms of conductivity.

10.39 Derive an expression for relaxation time in terms of electron mobility.

10.40 Derive an expression of thermal conductivity of a material.

10.41 State and prove Wiedemann–Franz law

10.42 What do you mean by Lorentz number? Derive an expression for it.

10.43 State and prove Ohm's law from the classical free electron theory of metals.

10.44 On the basis of the classical free electron theory of metals for pure metals, show that $\rho \propto T$.

10.45 Discuss the advantages and disadvantages of the classical free electron theory of metals.

10.46 What are the postulates of the quantum theory of free electrons?

10.47 Define the Fermi–Dirac distribution function.

10.48 Write the mathematical expression for the Fermi–Dirac distribution explaining all the symbols used.

10.49 Graphically represent the Fermi–Dirac distribution function for metals at different temperatures.

10.50 What are the characteristics of a Fermi level?

10.51 Derive an expression for density of energy states.

10.52 Derive an expression for the number of particles per unit volume at 0K

10.53 Derive an expression for the average energy of an electron at 0K

10.54 Derive an expression for the Fermi energy at 0 K

10.55 Derive an expression for the number of particles per unit volume N having energy from 0 to ∞

10.56 Derive an expression for the Fermi energy in terms of concentration of electrons at 0 K.

10.57 What do you mean by density of energy states? Derive an expression for it.

10.58 Derive an expression for energy density of fermions in terms of the Fermi–Dirac distribution function

10.59 Derive an expression for energy density of fermions at 0 K.

10.60 Derive an expression for average energy of fermions at 0 K.

10.61 Derive an expression for electronic specific heat in terms of the universal gas constant.

10.62 What is meant by effective mass of electrons?

10.63 Derive an expression for the Fermi energy at any temperature T.

10.64 Derive an expression for the average energy of one electron at any temperature T.

10.65 Derive an expression for the electronic molar specific heat at constant volume.

10.66 Derive an expression for the thermal conductivity of a material by applying Fermi–Dirac statistics.

10.67 Derive an expression for the Lorentz number by applying Fermi–Dirac statistics

10.68 Explain how quantum physics gives the correct value of the Lorentz number.

10.69 What is Hall effect?

10.70 What is Hall field?

10.71 Explain the origin of the Hall effect?

10.72 Explain the Hall effect when charge carriers are holes.

10.73 Explain the Hall effect when charge carriers are electrons.

10.74 Derive an expression for Hall field?

10.75 What is Hall coefficient? Obtain an expression for it.

10.76 Derive an expression for Hall coefficient when current conduction is by holes.

10.77 Derive an expression for Hall coefficient when current conduction is by electrons.

10.78 Mention a few applications of the Hall effect.

10.79 Describe an experiment to obtain Hall coefficient.

10.80 Describe an experiment to obtain Hall voltage.

Problems

10.1   Find the energy level in sodium for which probability of occupation is 0.25. Fermi energy of sodium is 3.13 eV at 300 K.   [Ans   3.16 eV]

10.2   Calculate the temperature at which there is a 10^{-6} probability that an energy state 0.55 eV above the Fermi level is occupied by an electron.   [Ans   462 K]

10.3   Given for copper that density is 8.96×10^3 kg/m^3, atomic mass 63.55 amu, and the relaxation time of electrons is 10^{-14} s, calculate the conductivity of copper.
                                                                [Ans 2.38×10^7 Ωm]

10.4   What should be the energy of the quantum state which has 19 for the sum of the squares of the quantum numbers n_x, n_y and n_z? What is the degeneracy of this state?

$$[Ans \quad \frac{19\pi^2\hbar^2}{2mL^2}, \ 3]$$

10.5   Calculate the Fermi energy in sodium at 0 K on the assumption that each sodium atom contributes one electron to the electron gas. The density and atomic mass of sodium is 0.971 g/cm^3 and 23 amu.   [Ans   3.15 eV]

10.6   Calculate the Fermi energy in zinc at 0 K. The density and atomic mass of zinc is 7.14 g/cm^3 and 65.38 amu.   [Ans   9.45 eV]

10.7   The Fermi energy of copper at 0 K is 7.04 eV. What is the average energy of free electrons in copper at 0 K?   [Ans   4.22 eV]

10.8   Evaluate the Fermi–Dirac distribution function for energy 2 kT above Fermi energy.
                                                                [Ans   0.119]

10.9   If the effective mass of electron in copper is m* = 1.01 m and the electrical conductivity of copper is 5.76×10^7 Ω^{-1}m^{-1}, calculate the relaxation time.

10.10 A current of 200 A is established in a rectangular slab of copper of width 2 cm and 1.0 mm thickness. A magnetic field of induction 1.5 T is applied perpendicular to both current and the plane of the slab. The concentration of free electrons in copper is 8.4×10^{28} m^{-3}. Calculate the Hall voltage across the slab. [Ans 2.59×10^{-5} V]

10.11 A current of 25 A is established in a rectangular slab of aluminum of width 5 cm and 0.5 cm thickness. A magnetic field of induction 2.5 T is applied perpendicular to both current and the plane of the slab. The concentration of free electrons in aluminum is 18×10^{28} m^{-3}. Calculate the Hall voltage across the slab. [Ans 1.04×10^{-7} V]

10.12 A current of 30 A is established in a rectangular slab of zinc of width 5 cm. A magnetic field of induction 1.0 T is applied perpendicular to both current and the plane of the slab. The Hall voltage developed across the slab is measured to be 3.36×10^{-8} V. Calculate the concentration of free electrons in copper. [Ans 13.15×10^{28} m^{-3}]

10.13 A current of 200 A is established in a rectangular slab of copper of width 2 cm and 1.0 mm thickness. A magnetic field of induction 1.5 T is applied perpendicular to both current and the plane of the slab. The concentration of free electrons in copper is 8.4×10^{28} m^{-3}. Calculate the Hall voltage across the slab. [Ans 2.59×10^{-5} V]

10.14 A current of 25 A is established in a rectangular slab of aluminum of width 5 cm and 0.5cm thickness. A magnetic field of induction 2.5 T is applied perpendicular to both current and the plane of the slab. The concentration of free electrons in aluminum is 18×10^{28} m^{-3}. Calculate the Hall voltage across the slab. [Ans 1.04×10^{-7} V]

10.15 A current of 30 A is established in a rectangular slab of zinc of width 5 cm. A magnetic field of induction 1.0 T is applied perpendicular to both current and the plane of the slab. The Hall voltage developed across the slab is measured to be 3.36×10^{-8} V. Calculate the concentration of free electrons in copper. [Ans 13.15×10^{28} m^{-3}]

Multiple Choice Questions

1. Which of the following properties is not possible to explain correctly by the classical free electron theory?

    (i)   Electrical conductivity

    (ii) Temperature dependence of electrical conductivity

    (iii) Thermal conductivity

    (iv) Magnetic susceptibility

2. To electron gas, which of the following statistics is applicable?

    (i)   Maxwell–Boltzmann   (ii) Fermi–Dirac

    (iii) Bose–Einstein   (iv) Stefan–Hawking

3. Which of the following statements are true?

    (i)   A solid is an insulator if it has $E_g \approx 1$ eV

    (ii) A solid is a semiconductor if it has $E_g > 3$ eV

    (iii) A solid is a metal if its valence band is partially filled

    (iv) A solid is a metal if its valence band and conduction band overlap

4. The relation between conductivity and absolute temperature is
    (i) $\sigma \propto T$ (ii) $\sigma \propto T^2$
    (iii) $\sigma \propto \sqrt{T}$ (iv) $\sigma \propto \dfrac{1}{T}$

5. Which of the following equation is not correct?
    (i) $\mu_e = \dfrac{e\tau}{m}$ (ii) $\tau = \dfrac{m\sigma}{Ne^2}$
    (iii) $\sigma = Ne\mu_e$ (iv) $I = Ne <v>$

6. Which of the following is not a unit of thermal conductivity?
    (i) WmK^{-1} (ii) $cal.cm.s^{-1}.K^{-1}$
    (iii) $erg.cm.s^{-1} K^{-1}$ (iv) WmK

7. The Lorentz number, according to the Drude–Lorentz theory is given by
    (i) $L = \dfrac{3k^2}{2e^2}$ (ii) $L = \dfrac{3k}{2e}$
    (iii) $L = \dfrac{3e^2}{2k^2}$ (iv) $L = \dfrac{3e}{2k}$

8. The Wiedemann–Franz law according to the Drude–Lorentz theory is given by
    (i) $\dfrac{K}{\sigma} = \dfrac{3}{2}\dfrac{k}{e}T$ (ii) $\dfrac{K}{\sigma} = \dfrac{3}{2}\dfrac{e^2}{k^2}T$
    (iii) $\dfrac{K}{\sigma} = \dfrac{3}{2}\dfrac{k^2}{e^2}T$ (iv) $\dfrac{K}{\sigma} = \dfrac{3}{2}\dfrac{e}{k}T$

9. The quantum mechanical form of the Wiedemann–Franz law is
    (i) $\dfrac{K}{\sigma} = \dfrac{\pi^2}{3}\dfrac{k}{e}T$ (ii) $\dfrac{K}{\sigma} = \dfrac{\pi}{3}\dfrac{e^2}{k^2}T$
    (iii) $\dfrac{K}{\sigma} = \dfrac{\pi^2}{3}\dfrac{k^2}{e^2}T$ (iv) $\dfrac{K}{\sigma} = \dfrac{\pi^2}{3}\dfrac{e}{k}T$

10. Fermi energy level is the highest energy state occupied by an electron at 0 K.
    (i) True   (ii) False

11. The probability of occupancy of an energy level below Fermi level at 0 K is
    (i) 0   (ii) ∞
    (iii) 1   (iv) None of the above

12. The probability of occupancy of an energy level above the Fermi level at 0 K is
    (i) 0   (ii) ∞
    (iii) 1   (iv) None of the above

13. The density of energy states between E and $E + dE$ is proportional to

    (i) $\dfrac{1}{E}$ (ii) E

    (iii) $E^{\frac{1}{2}}$ (iv) $E^{-\frac{1}{2}}$

14. Which of the following is correct for Fermi level in a metal?

    (i) Depends upon temperature   (ii) 1 at any temperature

    (iii) 0 at any temperature   (iv) 0.5 at any temperature

15. The electronic contribution to molar heat capacity is of the order of

    (i) $10^{-2}\,R$ (ii) $10^{-1}\,R$

    (iii) $10^{\circ}\,R$ (iv) $10^{1}\,R$

16. In an extrinsic semiconductor, donor atoms

    (i) Add holes to the valence band

    (ii) Add electrons to the valence band

    (iii) Add holes to the conduction band

    (iv) Add electrons to the conduction band

17. In a p-type semiconductor the Fermi level lies midway between the accepter level and the top level of the valence band

    (i) True   (ii) False

18. In Hall effect, only one surface becomes charged, whether current conduction is due to holes or electrons.

    (i) True   (ii) False

19. Which of the following is wrong?

    (i) Hall field is perpendicular to both magnetic induction and current

    (ii) Magnetic induction is perpendicular to both Hall field and current

    (iii) Current is perpendicular to both Hall field and magnetic induction

    (iv) None of the above

20. Hall coefficient is

    (i) Always positive

    (ii) Always negative

    (iii) May be positive or negative

    (iv) Independent of charge

Answers

1 (iv)	2 (ii)	3 (iii & iv)	4 (iv)	5 (iv)	6 (iv)	7 (i)	8 (iii)
9 (iii)	10 (i)	11 (iii)	12 (i)	13 (iii)	14 (iv)	15 (i)	16 (iv)
17 (i)	18 (i)	19 (iv)	20 (iii)				

11 Energy Bands in Solids

11.1 Introduction

The behaviour of electrons and their energies in solids is related to the behaviour of all other particles around them where as their behavior in free space is completely different. The electrons in free space can posses any specified energy. The certain ranges of energies possessed by the electrons in a solid are called allowed energy bands and certain ranges of energies that can not be possessed by the electrons are called forbidden energy bands. The band theory accounts for many of the electrical and thermal properties of solids and gives the basic concepts behind the technological advancements of materials. The energy band theory of solids plays the most vital role in the development of semicondutor devices. The energy changes of electrons in solids interacting with photons of light, energetic electrons, x-rays and the like confirm the general validity of the band theory and provide detailed information about allowed and forbidden energy bands.

11.2 Origin of Energy Bands

The concepts of energy bands in solids can be accounted for by two different approaches; one is the classical approach and the other is the quantum mechanical approach.

11.2.1 Origin of energy bands: The classical approach

In the classical approach, the band of energies permitted in a solid is related to the discrete allowed energies – the energy levels of single, isolated atoms. According to the Bohr's atomic model, the energy level diagram of a hydrogen atom is depicted in Fig. 11.1, for three principal quantum numbers. As illustrated in the Fig. 11.1, an electron cannot have any energy between –13.6 eV and –3.4 eV. These are the energy levels of Bohr's hydrogen atom for principal quantum numbers 1 and 2 respectively.

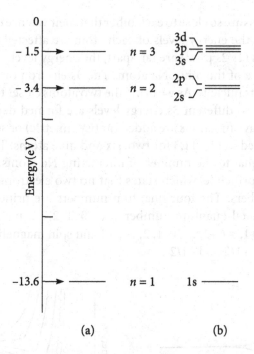

(a)   (b)

Figure 11.1 (a) The first three energy states of the hydrogen atom according to Bohr's model. (b) Electron energy states for the first three shells in the wave mechanical model of the hydrogen atom

The energy is quantized. The range between –13.6 eV and –3.4 eV is the forbidden energy gap. Now we shall see what happens to these energy levels when a solid is formed due to the close aggregation of individual atoms.

When two hydrogen atoms are brought together to form H_2^+, the 1s energy level of two hydrogen atoms is split into two as shown in Fig. 11.2.

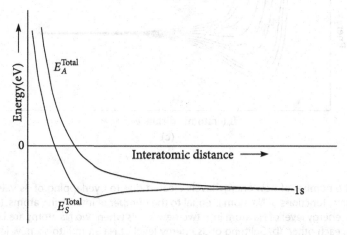

Figure 11.2 Total energy in H_2^+ as function of interatomic distance for symmetric and antisymmetric states. The antisymmetric state has no minimum in its total energy

In almost all solids atoms are so close to each other that their valence electron wave functions overlap each other i.e., the energy levels of each atom are affected by other atoms. When two identical Na atoms ($1s^22s^2p^63s^1$) are far apart, the energy levels of one Na atom are not affected by the presence of the other Na atom. The 3s electron of each atom has a single energy with respect to its nucleus. As we bring the two atoms closer together, decreasing the interatomic distance, two different 3s energy levels are formed depending upon whether two valence electron wave functions are added (antisymmetric) or subtracted (symmetric). This situation is depicted in Fig. 11.3 for two, six and nine atoms. The number of splitting of 3s energy level is equal to the number of interacting Na atoms. This is in accordance with Pauli's exclusion principle, which states that no two electrons can have the same set of four quantum numbers. The four quantum numbers are principal quantum number n (= 1, 2, 3, ..., n), orbital quantum number ℓ (= 0, 1, 2, ..., n – 1), magnetic quantum number m_ℓ (= $-\ell$, $-\ell+1$, $-\ell+2$, ..., 0, 1, 2, ..., ℓ) and spin magnetic quantum number m_s which for electrons are +1/2 and –1/2

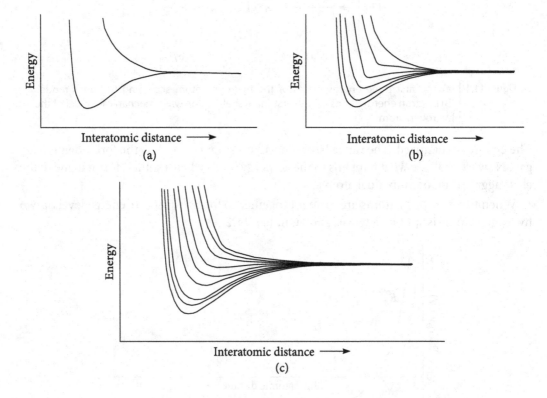

Figure 11.3 | The number of new energy levels created due to overlapping of 3s valence electron wave functions of Na atom is equal to the number of interacting atoms. (a) Splitting of 3s energy level of Na atom into two new levels when two Na atoms are brought closer to each other. (b) Splitting of 3s energy level of Na atom into six new levels when six Na atoms are brought closer to each other. (c) Splitting of 3s energy level of Na atom into nine new levels when nine Na atoms are brought closer to each other

When the Na atoms are far apart, all the 3s electrons of all the atoms have the same energy. As large numbers of atoms are brought closer to each other to form solid sodium, the electrons of the atoms are perturbed by the electrons and nuclei of the nearby atoms. As the interatomic distances among the coalescing atoms decrease, the electron wave functions of each energy level overlap each other splitting the individual energy levels of each atom into closely spaced new energy levels. The new energy levels are so close to each other that we are not in a position to distinguish them and they form an energy band consisting of a virtually continuous spread of permitted energies as shown in Fig. 11.4. The extent of splitting depends upon the interatomic distances among the coalescing atoms and starts from the outermost energy level. It is natural that splitting of inner most energy levels may not be possible due to the presence of nearby atoms. The gaps may exist between adjacent energy bands. The electrons cannot occupy these energy gaps because they cannot contain the energies lying within these energy gaps. These energy gaps are called forbidden energy bands as shown in Fig. 11.5. The energy bands identified with the 3 s energy level is called the 3 s energy band. The energy bands identified with the 3 p energy level is called the 3 p energy band. Figure 11.5 may be consulted. The energy bands of a solid, depending on the extent to which they are filled by the electrons, and the forbidden energy bands not only predict the electrical properties of the solid; they also have important bearings on the other properties of the solid.

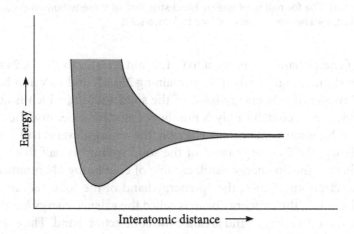

Figure 11.4 When the number of interacting atoms is very large as in solids, the new energy levels created due to the overlapping of electron wave functions of atoms are so close to each other that they appear continuos and form an energy band

Representation of energy bands

If the number of interacting atoms is N, then each energy band has a total of N individual levels. Each level, according to Pauli's exclusion principle, can accommodate $2(2\ell+1)$ number of electrons. Hence, each energy band, having N levels, can accommodate $2(2\ell+1)N$ number of electrons.

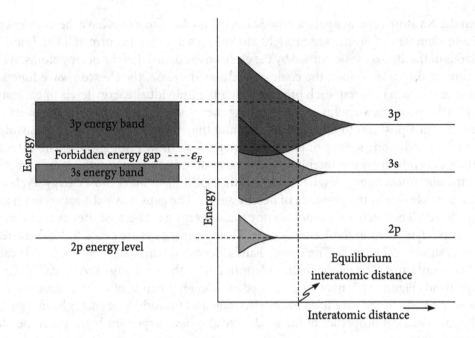

Figure 11.5 The typical plotting of electron energy versus interatomic distance of large number of atoms. The formation of energy band structure at the equilibrium distance has been illustrated here as atoms coal esce to form a solid

For s, p, d and f energy bands, ℓ is equal to 0, 1, 2 and 3 respectively. 1s, 2s and 2p energy bands of solid sodium is completely filled containing 2N, 6N and 10N number of electrons respectively. However, the 3s energy band of the solid sodium, which is able to contain 2N number of electrons, contains only N number of electrons; because the 3s energy level (valence level) of Na atom contains one electron, the energy level is half filled in the case of Na atom. Hence, the 3s energy band of the solid sodium is half filled. Above the 3s energy band, there is the 3p energy band, capable of containing 6N number of electrons, but contains no electrons. Hence, the 3p energy band of the solid sodium is completely empty. In solid sodium, the 3s energy band is called the valence energy band or simply the valence band and the 3p energy band is called the conduction band. The energy difference between the top of the valence band and the bottom of the conduction band is called the forbidden energy band or forbidden energy gap E_g. The situation described here is the ground state of solid sodium at 0 K and is depicted in Fig. 11.6.

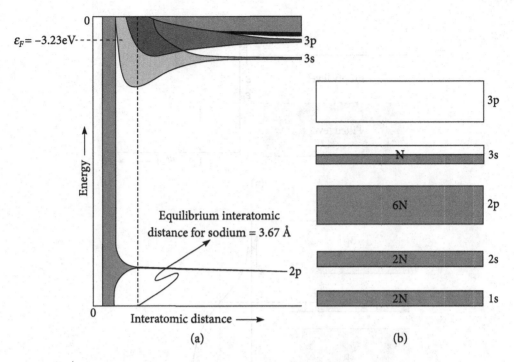

Figure 11.6 (a) The energy level of sodium atoms becomes energy bands as their interatomic distance decreases. The Fermi level is in the middle of the 3s band. The observed interatomic distance in solid sodium is 3.67 Å. (b) Representation of energy bands in solid sodium at 0 K

We can give electric energy to the material by connecting the solid sodium to an electric potential source or thermal energy by heating it. Since the 3s band is half empty, it contains filled energy states as well as unfilled energy states of equal amount. The electrons from 3s filled energy states may move either into 3s unfilled energy states or into the completely empty 3p band depending upon the amount of energy the electrons have absorbed. The 3s band in solid sodium is exactly half filled at 0 K. The Fermi level ε_F in this case will be at the middle of the 3s band. The energy band above the Fermi level ε_F is empty and below the Fermi level ε_F, the energy band is completely filled. Though in metals the Fermi level ε_F is approximately independent of temperature, the probability of occupation of energy bands by the electron depends strongly on temperature. As the temperature increases above 0 K, the energy level above the Fermi level ε_F is no longer empty. This has been depicted in Fig. 11.7 along with the Fermi–Dirac distribution function. Figure 11.7 shows that sodium is a good electrical conductor. By absorbing energy either from the electrical energy source or from the heat energy source, the electrons below the Fermi level ε_F can move up to the conduction band to conduct electric current.

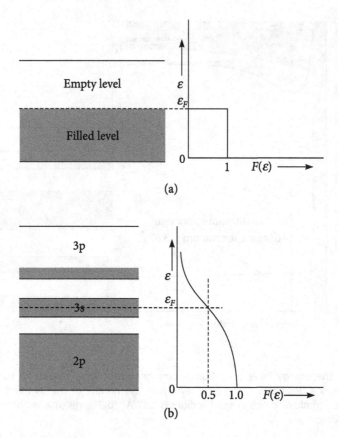

Figure 11.7 (a) The 3s energy bands of solid sodium at 0 K. It is exactly half filled (b) The energy bands of solid sodium above 0 K. In this case, The 2p band is neither completely full nor the 3p band is completely empty. In both cases, the corresponding Fermi–Dirac distribution function is plotted

The 3s valence band of magnesium ($1s^2 2s^2 2p^6 3s^2$) is completely filled. The forbidden energy band E_g between the 3s valence band and the 3p conduction band is large. However, magnesium is a metal and so is a good conductor of electricity. In reality, In solid magnesium, the 3s valence band and the 3p conduction band overlaps to produce a compound 3s+3p band at equilibrium interatomic distance as depicted in Fig. 11.8. The 3s and 3p bands can at best contain 2N and 6N electrons respectively. Hence, the compound 3s+3p band can accommodate 8N electrons. However, in magnesium, the 3s valence band contains 2 N electrons and the 3p conduction band is empty. Hence, the compound 3s+3p band contains only 2 N electrons, one-fourth of its maximum capacity. Therefore, as the compound 3s+3p band is only partially filled, it is a good conductor of electricity.

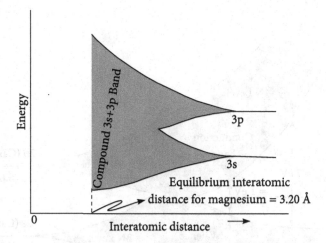

Figure 11.8 Energy bands in magnesium. The overlapping of 3s energy band and 3p energy band start as equilibrium interatomic distance is approached. At equilibrium distance 3s + 3p is a single energy band

Again diamond, a form of carbon ($1s^2 2s^2 2p^2$), has a partially filled valence band and is expected to be a good conductor. However, diamond is an insulator. During the formation of diamond, initially, the 2s band and the 2p band overlaps to form a compound 2s+2p band as carbon atoms are brought closer to each other. The 2s and 2p bands at best can contain 2 N and 6 N electrons respectively. Hence, the compound 2s+2p band of diamond can accommodate 8 N electrons. However, in carbon , the 2s valence band contains 2 N electrons and the 2p conduction band contains 2 N electrons and hence, in reality, the compound 2s+2p band of diamond contains only 4 N electrons. Later on, as the distance between coalescing carbon atoms approach the equilibrium interatomic distance, the compound band 2s+2p of diamond starts to separate out to form the valence band and the conduction band. At equilibrium interatomic distance, the forbidden energy gap between the valence band and the conduction band is around 7 eV and the only lower valence band is filled with all the 4 N electrons of the compound 2s+2p band making the conduction band empty. Here, the Fermi level ε_F lies on top of the valence band. Therefore, as the conduction band is completely empty, diamond is an insulaor of electricity. This is depicted in Fig. 11.9. The electrons in the valence band can acquire sufficient energy to move to the conduction band if an electric field of about 10^8 V/m is applied. This is just billion times greater than the electric field needed for current flow in a metal!

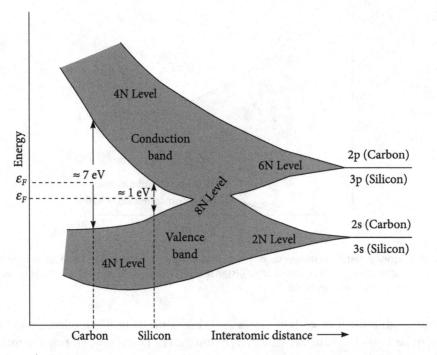

Figure 11.9 (a) The energy bands of diamond. The compound 2s+2p band splits as the interatomic distance approaches equilibrium distance. The lower valence band is filled and the upper conduction band is empty. The Fermi level ε_F lies above the completely filled valence band. The forbidden energy band E_g depends upon the interatomic distance and at equilibrium interatomic distance, it is around 7 eV which makes the diamond an insulator

Now let us apply the band theory to explain the properties of semiconductors. Silicon and germanium are semiconducting materials. Silicon ($1s^22s^22p^63s^23p^2$) and germanium ($1s^22s^22p^63s^23p^63d^{10}4s^24p^2$) have the same form of electronic configuration as that of carbon ($1s^22s^22p^2$). The outermost valence shells and number of valence electrons are the same in Si, Ge and C. They have similar crystal structure as well as similar band structure as shown in Fig. 11.10. However, to our astonishment, the electrical behaviour of silicon or germanium differs completely from that of carbon! As depicted in Fig. 11.10, the forbidden energy band in silicon and germanium is approximately 1 eV, which is very much less than that of carbon.

In carbon, the value of the forbidden energy band is around 7 eV, whereas in silicon and germanium, it is around 1 eV. In silicon, at room temperature, the few electrons in the top of the valence band can absorb 1 eV heat energy to jump the forbidden energy band and move into the conduction band by crossing the Fermi level. Here, the Fermi level lies approximately in the middle of the forbidden energy band. When a small electric field is applied, these few electrons in the conduction band constitute a small current in the silicon. Thus, the electrical conductivity of silicon lies in between that of the conductor and the insulator making it a semiconducting material or a semiconductor. Other solids having similar band structures are also semiconducting materials.

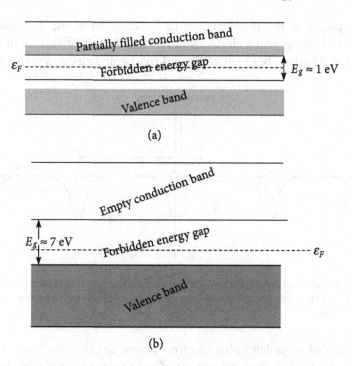

Figure 11.10 | (a) The energy bands of silicon or germanium at room temperature. The forbidden energy band in silicon or germanium is much less than that of an insulator. Here a small number of valence electrons near the top of the valence band can acquire enought thermal energy to jump the gap and enter into the conduction band. Hence, Fermi level e_F lies in the middle of the forbidden energy band of Si and Ge. (b) The energy band diagram of an insulator. Here the Fermi level ε_F lies on the top of the valence band. An electron in the valence band needs at least 7 eV to reach the empty conduction band

11.2.2 Origin of energy bands: The quantum mechanical approach

According to the quantum mechanical approach, which is more realistic and general, in solids, there exists periodic potential arising out of the positive ion cores at the lattice points. The formation of energy bands in solids can be well explained with the help of Schrödinger's equation, the basic equation of quantum physics. In the following, the phenomenon of formation of energy bands in solids is explained quantum mechanically.

The atoms in a perfect crystal are arranged, three-dimensionally, in a regular periodic pattern. The ion cores at the lattice points bear a net positive charge because the atoms are ionized in the metal, with valence electrons taken off to the conduction band. The electrostatic potential energy of an electron in the field of positive ions is negative so that the force between them is attractive as $\vec{F} = -\vec{\nabla}V$. Thus, conduction electrons move in a periodic potential produced by positive ions cores at lattice points. An electron moving in a periodic potential undergoes diffraction similar to that of a wave. As a result, the diffraction effects limit the electron to certain ranges of momenta that correspond to allowed energy bands. For simplicity of the mathematics, let us consider a one-dimensional lattice as shown in Fig. 11.11 along which electrons move. Due to the diffraction of electron waves in one-dimensional lattices, both left going and right going electron waves exist.

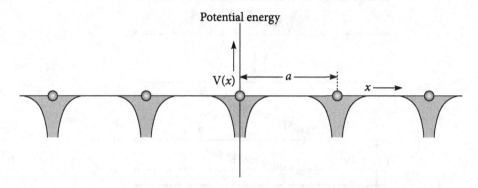

Potential energy

Figure 11.11 Variation of potential energy of a conduction electron in the field of positive ion cores in a one-dimensional crystal lattice. The potential energy is minimum at $x = 0, \pm a, \pm 2a, \pm 3a$...

The superposition of these left going electron waves, $\psi_-(x) = Ae^{-ikx}$ and the right going electron waves, $\psi_+(x) = Ae^{ikx}$ with magnitude of wave vector $k = n\pi/a$ electron waves gives rise to standing waves. The following two types of standing wave patterns are possible.

$$\psi_1(x) = \psi_+(x) + \psi_-(x) = 2A\cos\frac{n\pi}{a}x \qquad (11.1)$$

$$\psi_2(x) = \psi_+(x) - \psi_-(x) = 2iA\sin\frac{n\pi}{a}x \qquad (11.2)$$

The two standing waves $\psi_1(x)$ and $\psi_2(x)$ pile up electrons at different regions and therefore, the two waves have different values of potential energy. This is the origin of the energy gap.

The probability density ρ of a particle is the square of the wave function, i.e., $\rho = \psi\psi^* = |\psi|^2$. For the standing wave $\psi_1(x) = 2A\cos\frac{n\pi}{a}x$, the probability density $\rho_1(x)$ is calculated out to be

$$\rho_1(x) = 4A^2\cos^2\frac{n\pi}{a}x \qquad (11.3)$$

This function $\rho_1(x)$ piles up electrons on the positive ion centres at $x = 0, \pm a, \pm 2a, \pm 3a$ ..., where potential energy is minimum. For the other standing wave, $\psi_2(x) = 2iA\sin\frac{n\pi}{a}x$, the probability density $\rho_2(x)$ is calculated to be $\rho(x)$

$$\rho_2(x) = 4A^2 \sin^2 \frac{n\pi}{a} x \qquad (11.4)$$

This function $\rho_2(x)$ piles up electrons away from the positive ion cores at $x = \pm a/2, \pm 3a/2,$ $\pm 5a/2, \pm 7a/2, \ldots$. The functions $\rho_1(x)$ and $\rho_2(x)$ are shown graphically in Fig. 11.12 along with the one-dimensional lattice.

Figure 11.12 is important for explaining the origin of energy band in solids quantum mechanically. When we calculate the average or the expectation values of the potential energy over these three charge distributions, we find that the potential energy of $\rho_1(x)$ is lower than that of the travelling wave, whereas the potential energy of $\rho_2(x)$ is higher than that of the travelling wave. The potential energy difference due to $\rho_1(x)$ and $\rho_2(x)$ is the forbidden energy gap E_g.

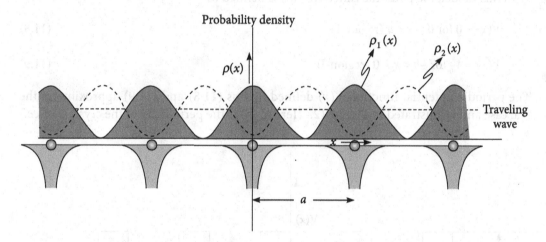

Figure 11.12	The distribution of probability density $\rho_1(x)$ and $\rho_2(x)$; and a travelling wave. The wave function $\psi_1(x) = 2A\cos\frac{n\pi}{a}x$ piles up electronic charge on the positive ion cores lowering the potential energy in comparison with the average potential energy of an electron defined by a travelling wave. The wave function $\psi_2(x) = 2iA\sin\frac{n\pi}{a}x$ piles up electronic charge in the region between positive ion cores raising the potential energy in comparison with the average potential energy of an electron defined by a travelling wave

Kronig–Penney model

The Kronig–Penney model assumed for periodic potential illustrates the behaviour of electrons in a crystalline solid. The Schrödinger equation for the motion of an electron in a crystalline solid is given by

$$\nabla^2\psi + \frac{2m}{\hbar^2}(E-V)\psi = 0 \qquad (11.5)$$

where V is the periodic potential due to lattice points. For simplicity, let us consider a one-dimensional crystal lattice. In one dimension, the Schrödinger equation for the motion of an electron is given by

$$\frac{d^2\psi}{dx^2} + \frac{2m}{\hbar^2}(E - V)\psi = 0, \text{ with } V(x + a) = V(x) \tag{11.6}$$

The solution of this equation is, according to Bloch's theorem[*], given by

$$\psi_k(x) = u_k(x)e^{ikx} \tag{11.7}$$

Here the function $u_k(x)$ known as the Bloch function in one-dimension satisfies the condition $u_k(x) = u_k(x + a)$, i.e., $u_k(x)$ is a periodic function with period a.

In this model, the potential function $V(x)$ is defined by

$$V(x) = 0 \text{ for } 0 < x < a \text{ (region I)} \tag{11.8}$$

$$V(x) = V_0 \text{ for } -b < x < 0 \text{ (region-II)} \tag{11.9}$$

The periodic potential function $V(x)$ defined by Eqs (11.8) and (11.9) approximates the real case and is illustrated in Fig. 11.12. Here (a+b) is the periodicity of the crystal lattice.

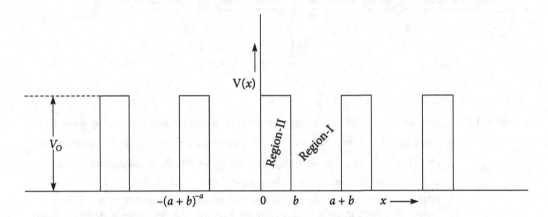

Figure 11.13 │ Kronig–Penney model of one-dimensional periodic potential of periodicity $(a + b)$

The Schrödinger Eq. (11.6) in region I and region II will become respectively

$$\frac{d^2\psi}{dx^2} + \frac{2mE}{\hbar^2}\psi = 0 \text{ region I} \tag{11.10}$$

---

[*] Bloch's theorem: The eigenfunctions of the wave equation for a periodic potential are the product of a plain wave $e^{i\vec{k}\cdot\vec{r}}$ times a periodic function $u_k(\vec{r})$ with the periodicity of the crystal.

and $\dfrac{d^2\psi}{dx^2} - \dfrac{2m}{\hbar^2}(V_0 - E)\psi = 0$ region II $\qquad\qquad$ (11.11)

When electrons move in the metal, the total energy of an electron is less than the potential energy of the moving electron, i.e., $E < V_0$. Substituting

$$k_O = \frac{\sqrt{2mE}}{\hbar} \qquad\qquad (11.12)$$

and $\quad k_1 = \dfrac{\sqrt{2m(V_0 - E)}}{\hbar} \qquad\qquad$ (11.13)

Equations (11.10) and (11.11) becomes respectively

$$\frac{d^2\psi}{dx^2} + k_0^2\psi = 0 \qquad\qquad (11.14)$$

$$\frac{d^2\psi}{dx^2} - k_1^2\psi = 0 \qquad\qquad (11.15)$$

According to Bloch's theorem, $\psi_k(x) = u_k(x)e^{ikx}$ is the solution of the Schrödinger equation for an electron moving in a one-dimensional periodic potential. Substituting $\psi_k(x) = u_k(x)e^{ikx}$ into Eqs (11.14) and (11.15), we have respectively

$$\frac{d^2u_k}{dx^2} + 2ik\frac{du_k}{dx} + \left(k_0^2 - k^2\right)u_k = 0 \qquad\qquad (11.16)$$

and $\quad \dfrac{d^2u_k}{dx^2} + 2ik\dfrac{du_k}{dx} - \left(k_1^2 + k^2\right)u_k = 0 \qquad\qquad$ (11.17)

The solutions of Eqs (11.16) and (11.17) are given respectively by

$$u_1 = Ae^{i(k_o - k)x} + Be^{-i(k_o + k)x} \qquad\qquad (11.18)$$

$$u_2 = Ce^{(k_1 - ik)x} + De^{-(k_1 + ik)x} \qquad\qquad (11.19)$$

Since $\psi_k(x) = u_k e^{ikx}$ is a solution of the Schrödinger equation, we have

$$u_1(x)\big|_{x=0} = u_2(x)\big|_{x=0} \text{ and } u_1(x)\big|_{x=a} = u_2(x)\big|_{x=-b} \tag{11.20}$$

$$\frac{du_1(x)}{dx}\bigg|_{x=0} = \frac{du_2(x)}{dx}\bigg|_{x=0} \text{ and } \frac{du_1(x)}{dx}\bigg|_{x=a} = \frac{du_2(x)}{dx}\bigg|_{x=-b} \tag{11.21}$$

Substituting Eqs (11.18) and (11.19) into Eqs (11.20) and (11.21), we get four simultaneous linear equations in A, B, C, D. The four equations will have non-trivial solutions only if the determinant of the coefficients vanishes. Simplifying the determinant and equating it to zero, we have

$$\left(\frac{k_1^2 - k_0^2}{2k_1 k_0}\right)\sin k_0 a \sin k_1 b + \cos k_0 a \cos k_1 b = \cos k(a+b) \tag{11.22}$$

This equation is a relation between energy E and $k\left(=\frac{2\pi}{\lambda}\right)$ and thus, is a dispersion relation.

The height of the potential barrier is very large, i.e., $V_0 \to \infty$ and at the same time, the width of the barrier $b \to 0$ in such a way that the product bV_0 remains finite. If $k_1 \gg k_0$ and $b \to 0$, Eq. (11.22) becomes

$$\left(\frac{k_1^2}{2k_1 k_0}\right)\frac{ab}{ab}\sin k_0 a \sin k_1 b + \cos k_0 a \cos k_1 b = \cos k(a+b)$$

or $\quad\left(\frac{k_1^2 ab}{2k_0 a}\right)\frac{\sin k_1 b}{k_1 b}\sin k_0 a + \cos k_0 a \cos k_1 b = \cos k(a+b)$

or $\quad\left(\frac{k_1^2 ab}{2k_0 a}\right)\sin k_0 a + \cos k_0 a = \cos ka$ since $\lim\limits_{k_1 b \to 0}\dfrac{\sin k_1 b}{k_1 b} = 1$

or $\quad P\dfrac{\sin k_0 a}{k_0 a} + \cos k_0 a = \cos ka \tag{11.23}$

where $P = \dfrac{k_1^2 ab}{2} = \dfrac{ab}{2}\dfrac{2m}{\hbar^2}(V_0 - E) \approx \dfrac{V_0 mab}{\hbar^2}$ \tag{11.24}

This equation shows that P is a measure of height of the potential barrier, i.e., it is a measure of the binding energy of the electrons.

Interpretation of Eq. (11.23)

The right-hand side of Eq. (11.23) is cos ka and it varies from a maximum value of +1 to a minimum value of –1. Figure 11.14 is a plot of

$$P\frac{\sin k_0 a}{k_0 a} + \cos k_0 a$$

versus $k_0 a$ for $P = 2\pi$ and range from -6π to $+6\pi$. The two extreme limits of cos ka are shown in Fig. 11.14 by two horizontal broken lines marked as +1 and –1. Since Eq. (11.23) has to be satisfied, those portions of the curve which lie beyond the limits of +1 and –1 are discarded and are shown shaded in Fig. 11.14. Hence, all the values of

$$k_0 a = \frac{\sqrt{2mE}}{\hbar} a$$

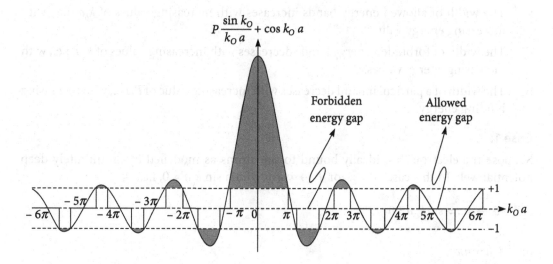

Figure 11.14 | Plot of function $P\dfrac{\sin k_0 a}{k_0 a} + \cos k_0 a$ versus $k_0 a$ for $P = 2\pi$ and range from -6π to $+6\pi$. The maximum and minimum values of the right-hand side of Eq. (11.23) (i.e., cos ka) are represented by two horizontal broken lines. The allowed values of energy are given by those ranges of $k_0 a = \dfrac{\sqrt{2mE}}{\hbar} a$ for which the function lies between +1 and –1. For other values of energy, there are no travelling waves or Bloch-like solutions to the wave equations, so that the forbidden energy gaps in the energy spectrum are formed

are not valid; only those values of

$$k_0 a = \frac{\sqrt{2mE}}{\hbar} a$$

are allowed for which the curve lies within the limits of +1 and −1. The forbidden values of $k_0 a$ are marked by dashed lines and the allowed values of

$$k_0 a = \frac{\sqrt{2mE}}{\hbar} a$$

are marked by continuous lines in Fig. 11.14 This leads to the very important conclusion that there are energy bands of allowed energy values separated by forbidden energy ranges.

Observations

From the previous discussions, the following observations were inferred from Fig. 11.14:

i. There exist a number of energy bands separated by forbidden energy ranges called energy gaps.

ii. The width of allowed energy bands increases with increasing values of $k_0 a$, i.e., with increasing energy values.

iii. The width of forbidden energy bands decreases with increasing values of $k_0 a$, i.e., with increasing energy values.

iv. The width of a particular band decreases with increasing value of P, i.e., with increasing binding energy.

Case 1:

Suppose the electrons are ideally bound to the atoms as modelled by an infinitely deep potential well. In this case, $P = \infty$ or $P \rightarrow \infty$ for which $\sin k_0 a = 0$, i.e.,

$$P \rightarrow \infty \Rightarrow \sin k_0 a = 0$$

or $\quad k_0 a = n\pi$

or $\quad \sqrt{2mE} = \frac{n\hbar\pi}{a}$

or $\quad E = \frac{n^2 \hbar^2 \pi^2}{2ma^2} = \frac{n^2 h^2}{8ma^2} \quad n = 1, 2, 3, 4, \dots$ $\qquad$ (11.25) which may be recognized as the energy levels of an electron in an infinitely deep potential well of width a. In this case, electrons are independent of each other and each electron is confined to one atom by an infinite potential barrier.

Case 2:

Suppose the electrons are completely free from the atoms as modelled by the free electron theory. In this case, $P = 0$ or $P \to 0$ for which $\cos ka = \cos k_o a$, i.e.,

$$P \to 0 \Rightarrow \cos ka = \cos k_o a$$

or $\quad k_o a = ka$

or $\quad \sqrt{2mE} = k\hbar$

or $\quad E = \dfrac{k^2 \hbar^2}{2m} = \dfrac{p^2}{2m}$ $\hfill (11.26)$

which may be recognized as the energy of a free electron. In this case, electrons are independent of each other and each one is free to move in the metal.

Thus, by varying the value of P from 0 to ∞, the motions of the electron from a completely free state to a completely bound state are discussed.

Case 3:

As we have discussed earlier, the forbidden energy values are obtained by setting

$$\cos ka = \pm 1$$

or $\quad ka = n\pi$

or $\quad k = \dfrac{n\pi}{a} \quad n = 1, 2, 3, \dots$ $\hfill (11.27)$

Equation (11.22) shows that the discontinuity in energy occurs at

$$k = \pm \frac{\pi}{a}, \pm \frac{2\pi}{a}, \pm \frac{3\pi}{a}, \pm \frac{4\pi}{a}, \pm \frac{5\pi}{a}, \dots\dots\dots$$

11.2.3 Dispersion curves

Using Eq. (11.23), it is possible to obtain the values of energy E as a function of k. The plotting of E versus k gives the dispersion curve as shown in Fig. 11.15.

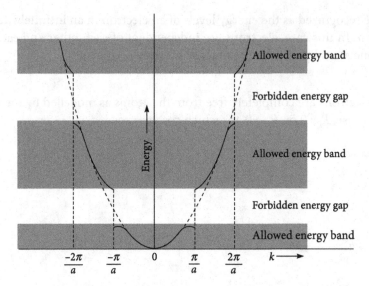

Figure 11.15 | The dispersion curve. It is a plot of energy E versus $k = 2\pi/\lambda$ of an electron moving in a one-dimensional periodic lattice. The dashed curve (a parabola) represents the same plot for a free electron [Eq. (11.26)]. The energy of a free electron is continuous, whereas the energy of an electron moving in a periodic lattice is not continuous

The curve has discontinuities at k values given by $k = \pm \dfrac{n\pi}{a}$, n = 1, 2, 3,........ The dispersion curve (i.e., plot between E and k) is of great importance as it determines the behaviour of electrons in the solid when electrons get accelerated. Since $p = \hbar k$, the variation of k signifies the variation of momentum of the electron. The energy values corresponding to the continuous k values form the allowed energy bands and the energy values between the discontinuities form the forbidden energy gaps.

11.2.4 Conclusions

The following conclusions may be drawn regarding the energy of conduction electrons in a periodic potential.

i.   There exist a number of energy bands separated by forbidden energy gaps.

ii.   The width of allowed energy bands increases with increasing values of $k_0 a$, i.e., with increasing energy values.

iii.   The width of a particular band decreases with increasing values of P, i.e., with increasing binding energy of the electrons.

11.3 Representation of Band Diagrams of Solids

According to the band theory, solids can be classified as conductors, insulators or semiconductors. From our discussions, we can now draw the band diagrams of conductors, insulators and semiconductors. Figure 11.16 shows the band diagrams of conductors and Fig. 11.17 shows the band diagrams of semiconductors and insulators.

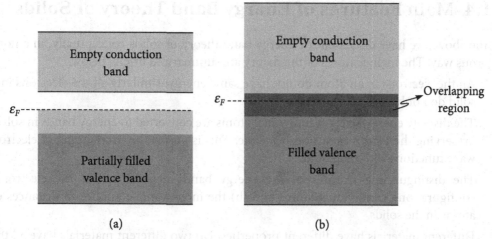

Figure 11.16 | The energy band diagrams along with the Fermi level ε_F of conductors at 0 K. (a) The energy band structures found in copper like metals. The Fermi level lies at the top of the partially valence band. The forbidden energy band is small. (b) The energy band structures found in magnesium like metals. The filled valence band overlaps the empty conduction band. The Fermi level lies in the overlapping region. The forbidden energy band is absent here

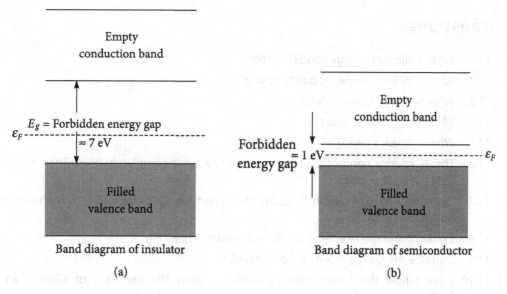

Figure 11.17 | The energy band diagrams of semiconductors and insulators along with the Fermi level ε_F at 0 K. (a) The energy band structures found in insulators. The forbidden energy band is large (>>1 eV) in comparison to conductors and semiconductors. The Fermi level lies on the top of the valence band. (b) The energy band structures found in semiconductors. The Fermi level lies in the middle of the forbidden energy band. The forbidden energy band is small ($\approx$ 1 eV) in comparison to insulators

11.4 Main Features of Energy Band Theory of Solids

In the above, we have discussed the energy band theory of solids conceptually, in a non-rigorous way. The main features of the theory are summarized briefly below.

i.   All the electrons in an atom do not have same energy; similarly all the electrons in a solid do not have same energy.

ii. The discrete energy levels of individual atoms are converted to energy bands in solids preserving the Pauli's exclusion principle. This is due to the overlapping of electron wave functions with each other.

iii. The distinguishing features of the energy bands depend on (a) the electronic configuration of individual atoms and (b) the interatomic equilibrium distances of atoms in the solids.

iv. Different materials have different properties. No two different materials have all the same properties. The variation of properties from solids to solids depends on (a) nature of energy bands, (b) occupation of energy bands by the electrons, (c) width of forbidden energy gaps, and (d) relative position of Fermi energy level in the energy bands.

v.   The electrical properties of almost all solids can be perfectly explained with help of the concepts of energy band theory.

Questions

11.1 What is allowed energy bands in solids?

11.2 What is forbidden energy bands in solids?

11.3 What is energy gap in solids?

11.4 What is conduction band?

11.5 What is valence band?

11.6 The 3s valence band of magnesium ($1s^2 2s^2 2p^6 3s^2$) is completely filled. Then explain how it is a conductor.

11.7 Diamond, a form of carbon ($1s^2 2s^2 2p^2$), has a partially filled valence band. Then explain how it is an insulator.

11.8 Explain the origin of energy bands quantum mechanically.

11.9 Describe the Kronig–Penney model of solids.

11.10 Explain how the Kronig–Penney model predicts the presence of allowed and forbidden energy bands in crystals.

11.11 What happens to the width of the allowed and forbidden bands with change in the strength of periodic potential?

11.12 Explain how the Kronig–Penney model predicts the presence of energy gaps in crystals.

11.13 Prove that the width of allowed energy bands increases with increasing energy values.

11.14 Prove that the width of forbidden energy bands decreases with increasing energy values.

11.15 Show that the width of a particular band decreases with increasing binding energy.

11.16 Based on the band theory of solids, how you will define a semiconductor?

11.17 What are the similarities and dissimilarities between the energy level band diagram of silicon and carbon?

11.18 Describe the energy band structure of the *n*-type semiconductor.

11.19 Describe the energy band structure of the *p*-type semiconductor.

11.20 Explain classically the origin of energy bands in solids.

Multiple Choice Questions

1. When a number of sodium atoms are brought together to form a solid, the number of splitting of 3s energy level is equal to

    (i)   Less than the number of interacting Na atoms

    (ii) The number of interacting Na atoms

    (iii) More than the number of interacting Na atoms

    (iv) Cannot be said

2. The conduction electrons move in a periodic potential produced by positive ions cores at lattice points.

    (i)   True

    (ii) False

3. When electrons move in the metal, the total energy of an electron is

    (i)   More than the potential energy of the moving electron

    (ii) Less than the potential energy of the moving electron

    (iii) Equal to the potential energy of the moving electron

    (iv) None of the above

4. The relation between total energy and propagation vector of magnitude $(2\pi/\lambda)$ is called

    (i)   Einstein's relation

    (ii) Newton's spectral relation

    (iii) Wein's displacement relation

    (iv) Dispersion relation

5. The statement, 'The eigenfunctions of the wave equation for a periodic potential are the product of a plain wave $e^{i\vec{k}\cdot\vec{r}}$ times a function $u_k(\vec{r})$ with the periodicity of the crystal.' is called
    (i)   Pythagoras' theorem   (ii) Nernst's theorem
    (iii) Bloch's theorem   (iv) Fleming's theorem

6. What do you mean by free electrons precisely?
    (i)   Electrons moving with energy
    (ii) Electrons for which potential energy gradient is zero
    (iii) Electrons for which kinetic energy gradient is zero
    (iv) Electrons moving randomly

7. The electrons in the conduction band are free to
    (i)   Transport impulses   (ii) Transport signals
    (iii) Transport vibrations   (iv) Transport charge

8. Which one has the greatest energy gap?
    (i)   Semi-metals   (ii) Semiconductors
    (iii) Insulators   (iv) Metals

9. Which one has the smallest energy or zero energy gap?
    (i)   Semi-metals   (ii) Semiconductors
    (iii) Insulators   (iv) Metals

10. The Kronig–Penney model is based on the assumption
    (i)   Electrons move in a periodic potential field
    (ii) Electrons move in a constant potential field
    (iii) Electrons move in a zero potential field
    (iv) Electrons move with constant potential energy

Answers

1 (ii)   2 (i)   3 (ii)   4 (iv)   5 (iii)   6 (ii)   7 (iv)   8 (iii)
9 (iv)   10 (i)

Semiconductors

12.1 Introduction

No property of solids varies as widely as their ability to conduct electric current. Copper, a good conductor, has a resistivity of 1.7×10^{-8} Ωm at room temperature, whereas quartz, a good insulator, has a resistivity of 7.5×10^{17} Ωm, more than 25 powers of ten greater. Thus, based on the ability of solids to carry electric currents, solids are classified into three categories:

i. Conductors

ii. Insulators

iii. Semiconductors

Materials which conduct electric current when a small potential difference is applied across them are known as good conductors. At room temperatures the resistivity of a good conductor is of the order of 10^{-8} Ωm. The valence band in conductors is completely filled, while the conduction band is partially filled. Therefore, when a small potential difference is applied to a conductor, it provides sufficient energy to the electrons in the valence band to jump to the conduction band where they result in currents in the conductor. The energy band diagrams of conductors have been discussed in chapter 11. Materials which do not conduct electric current under ordinary potential difference under normal conditions are known as insulators. At room temperature, the resistivity of an insulator is of the order of 10^{17} Ωm . In case of insulators, the valence band is completely filled, the conduction band is completely empty and the forbidden energy band is very large. In case of diamond, a good insulator, the forbidden energy gap is ≈ 7 eV and an enormous electric field of 10^8 V/m is required to make it conduct current. The energy band diagrams of insulators have already been discussed in chapter 11.

12.2 Semiconductors

Generally, the term semiconductor is applied to a class of materials having conductivity in between a conductor and an insulator. The forbidden energy band in semiconductors lies roughly in between 0.2 eV to 2.5 eV. Germanium and silicon are basic semiconductors. Early stage semiconductor electronic devices made of germanium have been replaced by silicon. Semiconductors possess the following characteristic properties:

i.   A pure semiconductor has a negative temperature coefficient of resistance [resistance decreases with increase in temperature]

ii. Semiconductors give high thermoelectric power with signs both positive and negative relative to a given metal.

iii. The junction between a *p*-type semiconductor and an *n*-type semiconductor shows rectification properties.

iv. Semiconductors are light sensitive, generating either a photo-voltage or a change in resistance upon irradiation by visible radiation.

12.3 Band Theory of Semiconductors

The band diagram of silicon and diamond are very much similar and has already been shown in Fig. 11.7. The atoms are so arranged that in diamond, the forbidden energy band is $\approx$ 7 eV and in silicon, it is only $\approx$ 1 eV. In case of a semiconductor, at room temperature, a small number of valence band electrons gain sufficient energy and jump to the conduction band. The energy band diagram of a semiconductor at 0K and at room temperature is shown in Fig. 12.1.

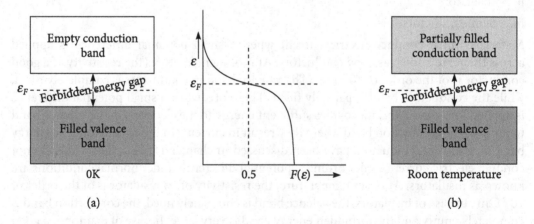

Figure 12.1 Energy band diagram of a semiconductor. (a) At 0 K, the valence band is completely filled and the conduction band is completely empty. (b) At room temperature, a small number of valence band electrons gain sufficient thermal energy and jump to the conduction band. Therefore, at room temperature, the conduction band is partially filled and the valence band is partially empty

In this figure, the conduction band, the valence band and the Fermi energy level ε_F have been shown. These electrons, small in number, allow a small current to flow in the semiconductors when an electric field is applied. Thus, these types of materials have resistivity in between conductors and insulators.

12.3.1 Intrinsic semiconductors

When the conductivity of a semiconductor is solely determined by thermally generated charge carriers, the semiconductor is called an intrinsic semiconductor. Intrinsic semiconductors are also called pure semiconductors. An intrinsic semiconductor exhibits a high degree of chemical purity (i.e., it has a ratio of impurities of only about one part in 10^{12}). The most commonly used intrinsic semiconductors are silicon ($_{14}Si^{28}$; 14 = $1s^2$, $2s^2$, $2p^6$, $3s^2$, $3p^2$) and germanium ($_{32}Ge^{74}$; 32= $1s^2$, $2s^2$, $2p^6$, $3s^2$, $3p^6$, $3d^{10}$, $4s^2$, $4p^2$). Both Si and Ge belong to the IV group of the periodic table. At 0 K, the valence band is completely filled and the conduction band is completely empty. Therefore, at 0 K, an intrinsic semiconductor behaves like an insulator. At room temperature, a small number of valence band electrons gain sufficient thermal energy and jump to the conduction band. Therefore, at room temperature, the conduction band becomes partially filled, the valence band becomes partially empty and the intrinsic semiconductor conducts electric current poorly.

The vacancies formed in the valence band of the semiconductor when some of the valence band electrons, gaining thermal energy, jump into the conduction band are termed as holes. The number of thermally generated holes per unit volume in the valence band p and the number of thermally generated electrons in the conduction band n are called intrinsic carrier concentration n_i. The number of holes in the valence band is equal to the number of thermally generated electrons in the conduction band because electro–hole pairs are produced simultaneously due to the breaking of covalent bonds in semiconductors. Thus, we have

$$p = n = n_i \tag{12.1}$$

Holes carry positive charge of magnitude equal to the charge of an electron. The electrons in the conduction band and the holes in the valence band are free and move within the crystal randomly due to thermal energy. An application of an external electric field directs all the electrons in one direction and all the holes in the opposite direction (to that of electrons) as shown in Fig. 12.2. When a semiconductor is connected to a battery, current is contributed by the electrons in the conduction band and by the holes in the valence band. The electrons move towards the positive terminal of the battery and the holes move towards the negative terminal of the battery.

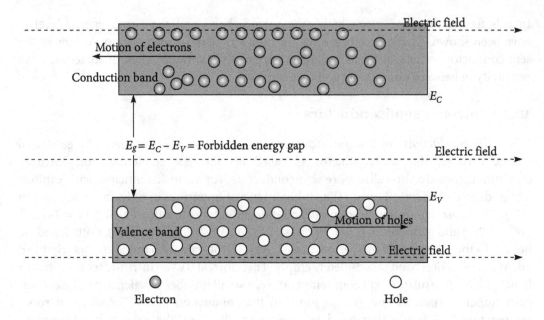

Figure 12.2 Motion of electrons and holes under the influence of an externally applied electric field. Black circles and empty circles represent electrons and holes respectively. Electrons move in a direction opposite to that of the electric field and holes move in the same direction as that of the electric field

Concentration of electrons in the conduction band

The number of electrons per unit volume in the conduction band is called concentration of electrons in the conduction band. The concentration of electrons having energy from 0 to ∞ will be given according to Eq. (10.73) as

$$n = \int_0^\infty F(\varepsilon)g(\varepsilon)d\varepsilon$$

where $F(\varepsilon) = \dfrac{1}{e^{\frac{\varepsilon - \varepsilon_F}{kT}} + 1}$

is the Fermi–Dirac distribution function and $g(\varepsilon)d\varepsilon = C\varepsilon^{\frac{1}{2}}d\varepsilon$ with $C = \dfrac{8\sqrt{2}\pi m_e^{*\frac{3}{2}}}{h^3}$, $g(\varepsilon)$ is called the density of energy states. m_e^* is called the average effective mass of the electron. If energy ε is measured with respect to the bottom level of the conduction band E_C, the density of the energy states $g(\varepsilon)$ will become

$$g(\varepsilon)d\varepsilon = C\left(\varepsilon - E_C\right)^{\frac{1}{2}}d\varepsilon \tag{12.2}$$

The concentration of electrons in the conduction band having energy from E_C to ∞ will be given by

$$n = \int_{E_C}^{\infty} F(\varepsilon)g(\varepsilon)d\varepsilon$$

At room temperature [300 K],

$$\varepsilon - \varepsilon_F \gg kT$$

or $\quad e^{\frac{\varepsilon - \varepsilon_F}{kT}} \gg 1$

Therefore, at room temperature, the Fermi–Dirac distribution function is given by

$$F(\varepsilon) = \frac{1}{e^{\frac{\varepsilon - \varepsilon_F}{kT}}} = e^{-\frac{\varepsilon - \varepsilon_F}{kT}}$$

The concentration of electrons in the conduction band having energy from E_C to ∞ will be given by

$$n = \int_{E_C}^{\infty} C(\varepsilon - E_C)^{\frac{1}{2}} e^{-\frac{\varepsilon - \varepsilon_F}{kT}} d\varepsilon = \int_{E_C}^{\infty} C(\varepsilon - E_C)^{\frac{1}{2}} e^{\frac{\varepsilon_F}{kT}} e^{-\frac{\varepsilon}{kT}} d\varepsilon = Ce^{\frac{\varepsilon_F}{kT}} \int_{E_C}^{\infty} (\varepsilon - E_C)^{\frac{1}{2}} e^{-\frac{\varepsilon}{kT}} d\varepsilon$$

or $\quad n = Ce^{\frac{\varepsilon_F}{kT}} \int_{E_C}^{\infty} (\varepsilon - E_C)^{\frac{1}{2}} e^{-\frac{\varepsilon}{kT}} d\varepsilon$ $\qquad\qquad$ (12.3)

Let

$$x = \frac{\varepsilon - E_C}{kT} \text{ or } \varepsilon = kTx + E_C, \text{ and } d\varepsilon = kTdx$$

When ε become E_C, x becomes 0 and when ε becomes ∞, x becomes ∞ and thus, the limits of integration E_C and ∞ become respectively 0 and ∞. Therefore, the concentration of electrons in the conduction band will be

$$n = Ce^{\frac{\varepsilon_F}{kT}} \int_{0}^{\infty} (kT)^{\frac{1}{2}} x^{\frac{1}{2}} e^{\frac{E_C}{kT}} e^{-x} kTdx = C(kT)^{\frac{3}{2}} e^{\frac{E_C}{kT}} e^{\frac{\varepsilon_F}{kT}} \int_{0}^{\infty} x^{\frac{1}{2}} e^{-x} dx$$

$$= C(kT)^{\frac{3}{2}} e^{-\frac{E_C}{kT}} e^{\frac{\varepsilon_F}{kT}} \int_0^\infty x^{\frac{3}{2}-1} e^{-x} dx$$

$$= C(kT)^{\frac{3}{2}} e^{-\frac{E_C}{kT}} e^{\frac{\varepsilon_F}{kT}} \Gamma\left(\frac{3}{2}\right) = \frac{1}{2}\sqrt{\pi}(kT)^{\frac{3}{2}} C e^{\frac{\varepsilon_F-E_C}{kT}}$$

Putting the value of C into this equation, we have

$$n = N_C e^{\frac{\varepsilon_F-E_C}{kT}}, \text{ where } N_C = 2\left(\frac{2\pi m_e^* kT}{h^2}\right)^{\frac{3}{2}} \tag{12.4}$$

This expression gives the concentration of electrons in the conduction band at any temperature T. Here N_C is called the effective density of states in the conduction band. It represents the number of electrons at energy level E_C when total conduction band is squeezed to E_C. At room temperature (300 K) for most of intrinsic semiconductors, it is of the order of 10^{19} cm^{-3}.

Concentration of holes in the valence band

If energy ε is measured with respect to the topmost level of the valence band E_V, the density of energy states in the valence band $g(\varepsilon)$ will become

$$g(\varepsilon)d\varepsilon = C\left(E_V - \varepsilon\right)^{\frac{1}{2}} d\varepsilon$$

where $C = \dfrac{8\sqrt{2}\pi m_h^{*\frac{3}{2}}}{h^3}$,

m_h^* is called the effective mass of the hole.

The Fermi–Dirac distribution function for holes in the valence band is given by

$$F_h(\varepsilon) = 1 - F_e(\varepsilon) = 1 - \frac{1}{e^{\frac{\varepsilon-\varepsilon_F}{kT}}+1} \approx e^{\frac{\varepsilon_F-\varepsilon}{kT}} \text{ for } \varepsilon_F - \varepsilon \gg kT$$

The concentration of holes in the valence band having energy from $-\infty$ to E_C will be given by

$$p = \int_{-\infty}^{E_V} C\left(E_V - \varepsilon\right)^{\frac{1}{2}} e^{\frac{\varepsilon_F-\varepsilon}{kT}} d\varepsilon = \int_{-\infty}^{E_V} C\left(E_V - \varepsilon\right)^{\frac{1}{2}} e^{\frac{\varepsilon_F}{kT}} e^{\frac{\varepsilon}{kT}} d\varepsilon$$

$$= Ce^{-\frac{\varepsilon_F}{kT}} \int_{-\infty}^{E_V} (E_V - \varepsilon)^{\frac{1}{2}} e^{\frac{\varepsilon}{kT}} d\varepsilon$$

Let

$$x = \frac{E_V - \varepsilon}{kT} \text{ ;or } \varepsilon = E_V - kTx \text{ and } d\varepsilon = -kTdx$$

When ε becomes E_V x becomes 0 and when ε becomes $-\infty$ x becomes ∞ and thus, the limits of integration $-\infty$ and E_V become ∞ and 0 respectively. Thus, we have

$$p = Ce^{-\frac{\varepsilon_F}{kT}} \int_{\infty}^{0} (ktx)^{\frac{1}{2}} e^{\frac{E_V - kTx}{kT}} (-kT)dx = -Ce^{-\frac{\varepsilon_F}{kT}} e^{\frac{E_V}{kT}} (kT)^{\frac{3}{2}} \int_{\infty}^{0} x^{\frac{1}{2}} e^{-x} dx$$

$$= Ce^{\frac{E_V - \varepsilon_F}{kT}} (kT)^{\frac{3}{2}} \int_{0}^{\infty} x^{\frac{1}{2}} e^{-x} dx = (kT)^{\frac{3}{2}} \Gamma(\frac{3}{2}) Ce^{\frac{E_V - \varepsilon_F}{kT}}$$

Putting the value of C into this equation, we have

$$p = N_V e^{\frac{\varepsilon_F - E_C}{kT}} \text{ where } N_V = 2\left(\frac{2\pi m_h^* kT}{h^2}\right)^{\frac{3}{2}} \tag{12.5}$$

This expression gives the concentration of holes in the valence band at any temperature T. Here N_V is called the effective density of states in the valence band. It represents the number of holes at energy level E_V when total valence band is squeezed to E_V. At room temperature (300 K) for most of intrinsic semiconductors, it is of the order of 10^{19} cm^{-3}.

Intrinsic carrier concentration

The number of holes in the valence band and the number of electrons in the conduction band are equal as explained earlier, i.e.,

$$n = p = n_i$$

n_i is called the intrinsic carrier concentration. The square of the intrinsic carrier concentration is obtained as

$$n_i^2 = pn$$

Putting the values of p and n into this equation, we get

$$n_i^2 = 2\left(\frac{2\pi m_e^* kT}{h^2}\right)^{\frac{3}{2}} e^{\frac{\varepsilon_F-E_C}{kT}} \times 2\left(\frac{2\pi m_h^* kT}{h^2}\right)^{\frac{3}{2}} e^{\frac{E_V-\varepsilon_F}{kT}}$$

$$= 4\left(\frac{2\pi kT}{h^2}\right)^{3}\left(m_e^* \times m_h^*\right)^{\frac{3}{2}} e^{-\frac{E_C-E_V}{kT}}$$

or $\quad n_i = Ae^{-\frac{E_g}{2kT}}$ where $A = 2\left(\frac{2\pi kT}{h^2}\right)^{\frac{3}{2}}\left(m_e^* \times m_h^*\right)^{\frac{3}{4}}$ (12.6)

This equation gives the intrinsic carrier concentration in intrinsic semiconductors.

Example 12.1

Calculate the intrinsic carrier concentration in an intrinsic semiconductor at 300 K using the following data: $\mu_e = 0.4$ m^2 V^{-1} s^{-1}, $\mu_h = 0.2$ m^2 V^{-1} s^{-1}, $E_g = 0.7$ eV, $m_e^* = 0.55 m_0$, $m_h^* = 0.37 m_0$.

Solution

The intrinsic carrier concentration in an intrinsic semiconductor is given by

$$n_i = 2\left(\frac{2\pi kT}{h^2}\right)^{\frac{3}{2}}\left(m_e^* \times m_h^*\right)^{\frac{3}{4}} e^{-\frac{E_C-E_V}{2kT}}$$

Putting the given data into this equation, we get

$$n_i = 2\left(\frac{2\pi \times 1.38\times 10^{-23} \times 300}{(6.626\times 10^{-34})^2}\right)^{\frac{3}{2}}\left(0.55\times 0.37 \times m_0^2\right)^{\frac{3}{4}} e^{-\frac{0.7\times 1.6\times 10^{-19}}{2\times 1.38\times 10^{-23}\times 300}} \text{m}^{-3} = 1.352\times 10^{13}\,\text{m}^{-3}$$

Example 12.2

Calculate the forbidden energy gap E_g of germanium at 300 K from the following data: intrinsic atom concentration for germanium is 5.56×10^{18} m^{-3}, $m_e^* = 0.55 m_0$ and $m_h^* = 0.37 m_0$

Solution

$$n_i = 2\left(\frac{2\pi kT}{h^2}\right)^{\frac{3}{2}}\left(m_e^* \times m_h^*\right)^{\frac{3}{4}} e^{-\frac{E_g}{2kT}}$$

or $e^{\frac{E_g}{2kT}} = \dfrac{2\left(\dfrac{2\pi kT}{h^2}\right)^{\frac{3}{2}}\left(m_e^* m_h^*\right)^{\frac{3}{4}}}{n_i}$

or $E_g = 2kT \times \ell n \dfrac{2\left(\dfrac{2\pi kT}{h^2}\right)^{\frac{3}{2}}\left(m_e^* m_h^*\right)^{\frac{3}{4}}}{n_i}$

Putting the given values into this equation, we obtain

$E_g = 2 \times 1.38 \times 10^{-23} \times 300$

$\times \ell n \dfrac{2\left(\dfrac{2\pi \times 1.38 \times 300}{\left(6.626\times 10^{-34}\right)^2}\right)^{\frac{3}{2}}\left(0.55 m_0 \times 0.37 m_0\right)^{\frac{3}{4}}}{5.56 \times 10^{18}}\text{eV} = 0.73\ \text{eV}$

Fermi level in an intrinsic semiconductor

In an intrinsic semiconductor, the concentration of holes and electrons are equal. Therefore, we have

$$m_e^{*3} e^{\frac{2(\varepsilon_F - E_C)}{kT}} = m_h^{*3} e^{\frac{2(E_V - \varepsilon_F)}{kT}}$$

or $\dfrac{m_h^{*3}}{m_e^{*3}} = e^{\frac{2(\varepsilon_F - E_C)}{kT}} e^{\frac{-2(E_V - \varepsilon_F)}{kT}} = e^{\frac{2(2\varepsilon_F - E_C - E_V)}{kT}}$

Taking the logarithm of both sides, we have

$$3\ell n \dfrac{m_h^*}{m_e^*} = \dfrac{2\left(2\varepsilon_F - E_C - E_V\right)}{kT}$$

or $\quad \varepsilon_F = \dfrac{3}{4} kT \ell n \dfrac{m_h^*}{m_e^*} + \dfrac{1}{2}\left(E_V + E_C\right)$ (12.7)

If at 0 K, $m_h^* = m_e^*$, then $\ln\dfrac{m_h^*}{m_e^*} = 0$ and

$$\varepsilon_F = \frac{1}{2}\left(E_V + E_C\right),$$

i.e., the Fermi level ε_F lies in the middle of the forbidden energy band. However, $\ln\dfrac{m_h^*}{m_e^*}$ is

more than 1 and hence, the Fermi level ε_F lies just above the middle of the forbidden energy band. Again since the first term of the expression for the Fermi level contains temperature, the Fermi level ε_F increases linearly as temperature increases.

Intrinsic electrical conductivity

The electric current density J in terms of carrier density N as we know is given by

$$J = Nvq$$

However, $J = \sigma E$, σ = electrical conductivity

Therefore, $\sigma E = Nqv$

or $\sigma = Nq\dfrac{v}{E} = Nq\mu$ (12.8)

where $\mu = \dfrac{v}{E}$

is called the mobility of the charge carrier and depends weakly on temperature.

In an intrinsic semiconductor, electron and holes both are charge carriers and charge of both carriers are equal to e, the charge of an electron. Hence, the electrical conductivity of an intrinsic semiconductor will be given by

$$\sigma_i = ne\mu_n + pe\mu_p \qquad (12.9)$$

where μ_e and μ_h are the mobilities of the electrons and holes respectively. Since $n = p = n_i$, this expression becomes

$$\sigma_i = n_i e\left(\mu_n + \mu_p\right)$$

Putting the value of n_i from Eq. (12.6) into this equation, we have

$$\sigma_i = Ae\left(\mu_n + \mu_p\right)T^{\frac{3}{2}}e^{-\frac{E_g}{2kT}} = \sigma_0 e^{-\frac{E_g}{2kT}} \qquad (12.10)$$

Here $\sigma_0 = Ae(\mu_n + \mu_p)$ is nearly independent of temperature and is a characteristic parameter of the semiconductor. Equation (12.10) shows the temperature dependence of σ_i.

Example 12.3

Calculate the current produced in germanium crystal having cross-sectional area 2 cm² and length 0.2 mm when a potential difference of 2 volt is applied. Given: concentration of free electrons in germanium crystal = 2×10^{19} m⁻³, $\mu_n = 0.36$ m³V⁻¹s⁻¹, $\mu_p = 0.17$ m³V⁻¹s⁻¹.

Solution

The data given are $A = 2$ cm² $= 2 \times 10^{-4}$ m², $L = 0.2$ mm $= 0.2 \times 10^{-3}$, $V = 2$ V, $n_i = 2 \times 10^{19}$ m⁻³, $\mu_n = 0.36$ m³V⁻¹s⁻¹, $\mu_p = 0.17$ m³V⁻¹s⁻¹.

The current produced in germanium crystal is given by

$$I = JA = \sigma EA = n_i e \left(\mu_n + \mu_p \right) \frac{AV}{L} = 2 \times 10^{19} \times 1.6 \times 10^{-19} (0.36 + 0.17) \frac{2 \times 10^{-4} \times 2}{0.2 \times 10^{-3}} A = 3.39 \, A.$$

12.3.2 Extrinsic semiconductors

Electron and holes may be generated inside a semiconductor by the action of heat and light as in the case of intrinsic semiconductors. However, the most efficient and convenient method of generating electron and hole pairs is to add a small amount of selected impurities into the intrinsic semiconductor. The process of addition of selected impurities into an intrinsic semiconductor to increase its conductivity is called doping. The impurity that is added is known as dopant. The semiconductor containing selected impurities is called an extrinsic semiconductor.

If donor atoms are doped at the rate of one part in 10⁶ (i.e., to 10⁶ intrinsic atoms, one dopant is added) and N is the intrinsic atom concentration, then the concentration of donor atoms N_d will be

$$N_d = \frac{N}{10^6} \tag{12.11}$$

Example 12.4

A silicon crystal is doped with arsenic with a doping concentration of 1 in 10⁹. Calculate the concentration of donor atoms if the intrinsic atom concentration for silicon is 5×10^{28} m⁻³.

Solution

According to the question impurities are added at the rate of 1 part in 10⁹ i.e.,
To 10⁹ silicon atoms, the number of arsenic atom(s) doped is 1.

or   To 1 silicon atom, the number of arsenic atom(s) doped will be $\dfrac{1}{10^9}$

or   to 5×10^{28} atoms/m^3 of silicon, the number of arsenic atom(s) doped will be

$$\frac{5 \times 10^{28}}{10^9} \text{atoms/m}^3 = 5 \times 10^{19} \text{atoms/m}^3$$

Thus, the concentration of donor atoms (arsenic) is calculated to be 5×10^{19} atoms/m^3.

The conductivity of an extrinsic semiconductor is solely determined by the excess of electrons or holes due to impurity atoms. The impurities frequently employed for Si and Ge [group IV elements] are the elements of group III and group V of the periodic table. They are generally boron, gallium, indium and aluminum from group III and phosphorous, antimony and arsenic from group IV. Depending upon the type of impurity present, the extrinsic semiconductors are of two types: (i) *n*-type and (ii) *p*-type.

n-type semiconductors

The *n*-type semiconductor results when pentavalent impurities are added to an intrinsic semiconductor. When a pentavalent atom [arsenic, antimony, phosphorous, bismuth] is added as dopant in a tetravalent silicon crystal, the pentavalent atom will occupy one site of the silicon atom. Out of five valence electrons of the dopant, four electrons form covalent bonds with four valence electrons of the silicon atom and the fifth electron of the pentavalent atom is loosely attached to it as shown in Fig. 12.3(a). This fifth electron revolves around the positively charged pentavalent atom inside the dielectric medium of silicon [of dielectric constant ε_r] in the same way as the 1s electron around the hydrogen atom. Following the quantum mechanics of hydrogen atom, the radius of revolution of the fifth electron around the pentavalent atom will be

$$r_d = \frac{4\pi\varepsilon\hbar^2}{m_e^* e^2} = \frac{4\pi\varepsilon_r\varepsilon_0\hbar^2}{m_e^* e^2}$$

For silicon, $m_e^* = 0.2m_e$ and $\varepsilon_r = 11.7$. Putting these values into this equation, we get

$$r_d = \frac{4\pi \times 11.7 \times 8.85 \times 10^{-12} \left(1.054 \times 10^{-34}\right)^2}{0.2 \times 9.11 \times 10^{-31} \left(1.6 \times 10^{-19}\right)^2} \text{m} = 31\text{Å}$$

The ionization energy [the minimum amount of energy to be given so that the outermost electron becomes completely free] of the donor atom E_d following the quantum mechanics of hydrogen atom is given by

$$E_d = \frac{m_e^* e^4}{2\left(4\pi\varepsilon\hbar\right)^2} = \frac{m_e^* e^4}{2\left(4\pi\varepsilon_0\varepsilon_r\hbar\right)^2} \tag{12.12}$$

Putting the values $m_e^* = 0.2 m_e$ and $\varepsilon_r = 11.7$. into this equation, we get

$$E_d = \frac{(9 \times 10^9)^2 0.2 \times 9.11 \times 10^{-31}(1.6 \times 10^{-19})^4}{2(11.7 \times 1.054 \times 10^{-34})^2} J = 0.02 \text{ eV}$$

Thus, the ionization energy of the donor atom is very low as a result of which it is in an energy level very close to the conduction band E_C. Hence, a very small amount of energy is required to make this fifth electron completely free; thermal energy of the material is enough to shift this fifth electron to the conduction band. At room temperature, $kT = 0.0259$ eV. The energy level of the fifth electron is called the donor level and it lies just below the conduction band. Each pentavalent atom contributes one free electron to the silicon crystal and is called a donor atom or donor impurity. The silicon crystal (group IV) doped with pentavalent atoms results in n-type semiconductors. Here n stands for negative charge carriers, i.e., electrons. In an n-type semiconductor, electrons are majority carriers and holes are minority carriers. The formation of covalent bonds and the energy band picture of an n-type semiconductor is shown in Figs 12.3(a) and (b) respectively.

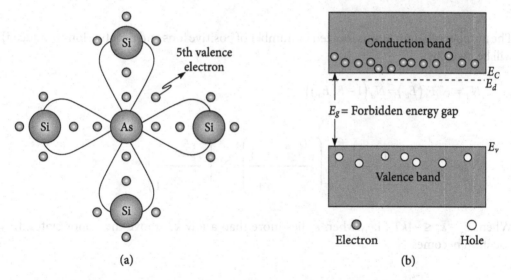

(a)   (b)

Figure 12.3 | (a) Formation of covalent bonds in an n-type semiconductor. The arsenic atom (impurity atom) is surrounded by four silicon atoms. The four silicon atoms will form covalent bonds with four of the five electrons of the arsenic atom. The fifth valence electron of the arsenic atom is loosely bound to its nucleus and so can be detached and made free by the expenditure of energy which is much less than that required for breaking a covalent bond. The fifth electron gets this small amount of energy easily from thermal agitation of the crystal. (b) Band diagram of an n-type semiconductor. Donor level E_d is very close to the conduction band E_C. Number of electrons in the conduction band is more than that of holes in the valence band. Electrons are majority carriers and holes are minority carriers in n-type semiconductors

Concentration of majority carriers

Let

N_d = density of donor atoms

N_d^+ = density of donor atoms ionized = concentration of electrons in the conduction band

The number of electrons [= number of quantum states] per unit volume in the conduction band of an intrinsic semiconductor is given by (from Eq. (12.3))

$$n = Ce^{\frac{\varepsilon_F}{kT}} \int_{E_C}^{\infty} (\varepsilon - E_C)^{\frac{1}{2}} e^{-\frac{\varepsilon}{kT}} d\varepsilon$$

This number must be exactly equal to the number of vacant or ionized donor states. The Fermi–Dirac distribution function for the vacant states [missing electrons] at the donor level will be

$$F_d(\varepsilon_d) = 1 - F(E_d)$$

The number of donor atoms ionized [= number of positive ions and negative ions produced] will be given by

$$N_d^+ = N_d F_d(E_d) = N_d\left(1 - F(E_d)\right)$$

or $\quad N_d^+ = N_d\left(1 - \dfrac{1}{e^{\frac{E_d-\varepsilon_F}{kT}}+1}\right) = N_d\left(\dfrac{e^{\frac{E_d-\varepsilon_F}{kT}}+1-1}{e^{\frac{E_d-\varepsilon_F}{kT}}+1}\right) = N_d\left(\dfrac{e^{\frac{E_d-\varepsilon_F}{kT}}}{e^{\frac{E_d-\varepsilon_F}{kT}}+1}\right)$

When $E_d - \varepsilon_F \leq -4kT$, i.e., when ε_F lies more than a few kT above the donor states, this equation becomes

$$N_d^+ \approx N_d e^{\frac{E_d-\varepsilon_F}{kT}}$$

The number of occupied conduction states and the number of empty donor states are equal. Therefore, we have

$$2\left(\frac{2\pi m_e^* kT}{h^2}\right)^{\frac{3}{2}} e^{\frac{\varepsilon_F - E_C}{kT}} = N_d e^{\frac{E_d-\varepsilon_F}{kT}} \tag{12.13}$$

or $\quad e^{\frac{\varepsilon_F - E_C - E_d + \varepsilon_F}{kT}} = \dfrac{N_d}{2\left(\dfrac{2\pi m_e^* kT}{h^2}\right)^{\frac{3}{2}}}$

Taking the natural logarithm of both sides, we get

$$\dfrac{2\varepsilon_F - E_c - E_d}{kT} = \ell n \dfrac{N_d}{2\left(\dfrac{2\pi m_e^* kT}{h^2}\right)^{\frac{3}{2}}}$$

or $\quad \varepsilon_F = \dfrac{1}{2}\left(E_c + E_d\right) + \dfrac{kT}{2}\ell n \dfrac{N_d}{2\left(\dfrac{2\pi m_e^* kT}{h^2}\right)^{\frac{3}{2}}}$ $\hspace{2cm}$ (12.14)

This equation shows that at $T = 0$ K

$$\varepsilon_F = \dfrac{1}{2}\left(E_C + E_d\right) \hspace{2cm} (12.15)$$

Thus, at absolute zero, the Fermi level lies exactly half way between donor levels and the bottom of the conduction band. As T increases, the Fermi level decreases as shown in Fig. 12.4.

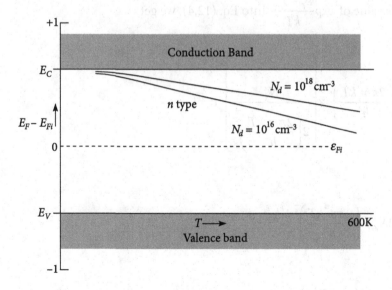

Figure 12.4 | Variation of Fermi energy with temperature for different concentrations in case of n-type semiconductors

From Eq. (12.14), we have

$$\frac{\varepsilon_F - E_C}{kT} = \frac{E_d - E_c}{2kT} + \frac{1}{2}\ln\frac{N_d}{2\left(\dfrac{2\pi m_e^* kT}{h^2}\right)^{\frac{3}{2}}}$$

or $$\exp\frac{\varepsilon_F - E_C}{kT} = \exp\frac{E_d - E_c}{2kT} \times \exp\left[\ln\left(\frac{N_d}{2\left(\dfrac{2\pi m_e^* kT}{h^2}\right)^{\frac{3}{2}}}\right)^{\frac{1}{2}}\right]$$

or $$\exp\frac{\varepsilon_F - E_C}{kT} = \exp\frac{E_d - E_c}{2kT} \times \left(\frac{N_d}{2\left(\dfrac{2\pi m_e^* kT}{h^2}\right)^{\frac{3}{2}}}\right)^{\frac{1}{2}}$$

Putting this value of $\exp\dfrac{\varepsilon_F - E_C}{kT}$ into Eq. (12.4), we get

$$n = 2\left(\frac{2\pi m_e^* kT}{h^2}\right)^{\frac{3}{2}} \times \left(\frac{N_d}{2\left(\dfrac{2\pi m_e^* kT}{h^2}\right)^{\frac{3}{2}}}\right)^{\frac{1}{2}} e^{\frac{E_d - E_C}{2kT}}$$

or $$n = \left(2N_d\right)^{\frac{1}{2}}\left(\frac{2\pi m_e^* kT}{h^2}\right)^{\frac{3}{4}} e^{\frac{E_d - E_C}{2kT}}$$

or $$n = \left(2N_d\right)^{\frac{1}{2}}\left(\frac{2\pi m_e^* kT}{h^2}\right)^{\frac{3}{4}} e^{-\frac{\Delta E}{2kT}} \tag{12.16}$$

In these equations, $\Delta E = E_C - E_d$ represents the ionization energy of the donor atoms. We can conclude the following from the equations:

i.   The density of electrons n in the conduction band is proportional to the square root of donor concentration N_d.

ii.   As temperature increases, the Fermi level falls below the donor level and approaches the centre of the forbidden gap which makes the substance an intrinsic semiconductor. The conductivity of an intrinsic semiconductor is less than that of an the extrinsic semiconductor. This places a limit on the operating temperature of a semiconducting device.

Example 12.5

All the 10^{19} donor atoms in an n-type semiconductor are ionized at 200 K. If the conductivity at 300 K is 26 $\Omega^{-1}m^{-1}$, calculate the conductivity at 200 K. Given: mobility at 200 K and 300 K are $\mu_n = 0.39$ m^{2}V^{-1}s^{-1}, $\mu_n = 0.41$ m^{2}V^{-1}s^{-1} respectively.

Solution

The data given are $N^+ = 10^{19}$, $\mu_{n200} = 0.39$ m^{2}V^{-1}s^{-1}, $\mu_{n300} = 0.41$ m^{2}V^{-1}s^{-1}, $\sigma_{300} = 0.26$ $\Omega^{-1}m^{-1}$. The conductivity of an n-type semiconductor is given by $\sigma_n = N^+ e\mu_n$. Hence, we have

$$\sigma_{n200} = N^+ e\mu_{n200} \text{ and } \sigma_{n300} = N^+ e\mu_{n300}$$

or $$\frac{\sigma_{n200}}{\sigma_{n300}} = \frac{N^+ e\mu_{n200}}{N^+ e\mu_{n300}} = \frac{\mu_{n200}}{\mu_{n300}} = \frac{0.39}{0.41}$$

or $$\sigma_{n200} = \frac{0.39}{0.41} \times \sigma_{n300} = \frac{0.39}{0.41} \times 26 = 24.73 \Omega m$$

Example 12.6

An n-type silicon wafer has been doped uniformly with antimony and the doped silicon has a donor concentration of 10^{14} cm^{-3}. Calculate the Fermi energy with respect to the Fermi energy in an intrinsic semiconductor at 300 K. The intrinsic carrier concentration of silicon at 300 K is $n_i = 1.45 \times 10^{10}$ cm^{-3}.

Solution

The Fermi energy with respect to the Fermi energy in an intrinsic semiconductor is given by $\varepsilon_{Fe} - \varepsilon_{Fi}$. We can derive an expression for $\varepsilon_{Fe} - \varepsilon_{Fi}$ in the following way. By using Eq. (12.4), the intrinsic carrier concentration n_i in the conduction band can be given by

$$N_i = 2\left(\frac{2\pi m_e^* kT}{h^2}\right)^{\frac{3}{2}} e^{\frac{\varepsilon_{Fi} - E_C}{kT}},$$

ε_{Fi} = Fermi energy in an intrinsic semiconductor

By using Eq. (12.13), donor concentration in the conduction band N_d^+ can be given by

$$N_d^+ = 2\left(\frac{2\pi m_e^* kT}{h^2}\right)^{\frac{3}{2}} e^{\frac{\varepsilon_F - E_C}{kT}},$$

ε_{Fe} = Fermi energy in an extrinsic semiconductor

Dividing the equation for N_d^+ by the equation for N_i, we have

$$\frac{N_d^+}{N_i} = \frac{e^{\frac{\varepsilon_{Fe} - E_C}{kT}}}{e^{\frac{\varepsilon_{Fi} - E_C}{kT}}} = e^{\frac{\varepsilon_{Fe} - E_C}{kT}} e^{\frac{E_C - \varepsilon_{Fi}}{kT}} = e^{\frac{\varepsilon_{Fe} - \varepsilon_{Fi}}{kT}}$$

or $\varepsilon_{Fe} - \varepsilon_{Fi} = kT \ell n\left(\dfrac{N_d^+}{N_1}\right)$

Putting the given data into this equation, we get

$$\varepsilon_{Fe} - \varepsilon_{Fi} = kT \ell n\left(\frac{N_d^+}{N_1}\right) = 0.229 \text{ eV}$$

Example 12.7

In an n-type semiconductor, the Fermi level lies 0.3 eV below the conduction band at 300 K. If the temperature is increased to 330 K, find the new position of the Fermi level assuming that the concentration of carriers does not change with temperature.

Solution

By using Eq. (12.4), the intrinsic carrier concentration n_i in the conduction band can be given by

$$n_i = 2\left(\frac{2\pi m_e^* kT}{h^2}\right)^{\frac{3}{2}} e^{\frac{\varepsilon_F - E_C}{kT}}.$$

For temperatures 300 K and 330 K, this equation becomes

$$n_{i300} = 2\left(\frac{2\pi m_e^* k \times 300}{h^2}\right)^{\frac{3}{2}} e^{\frac{\varepsilon_{F300} - E_C}{k \times 300}} \text{ and } n_{i330} = 2\left(\frac{2\pi m_e^* k \times 330}{h^2}\right)^{\frac{3}{2}} e^{\frac{\varepsilon_{F330} - E_C}{k \times 330}}. \text{ Since } n_{i300} = n_{i330},$$

we have

$$2\left(\frac{2\pi m_e^* k \times 300}{h^2}\right)^{\frac{3}{2}} e^{\frac{\varepsilon_{F300}-E_C}{k\times 300}} = 2\left(\frac{2\pi m_e^* k \times 330}{h^2}\right)^{\frac{3}{2}} e^{\frac{\varepsilon_{F330}-E_C}{k\times 330}}$$

or $\quad e^{\frac{2(\varepsilon_{F300}-E_C)}{3k\times 300}} = 1.1 \times e^{\frac{2(\varepsilon_{F330}-E_C)}{3k\times 330}}$

or $\quad e^{\frac{2(\varepsilon_{F330}-E_C)}{3k\times 330}} = \frac{300}{330} \times e^{\frac{2(\varepsilon_{F300}-E_C)}{3k\times 300}}$

or $\quad \varepsilon_{F330} - E_C = \frac{3k\times 330}{2}\ell n\frac{300}{330} + \frac{3k\times 330}{2}\frac{2\left(\varepsilon_{F300}-E_C\right)}{3k\times 300}$

or $\quad \varepsilon_{F330} - E_C = 495k \times \ell n\frac{300}{330} + \frac{33}{30} \times \left(\varepsilon_{F300}-E_E\right)$

According to the question, the Fermi level lies 0.3eV below the conduction band at 300 K, i.e.,

$E_C - \varepsilon_{F300} = 0.3$ eV. Putting $r_a = \dfrac{4\pi\varepsilon_r\varepsilon_0\hbar^2}{m_h^*e^2}$ into this equation, we have

$$\varepsilon_{F330} - E_C = 495k \times \ell n\frac{300}{330} + \frac{33}{30} \times 0.3$$

The first term in the RHS of this equation is negligibly small due to Boltzmann constant. Hence, we have

$$\varepsilon_{F330} - E_C = \frac{33}{30} \times 0.3 = 0.33 \text{ eV}$$

p-type semiconductors

The p-type semiconductor results when trivalent impurities are added to an intrinsic semiconductor. When a trivalent atom [boron, aluminum, gallium] is added as dopant in a tetravalent silicon crystal, the trivalent atom will occupy one site of the silicon atom. Three valence electrons of the trivalent atom form three covalent bonds with three out of the four valence electrons of the silicon atoms and the fourth electron goes to the trivalent atom, making it negatively charged to form a covalent bond as shown in Fig. 12.5. However, this fourth covalent bond is devoid of one electron. Devoid of one electron implies the presence of a hole. Thus, addition of trivalent impurity to silicon creates holes in the valence bands. This hole revolves around the negatively charged trivalent atom inside the dielectric medium of silicon [of dielectric constant ε_r] in the same way as the 1s electron around the hydrogen atom. Following the quantum mechanics of hydrogen atom, the radius of revolution of the hole around the trivalent atom will be

$$r_a = \frac{4\pi\varepsilon_r\varepsilon_0\hbar^2}{m_h^*e^2} \; [\, m_h \text{ is replaced by } m_h^*\, !]$$

The ionization energy of the accepter atom E_a following the quantum mechanics of hydrogen atom is given by

$$E_a = \frac{m_h^* e^4}{2(4\pi\varepsilon_0\varepsilon_r\hbar)^2}$$

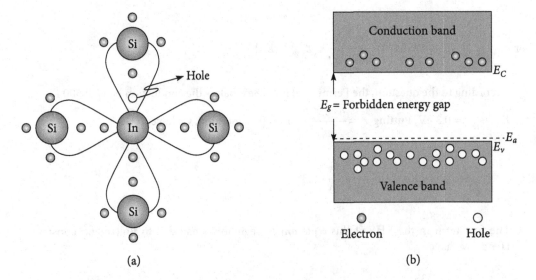

(a)   (b)

Figure 12.5 (a) Formation of covalent bonds in a p-type semiconductor. The indium atom (impurity atom) is surrounded by four silicon atoms. Three valence electrons of the indium atom form three covalent bonds with three out of four valence electrons of the silicon atoms and the fourth electron of the silicon atom goes to the indium atom making it negativelycharged to form a covalent bond. However, this fourth covalent bond is devoid of one electron. Devoid of one electron implies the presence of a hole. This hole is almost free and so can be detached and made free by the expenditure of energy which is much less han that required for breaking a covalent bond. The hole gets this small amount of energy easily from thermal agitation of the crystal (b) Band diagram of a p-type semiconductor. Accepter level E_a is very close to the valence band E_v. The number of holes in the valence band is more than that of electrons in the conduction band. Holes are majority carriers and electrons are minority carriers in p-type semiconductors

Thus, ionization energy of the accepter atom is very low as a result of which it is in an energy level very close to the valence band. Hence, a very small amount of energy is required to make this hole completely free; the thermal energy of the material is enough to shift this hole to the valence band. The energy level of the hole is called the accepter level and lies very close to the valence band E_v. Each trivalent atom accepts one electron from the silicon crystal and is called an accepter atom or accepter impurity. The silicon crystal (group IV) doped with trivalent atoms results in a p-type semiconductor. The p stands for positive charge carriers, i.e., holes. In a p-type semiconductor, holes are majority carriers and electrons are minority carriers. The formation of covalent bonds and the energy band picture of a p-type semiconductor is shown in Figs 12.5(a) and (b) respectively.

Concentration of majority carriers

Let

N_a = density of acceptor atoms

N_a^+ = density of accepter atoms ionized = concentration of holes in valence band

The number of electrons per unit volume in the conduction band of the intrinsic semiconductor is given by from Eq. (12.3)

$$n = Ce^{\frac{\varepsilon_F}{kT}} \int_{E_C}^{\infty} (\varepsilon - E_C)^{\frac{1}{2}} e^{-\frac{\varepsilon}{kT}} d\varepsilon$$

This number must be exactly equal to the number of ionized accepter states. The number of accepter atoms ionized will be given by

$$N_a^- = N_a F(E_a)$$

or $\quad N_a^- = N_a \dfrac{1}{e^{\frac{E_a - \varepsilon_F}{kT}} + 1}$

When $E_d - \varepsilon_F \le -4kT$, i.e., when ε lies more than a few kT below the accepter states, this equation becomes

$$N_a^- \approx N_a e^{-\frac{E_a - \varepsilon_F}{kT}}$$

Equating the number of occupied valence states to the number of accepter states, we have

$$2\left(\frac{2\pi m_h^* kT}{h^2}\right)^{\frac{3}{2}} e^{\frac{E_v - \varepsilon_F}{kT}} = N_a e^{-\frac{E_a - \varepsilon_F}{kT}} \tag{12.17}$$

or $\quad e^{\frac{E_a + E_v - 2\varepsilon_F}{kT}} = \dfrac{N_a}{2\left(\dfrac{2\pi m_h^* kT}{h^2}\right)^{\frac{3}{2}}}$

Taking the natural logarithm of both sides, we get

$$\frac{E_a + E_v - 2\varepsilon_F}{kT} = \ell n \frac{N_a}{2\left(\dfrac{2\pi m_h^* kT}{h^2}\right)^{\frac{3}{2}}}$$

or $\quad \varepsilon_F = \dfrac{1}{2}\left(E_a + E_v\right) - \dfrac{kT}{2}\ell n \dfrac{N_a}{2\left(\dfrac{2\pi m_e^* kT}{h^2}\right)^{\frac{3}{2}}}$

or $\quad \varepsilon_F = \dfrac{1}{2}\left(E_a + E_v\right) + \dfrac{kT}{2}\ell n \dfrac{2\left(\dfrac{2\pi m_h^* kT}{h^2}\right)^{\frac{3}{2}}}{N_a}$ (12.18)

This equation shows that at $T = 0$ K,

$$\varepsilon_F = \frac{1}{2}\left(E_a + E_v\right)$$ (12.19)

Thus, at absolute zero, the Fermi level lies exactly half way between accepter levels and top of the valence band. As T increases, the Fermi level increases as shown in Fig. 12.6.

From Eq. (12.14), we have

$$\frac{\varepsilon_F - E_v}{kT} = \frac{E_a - E_v}{2kT} - \frac{1}{2}\ell n \frac{N_a}{2\left(\dfrac{2\pi m_h^* kT}{h^2}\right)^{\frac{3}{2}}}$$

or $\quad \exp\dfrac{\varepsilon_F - E_v}{kT} = \exp\dfrac{E_a - E_v}{2kT} \times \exp\left[\ell n \left(\dfrac{N_a}{2\left(\dfrac{2\pi m_h^* kT}{h^2}\right)^{\frac{3}{2}}}\right)^{-\frac{1}{2}}\right]$

or $\quad \exp\dfrac{\varepsilon_F - E_v}{kT} = \exp\dfrac{E_a - E_v}{2kT} \times \left(\dfrac{N_a}{2\left(\dfrac{2\pi m_h^* kT}{h^2}\right)^{\frac{3}{2}}} \right)^{-\frac{1}{2}}$

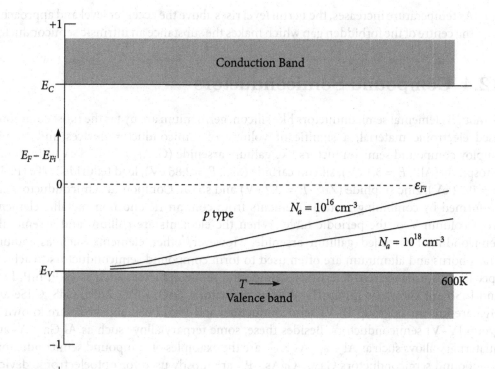

Figure 12.6 | Variation of Fermi energy with temperature for different concentrations in case of p-type semiconductors

Putting this value of $\exp\dfrac{\varepsilon_F - E_v}{kT}$ into the Eq. (12.5), we get

$$p = 2\left(\frac{2\pi m_h^* kT}{h^2}\right)^{\frac{3}{2}} \times \left(\frac{N_a}{2\left(\dfrac{2\pi m_h^* kT}{h^2}\right)^{\frac{3}{2}}} \right)^{-\frac{1}{2}} e^{\frac{E_a - E_v}{2kT}}$$

or $\quad p = \left(2N_a\right)^{\frac{1}{2}} \left(\dfrac{2\pi m_h^* kT}{h^2}\right)^{\frac{3}{4}} e^{-\frac{\Delta E}{2kT}}$ $\qquad\qquad\qquad\qquad$ (12.20)

In these equations, $\Delta E = E_v - E_a$ represents the ionization energy of the accepter atoms. We can conclude the following from these equations:

i.　The density of holes p in the valence band is proportional to the square root of accepter concentration N_a

ii.　As temperature increases, the Fermi level rises above the accepter level and approaches the centre of the forbidden gap which makes the substance an intrinsic semiconductor.

12.4　Compound Semiconductors

Although elemental semiconductors like silicon, germanium are by far the most commonly used electronic material, a significant volume of semiconductor devices and circuits employ compound semiconductors like gallium arsenide (GaAs, E_g = 1.38 eV), aluminum phosphide (AlP, E_g = 3.1 eV), silicon carbide (SiC, E_g = 2.86 eV), lead telluride PbTe (PbTe, E_g = 0.32 eV), zinc sulphide (ZnS, E_g = 3.7 eV) and so on. Compound semiconductors can be formed by combining metallic elements from column III and non-metallic elements from column V of the periodic table. When the elements are gallium and arsenic, the semiconductor is called gallium arsenide. However, other elements such as indium, phosphorus and aluminum are often used to form compound semiconductors to achieve specific performance characteristics. Compound semiconductors such as GaP, InP, InAs and InSb are known as group III–V semiconductors; ZnO, ZnS , ZnSe, CdS, CdSe and HgS are known as group II–VI semiconductors; and PbS, PbSe and PbTe are known as group IV–VI semiconductors. Besides these, some ternary alloys such as $Al_xGa_{1-x}As$ and quaternary alloys such as $Al_xGa_{1-x}As_ySb_{1-y}$ are the examples of compound semiconductors. Compound semiconductors GaAs, $GaAs_{1-x}P_x$ are mostly used for optoelectronic devices such as LEDs and the compound semiconductor $AlGa_{1-x}As$ is used in modulation doped field effect transistors. Compound semiconductors are, in general, ionic in character.

For electronic applications, compound semiconductors offer the basic advantage of higher electron mobility, which translates into higher operating speeds. In addition, devices made with compound semiconductors provide lower voltage operation for specific functions, radiation hardness (especially important for satellites and space vehicles) and semi-insulating substrates (avoiding the presence of parasitic capacitance in switching devices).

Compound semiconductors are more difficult to handle than silicon and a compound semiconductor wafer or substrate is usually less than half the size of a silicon wafer. In addition, a gallium arsenide wafer entering the processing facility can be expected to cost

10 to 20 times as much as a silicon wafer, although that cost difference narrows somewhat after fabrication, packaging and testing. Nevertheless, there is one major characteristic of compound semiconductors with which silicon cannot compete: compound semiconductors can be tailored to generate or detect photons of a specific wavelength. For example, an indium gallium arsenide phosphide (InGaAsP) laser can generate radiation at 1.55 micrometres to carry digitally coded information streams. This means that compound semiconductors can carry out both electronic and photonic functions in the same integrated circuit.

12.4.1 Semiconducting properties

Silicon, carbon, germanium, and a few other elements form covalently bonded solids. In these elements, there are four electrons in the outer sp shell, which is half filled (The sp shell is a hybrid formed from one s and one p sub-shell.). In the covalent bond, an atom shares one valence (outer shell) electron with each of its four nearest neighbour atoms. The bonds are highly directional and prefer a tetrahedral arrangement.

Besides the elemental semiconductors, such as silicon and germanium, some binary crystals are covalently bonded. Gallium has three electrons in the outer shell, while arsenic lacks three. Gallium arsenide (GaAs) could be formed as an insulator by transferring three electrons from gallium to arsenic; however, this does not occur. Instead, the bonding is more covalent, and gallium arsenide is a covalent semiconductor. The outer shells of the gallium atoms contribute three electrons, and those of the arsenic atoms contribute five, providing the eight electrons needed for four covalent bonds. The centres of the bonds are not at the midpoint between the ions but are shifted slightly towards arsenic. Such bonding is typical of the III–V semiconductors i.e., those consisting of one element from the third column of the periodic table and one from the fifth column. Elements from the third column (boron, aluminum, gallium and indium) contribute three electrons, while the fifth column elements (nitrogen, phosphorus, arsenic and antimony) contribute five electrons. All III–V semiconductors are covalently bonded and typically have the zinc blende structure with four neighbours per atom. Most common semiconductors favour this arrangement.

The factor that determines whether a binary crystal will act as an insulator or a semiconductor is the valence of its constituent atoms. Ions that donate or accept one or two valence electrons form insulators. Those that have three to five valence electrons tend to have covalent bonds and form semiconductors. There are exceptions to these rules, however, as is the case with the IV–VI semiconductors such as lead sulphide. Heavier elements from the fourth column of the periodic table (germanium, tin and lead) combine with the chalcogenides from the sixth row to form good binary semiconductors such as germanium telluride (GeTe) or tin sulphide (SnS). They have the sodium chloride structure, where each atom has six neighbours. Although not tetrahedrally bonded, they are good semiconductors.

Questions

12.1   Based on the ability of solids to carry electric currents, what are the different types of solids?

12.2   What is an insulator? Based on the band theory of solids, how will you define an insulator?

12.3   What is a conductor? Based on the band theory of solids, how will you define a conductor?

12.4   What is a semiconductor? Based on the band theory of solids, how will you define a semiconductor?

12.5   What are the similarities and dissimilarities between the energy level band diagram of silicon and carbon?

12.6   What is an intrinsic semiconductor?

12.7   Why is pure silicon an insulator at 0 K, whereas at room temperature it is a semiconductor?

12.8   What is an extrinsic semiconductor?

12.9   What is a donor atom?

12.10 What is the donor level?

12.11 What is an accepter atom?

12.12 What is the accepter level?

12.13 Give the band diagram of an n-type semiconductor and a p-type semiconductor.

12.14 What are the properties of a semiconductor?

12.15 Explain how an intrinsic semiconductor carries currents.

12.16 What do you mean by concentration of holes in the valence band? Write an expression for it in terms of the Fermi–Dirac distribution function.

12.17 Derive an expression for concentration of holes in a valence band of an intrinsic semiconductor.

12.18 Derive an expression for concentration of electrons in a conduction band of an intrinsic semiconductor

12.19 Derive an expression for the intrinsic carrier concentration in intrinsic semiconductors.

12.20 Derive an expression for the Fermi energy level in intrinsic semiconductors in terms of the conduction band and the valence band.

12.21 Derive an expression for intrinsic electrical conductivity in case of intrinsic semiconductors.

12.22 What do you mean by concentration of electrons in the conduction band? Write an expression for it in terms of the Fermi–Dirac distribution function for an intrinsic semiconductor.

12.23 Show that the Fermi level in an intrinsic semiconductor is a function of temperature.

12.24 Derive an expression for the Fermi level in an intrinsic semiconductor.

12.25 Derive an expression for the electrical conductivity of an intrinsic semiconductor.

12.26 Show that the electrical conductivity of intrinsic semiconductors is temperature dependent.

12.27 Show that the electrical conductivity of intrinsic semiconductors varies exponentially with temperature.

12.28 What are different types of impurities?

12.29 Describe the energy band structure of an *n*-type semiconductor.

12.30 Describe the formation of covalent bonds in *n*-type semiconductors.

12.31 Show that the density of electrons in the conduction band is proportional to the square root of donor concentration

12.32 Derive an expression for the concentration of majority carriers in *n*-type semiconductors.

12.33 Explain how the concentration of majority carriers in *n*-type semiconductors varies with temperature.

12.34 Derive an expression to show the temperature variation of the Fermi level in *n*-type semiconductors.

12.35 Calculate the position of the Fermi level in an *n*-type semiconductor at very low temperatures.

12.36 Explain how the addition of trivalent impurity to silicon creates holes in the valence bands.

12.37 Explain how the number of electrons in the conduction band is more than that of holes in the valence band in an *n*-type semiconductor at room temperature.

12.38 Describe the energy band structure of a *p*-type semiconductor.

12.39 Describe the formation of covalent bonds in *p*-type semiconductors.

12.40 Explain how the number of holes in the valence band is more than that of electrons in the conduction band.

12.41 Show that the density of holes in the valence band is proportional to the square root of accepter concentration.

12.42 Derive an expression for the concentration of majority carriers in a *p*-type semiconductor

12.43 Explain how the concentration of majority carriers in a *p*-type semiconductor varies with temperature.

12.44 Derive an expression to show the temperature variation of the Fermi level in a *p*-type semiconductor.

12.45 Calculate the position of the Fermi level in a *p*-type semiconductor at very low temperatures.

Problems

12.1 A silicon crystal is doped with arsenic with a doping concentration of 1 in 10^7. Calculate the concentration of donor atoms if the intrinsic atom concentration for silicon is 5×10^{28} m^{-3}.   [Ans 5×10^{21} m^{-3}]

12.2   Calculate the forbidden energy gap E_g of a semiconductor at 250 K from the following data: intrinsic atom concentration for the semiconductor is 0.6×10^{19} met^{-3}, $m_e^* = 0.55m_e$ and $m_h^* = 0.37m_p$.   [Ans   0.59 eV]

12.3   Calculate the current produced in a germanium crystal having cross-sectional area 2 cm^2 and length 0.4 mm when a potential difference of 1.5 volt is applied. Given: concentration of free electrons in germanium crystal = 2×10^{19} m^{-3}, $\mu_e = 0.36$ m^{3}V^{-1}s^{-1}, $\mu_p = 0.17$ m^{3}V^{-1}s^{-1}.   [Ans   1.27 Amp]

12.4   Calculate the current produced in a germanium crystal having cross-sectional area 1 cm^2 and length 0.3 mm when a potential difference of 2 volt is applied. Given: concentration of free electrons in germanium crystal = 2×10^{19} m^{-3}, $\mu_e = 0.36$ m^{3}V^{-1}s^{-1}, $\mu_p = 0.17$ m^{3}V^{-1}s^{-1}.   [Ans   1.13 A]

12.5   All the 10^{20} donor atoms in an n-type semiconductor are ionized at 200 K. If the conductivity at 300 K is 24 Ω^{-1}m^{-1}, calculate the conductivity at 200 K. Given: mobility at 250 K and 300 K are $\mu_e = 0.37$ m^{2}V^{-1}s^{-1}, $\mu_e = 0.40$ m^{2}V^{-1}s^{-1} respectively.   [Ans   22.2 Ωm]

12.6   All the 10^{20} donor atoms in an n-type semiconductor are ionized at 250 K. If the conductivity at 300 K is 24 Ω^{-1}m^{-1}, calculate the conductivity at 200 K. Given: mobility at 250 K and 300 K are $\mu_e = 0.39$ m^{2}V^{-1}s^{-1}, $\mu_e = 0.43$ m^{2}V^{-1}s^{-1} respectively.   [Ans   21.77 Ω^{-1}m^{-1}]

12.7   Calculate the intrinsic carrier concentration in an intrinsic semiconductor at 250 K, using the following data: $\mu_e = 0.4$ m^{2}V^{-1}s^{-1}, $\mu_h = 0.2$ m^{2}V^{-1}s^{-1}, $E_g = 0.7$ eV, $m_e^* = 0.55m_e$, $m_h^* = 0.37m_p$.   [Ans 1.028×10^{13} m^{-3}]

12.8   In an n-type semiconductor, the Fermi level lies 0.2 eV below the conduction band at 300 K. If the temperature is increased to 320 K, find the new position of the Fermi level assuming that the concentration of carriers does not change with temperature.   [Ans   0.213 eV]

12.9   An n-type silicon wafer has been doped uniformly with antimony and the doped silicon has a donor concentration of 10^{16} cm^{-3}. The intrinsic carrier concentration of silicon at 300 K is $n_i = 1.45 \times 10^{10}$ cm^{-3}. Calculate the Fermi energy with respect to the Fermi energy in an intrinsic semiconductor at 300 K.   [Ans $\varepsilon_{Fd} - \varepsilon_{Fi} = 0.348$]

12.10   In a semiconductor, the electron and hole mobilities are 0.85 and 0.04 m^{2}V^{-1}s^{-1} respectively and the effective masses of electron and hole respectively are 0.068 and 0.50 times the electron mass. The energy band gap is 1.43 eV. Calculate the intrinsic carrier density and conductivity at 300 K.   [Ans 2×10^6 cm^{-3}, 2.85×10^{-7} Ω^{-1}m^{-1}]

Multiple Choice Questions

1. Which of the following property of solids has maximum range of variation?

    (i)   Electrical

    (ii) Mechanical

    (iii) Optical

    (iv) Chemical

2. If ρ_c, ρ_i, and ρ_s are the resistivity of conductors, insulators and semiconductors respectively, which of the following relation is correct?

    (i) $\rho_c < \rho_i < \rho_s$

    (ii) $\rho_s < \rho_i < \rho_c$

    (iii) $\rho_c < \rho_s < \rho_i$

    (iv) $\rho_i < \rho_s < \rho_c$

3. Which of the following statements are true?

    (i)   A solid is an insulator if it has $E_g \approx 1$ eV

    (ii) A solid is a semiconductor if it has $E_g > 3$ eV

    (iii) A solid is a metal if its valence band is partially filled

    (iv) A solid is a metal if its valence band and conduction band overlap

4. In a material, the conduction band and valence band overlaps. The materials is

    (i)   Conductor

    (ii) Insulator

    (iii) Semiconductor

    (iv) The case is not possible.

5. In a material, the gap between the conduction band and the valence band is ≈ 1 eV. The material is

    (i)   Conductor

    (ii) Insulator

    (iii) Semiconductor

    (iv) The case is not possible

6. In a material, the gap between the conduction band and the valence band is ≈ 5 eV. The material is

    (i)   Conductor

    (ii) Insulator

    (iii) Semiconductor

    (iv) The case is not possible.

7. The temperature coefficient of resistance of a pure semiconductor is

    (i)   Negative

    (ii) Positive

    (iii) 0

    (iv) Depends on the applied voltage.

8. What is the relation between the number of holes in the valence band N_h and the number of thermally generated electrons in the conduction band N_e

    (i) $N_h < N_e$

    (ii) $N_h > N_e$

    (iii) $N_h = N_e$

    (iv) none of the above

9. When an electric field is applied to an intrinsic semiconductor
    (i) Both electrons and holes move in the same direction as the electric field
    (ii) Both electrons and holes move in the opposite direction as the electric field
    (iii) Electrons move in the direction of the field and holes move in the opposite direction
    (iv) Electrons move in the opposite direction of the field and holes move in the same direction as the field.

10. Which of the following is the Fermi–Dirac distribution function?

    (i) $F(\varepsilon) = \dfrac{1}{e^{\frac{\varepsilon - \varepsilon_F}{kT}} - 1}$

    (ii) $F(\varepsilon) = \dfrac{1}{e^{\frac{\varepsilon - \varepsilon_F}{kT}} + 1}$

    (iii) $F(\varepsilon) = \dfrac{1}{e^{\frac{\varepsilon - \varepsilon_F}{kT}}}$

    (iv) $F(\varepsilon) = \dfrac{1}{e^{\frac{\varepsilon + \varepsilon_F}{kT}} + 1}$

11. Which distribution function does electrons obey?
    (i) Maxwell–Boltzmann
    (ii) Bose–Einstein
    (iii) Fermi–Dirac
    (iv) All the distribution functions

12. Which of the following is correct for intrinsic semiconductors? At room temperature,
    (i) Concentration of electrons in the conduction band is more than the concentration of holes in the valence band
    (ii) Concentration of electrons in the conduction band is less than the concentration of holes in the valence band
    (iii) Concentration of electrons in the conduction band is equal to the concentration of holes in the valence band
    (iv) Electrons do not exist in the conduction band and holes do not exist in the valence band

13. Which of the following is correct for intrinsic semiconductors? At room temperature,
    (i) Electrons cannot exist in the conduction band and holes cannot exist in the valence band
    (ii) Electrons exist in the conduction band and holes exist in the valence band
    (iii) Electrons can only exist in the valence band and holes can only exist in the conduction band

14. In intrinsic semiconductors, the Fermi level is independent of temperature
    (i) True
    (ii) False

15. The extrinsic semiconductor is an insulator at 0K.
    (i) True
    (ii) False

16. When pentavalent impurities are added to silicon, its conductivity
    (i) Does not change
    (ii) Increases
    (iii) Decreases
    (iv) May increase or may decrease

17. In an *n*-type semiconductor,
    (i)   Holes are majority carriers   (ii) Holes are minority carriers
    (iii) Electrons are majority carriers (iv) Electrons are minority carriers

18. The conductivity of an intrinsic semiconductor is
    (i)   Less than that of extrinsic semiconductors
    (ii) More than that of extrinsic semiconductors
    (iii) Equal to that of the extrinsic semiconductors
    (iv) No such relation exists

19. When trivalent impurities are added to silicon, its conductivity
    (i)   Does not chang   (ii) Increases
    (iii) Decreases   (iv) May increase or may decrease

20. In *p*-type semiconductors
    (i)   Holes are majority carriers
    (ii) Holes are minority carriers
    (iii) Electrons are majority carriers
    (iv) Electrons are minority carriers

21. In an extrinsic semiconductor, donor atoms
    (i)   Add holes to the valence band
    (ii) Add electrons to the valence band
    (iii) Add holes to the conduction band
    (iv) Add electrons to the conduction band

22. In *p*-type semiconductors the Fermi level lies midway between the accepter level and the top level of the valence band.
    (i)   True
    (ii) False

23. When an electron jumps from the valence shell to the conduction band, it leaves a gap. What is this gap called?
    (i)   Electron–hole pair gap   (ii) Recombination gap
    (iii) Hole   (iv) Energy gap

24. The conduction band is
    (i)   Same as the forbidden energy gap
    (ii) Generally located on the top of the crystal
    (iii) Generally located on the bottom of the crystal
    (iv) A range of energies corresponding to the energies of free electron

25.   The type of atomic bonding most common in semiconductors is
   (i)   Metallic   (ii) Covalent
   (iii) Ionic   (iv) Chemical

Answers

1 (i)	2 (iii)	3 (iii & iv)	4 (i)	5 (iii)	6 (ii)	7 (i)	8 (iii)
9 (iv)	10 (ii)	11 (iii)	12 (iii)	13 (ii)	14 (ii)	15 (ii)	16 (ii)
17 (iii)	18 (i)	19 (ii)	20 (i)	21 (iv)	22 (i)	23 (iii)	24 (iv)
25 (ii)							

13 Nano Structures and Thin Films

13.1 Introduction

The concept of nano structures and nano technology dates back to the history of the Nobel Laureate Richard Feynman's famous lecture in 1959 where he speculated the possibility of manoeuvring things atom by atom. He proudly said "There is plenty of room at the bottom". By this sentence, he strongly pointed out that there is a lot of scope and application of materials at very small scale, i.e., at atomic or molecular level. Why cannot we write the entire 24 volumes of the Encyclopaedia Britannica on the head of a pin? One may wonder what it is so special at the nano scale? The answer to these questions is against the very notion that specific physical properties of a given material are the characteristic of the material itself and are irrespective of their size. This is no longer valid at the nano scale. There will be a drastic change in properties which are counter-intuitive and unbelievable. Yellow metal gold is no more yellow in any size or dimension. It may be purple red and orange. Cds under colloidal solution may yield a rainbow spectrum of colour. Ceramic oxides are not always brittle. Nanoceramics are ductile and malleable. What sounded as scientific fantasy or myth at that time, today has turned into reality.

Nano science and technology cut across disciplines without boundary. For physicists and chemists, nano means in the range of dimension of a few atoms and molecules. For biologists, the visualization is to the dimension of the size of a DNA or to the scale of a cell. Scientists have now devised techniques to prepare tiny nano-particles which have peculiar properties. Nano technology is now a portal opening onto a new world. This chapter will provide the readers a brief look at: nano scale and its visualization, i.e., what nano is; nano science and nanotechnology, i.e., why one should care about nano; surface to volume ratio and quantum confinement, i.e., what makes nano so important; nano cluster and nano fabrication, i.e., how we prepare nano.

13.2 Nano Scale and its Visualization

The word "nano" has its origin from the Greek word meaning dwarf or something small. But how small is this small? One micrometer, the so-called micron (1 μm = 10^{-6}m) is one millionth of a metre and one nanometer (nm) is one billionth of a unit meter (1 nm = 10^{-9} m = 1/1,000,000,000 of a meter). One micron is 1000 nanometer (nm). Our mind is preoccupied with the average size of bodies (cm, meter, feet, etc) used in our daily life. Visualization of the nanometer (nm) needs some examples as references.

One nanometer is approximately 50,000 times smaller than the diameter of an average human hair. In the unit of size of the atom, nano materials of approximately 100 nanometers size contain no fewer than tens of thousands atom, whereas bulk materials that are of micron to millimeter size contain several billion atoms.

Other examples that will help to get a sense of the nano scale are the following. The smallest thing one can see with the naked eye is 10,000 nm. 10 water molecules in a line will make one nm. The width of a DNA molecule is one nm. One red blood cell is 10 μm. 3 nm diameter clusters contain approximately 900 atoms. Its size is million times smaller than the tip of the needle.

A physical three-dimensional body with at least one dimension roughly between 1 and 100 nanometres will fall into the category of nano science.

One angstrom =1 Å= 0.1 nanometer (nm) = 1×10^{-10} m

100 angstrom= 100 Å = 10 nanometer (nm) or 1 nm = 10 Å.

13.3 Nano Science and Nanotechnology

Nano science is the study of the phenomenon and manipulation of materials at nano scale where properties differ significantly from those at larger scale. A ten centimeter piece of any material, say, copper would have similar properties as a one centimeter piece, or even a one millimeter piece. However, at the other extreme, one or two isolated atoms of copper would show quite different properties. Some bulk properties of matter lose their identity while others show a drastic change in the context of atoms and molecules. The properties of nano particles are therefore strongly dependent on the number of atoms or size of the cluster.

Nanotechnology deals with the design, characterization, production and application of structures, devices by controlling the shape and size of the material at nanometer scale. Size effects constitute a peculiar and fascinating aspect of nanomaterials. Nanoscience and nanotechnology explore and benefit from quantum phenomenology in the ultimate limit of miniaturization.

At length scales comparable to atoms and molecules, quantum effects strongly modify properties of matter like "colour", reactivity, magnetic or electrical and other physical properties. The drastic change in properties make nano a thrust area or driving force for developing high performance, low cost, miniaturized solid-state devices and the study of size effect on physical properties important in materials science. Nature also mimics nano science, for example, the minute structures that make up the wings of a butterfly, a peacock feather, spider thread and a lotus leaf.

To be more specific, on the properties front, we have the following amazing properties:

i.   Six nm copper is 5 times as hard as the bulk copper

ii. Nanoceramics are ductile and malleable whereas bulk ceramics are rigid.

iii. Nano size titanium sinters at lower temperature (600°C), whereas bulk titanium sinters at 1400°C

iv. Rutiles becomes ductile in the nano form

v.   Particles of size 10–30 nm cannot scatter visible light, hence all particles of this size, no matter what they are made of, are transparent

vi. Different sized nano CdSe exhibit different colours – a boon for band gap engineers; this is due to the quantum size effect.

vii. Electrical insulators at nano size conduct due to tunnelling current.

viii. Three nm CdS melts at 700 K compared to bulk CdS at 1678 K.

Other wonders of nano science include: (i) scratch resistance optical glass, (ii) dirt repellent trains, (iii) ceramic motor car engines and tools, (iv) cancer treatment due to drug delivery and (v) corrosion protecting surface.

The widespread use of nano technology is due to (i) small scale miniaturization (ii) faster and cheaper manufacture, (iii) higher reliability, (iv) lower cost nano electronics and many others.

13.4 Surface to Volume Ratio

It is well known that finer particles are more reactive than larger particles. Most common examples are: fine sugar dissolves faster, fine coal particles burn stronger than bigger size. This may be the reason why many Ayurvedic medicines are prepared after extensive grinding and compounding. Homeopathic medicines are taken in the form of micro fine globules. The high reactivity results from the availability of a greater surface area by reducing the size either by grinding, pounding, milling and so on. In estimation, it has been proved that if the size of the particle is reduced to half, the area increases two times. As a result, the number of particles on the surface will be eight times larger. Few hours of milling increases the surface area from around 0.5 m²/g to 20 m²/gm. Another approximate calculation showing more number of surface atoms may be as follows.

The volume density and surface density of a typical material are 10^{23} atom/cm³ and 10^{15} atoms/cm² respectively. Consider a cube of 1 nm (10^{-7} cm).

The total number of atoms will be $10^{23} \times (10^{-7})^3 = 100$. The total number of surface atoms will be $10^{15} \times (10^{-7})^2 = 60$. This shows that 60% of the atoms will sit on the surface. Since the surface atom possesses more energy than the bulk atoms, they are chemically more active.

Particles on surfaces are very sensitive to the environment and physical chemical properties get modified. The new surface is more reactive and acquires special features.

Minerals get activated on grinding. Grinding and pounding of solids increase dissolution rates not only because of increase in surface area; it also supports structural change, defects formation and accumulation of mechanical energy that is released during dissolution. As the size becomes smaller, the surface to volume ratio increases and this increase begins to go up rapidly as one reaches a smaller size range.

It is therefore clear that nano materials possess size dependent properties. This can be categorized as under.

i.   Thermal properties (melting point, sintering temperature, etc)

ii.   Chemical properties (reactions, catalytic properties, activation energy and dissolution)

iii.   Mechanical properties (strength, adhesion, hardness, scratch resistance)

iv.   Optical properties (absorption, scattering, band gap)

v.   Electrical properties (transport, tunnelling, resistance)

vi.   Magnetic (nano magnetism, superparamagnetism)

The list is not exhaustive. Other properties make them useful as sensors, transducers, biochemicals and the like.

13.5 Quantum Confinement

The drastic change in surface to volume ratio at nano range results in change of the atomic and electronic structure as a result, electrical, optical, magnetic, and transport properties changes. If the size is in the nano range only in one dimension, it leads to a thin film, i.e., 2-D structures and 1-D confinement; for example, graphene. The basic idea behind quantum confinement is keeping an electron trapped in a small area. It is due to the fact that the energy available to the electron is less than the potential barrier it has to cross. This is similar to a particle in a well in quantum mechanics, known as quantum well. The second arrangement is a nano rod or nano tube (with 1-D structures and 2-D confinement), popularly known as a quantum wire. An example of a quantum wire is the carbon nano tube (CNT). Three-dimensional graphite has a 3-D structure and 0D confinement, whereas fullerene has a 0-D structure and 3-D confinement.

The most important finding of quantum confinement is the quantum dot. The quantum dot is treated as an artificial atom or superatom having three-dimensional confinements and zero dimensional structures. Like a particle in a box, the allowed energy levels can be varied at the will and pleasure of the experimenter by changing the size of the quantum dot cluster. This in turn changes the material properties such as electrical and non-linear optical properties and so on. On excitation, a smaller dot emits light of higher energy and intensity or smaller wavelength. Hence, the emitted light is blue. As the dot enlarges in size, the emitted light is of lower energy or larger wavelength, moving towards red. This can be verified experimentally by taking a colloidal suspension of CdSe core with ZnS shell quantum dots of varying sizes. Larger and smaller quantum dots exhibit red and blue colours, respectively. Cadmium sulphide, which is orange in colour in its bulk form, for example, keeps on changing colour through different shades of yellow and becomes white as its cluster size changes. Optical property depends strongly upon the particle shape and size. Bulk gold (Au) is yellowish in reflected light, but a thin gold film appears blue in transmission. Further, this blue colour changes to purple, red and finally orange as the particle size is reduced down to 3 nm.

13.6 Nano Cluster

Nano materials can also exist in the form of clusters, as independent, separate particles. Such clusters, when brought together, coalesce and form bigger particles. A cluster is a special type of nano material. It is an independent particle whose properties are determined by the number of atoms in it. Clusters are also variously called nano particles, quantum dots, Q-particles, coulomb islands, artificial atoms, and so on. All these refer to an aggregate of atoms ranging from a few atoms to a few thousand atoms that are bound together. The diameter of a cluster usually ranges from 1 to 100 nanometers. The clusters are too large to be considered as molecules and too small to be treated as bulk material. That is why they form a new and different class of their own. The colour of tinted glasses is due to clusters of transition metal ions. A major breakthrough was made with the discovery of fullerenes.

Interesting things happen while the clusters grow. The atoms re-arrange themselves into a larger cluster, thus forming clusters of different sizes, shapes, strength, internal bonding and electronic properties. On the basis of size, clusters may be of large (one nm to 50 nm), medium where there is systematic variation of properties with constituent atoms and small, ranging from a few atoms to a few hundred atoms. A cluster may be homogeneous consisting of atoms of the same element and heterogeneous with atoms of different elements.

A very interesting cluster is a stable cluster of carbon consisting of sixty atoms having the shape of a football called a fullerene. In such a fullerene, atoms are arranged on 12 pentagons which have one side common with a hexagon. The next stable structures have 72 atoms, 78 atoms, and so on.

It is now known that only clusters with a certain number of atoms are stable. These numbers are known as magic numbers. In close similarity with the atomic structure, i.e., atoms with atomic number 2, 10, 18, 36, that is helium, neon, argon, krypton are more stable, there is, as mentioned before, a magic number which decides the stability of cluster. Here atomic and electronic shell closure determines stability. Thus, for stable carbon clusters, the magic numbers are 3, 5, 7, ... for smaller clusters and 60, 72, ... for larger clusters. This also resembles the magic number in nuclear physics, where for stable nucleus, the nucleon number (i.e., no of protons and neutrons) should be 2, 8, 20, 50, 82, 126, and so on. Nuclear fission is analogous to cluster fragmentation. The source of the cluster may be smoke or laser ablation and the most common deposition technique is low energy cluster beam deposition (LECBD).

13.7 Nano Fabrication

There are various ways to categorize the synthesis method of nano structured materials. Depending on the phase of the starting material, there are two classes: via gas phase and via condensed phase.

Alternate classifications may be physical or chemical and top down or bottom up approaches. Further, the crystallographic structure of the precursor also decides the method of fabrication. Amorphous materials on crystallization yield 3D nano structured materials. Poly-crystals on ball milling undergo severe plastic deformation and give rise to nano materials. Single atoms may be electrodeposited in various ways or condensed to give clusters and subsequently, on compaction yield nano particles.

It should be noted that there exists certain overlap between the different methods. The division is not sharp.

13.7.1 Gas phase or condensed phase classification

Gas phase synthesis is sophisticated and expensive. This method gives more insight into understanding the basic aspects of the nano structure.

Condensed phase synthesis deals with chemical vapour deposition (CVD) and physical vapour deposition (PVD) from atomic or molecular precursors. Bulk precursors under mechanical attrition, grinding and pounding, laser ablation, evaporation and so on, supports nano fabrication

13.7.2 Gas phase evaporation method

This method is based on the evaporation or sublimation of materials under pressure into a static inert gas followed by condensation. Vaporization can be achieved by resistive heating, i.e., thermal evaporation, electron beam evaporation, laser vaporization and so on. Cluster growth occurs in the gas phase and depends on the gas temperature and pressure. The clusters so formed are condensed onto a suitable substrate. In laser vaporization, the common techniques are pulsed laser deposition (PLD) and molecular beam epitaxy (MBE)

Evaporation is a thermally activated process. Appreciable amount of evaporating atoms are possible at high temperature. For conducting film, evaporation occurs from the surface. For dielectric and insulating film, evaporation may take place below the layer. Materials having high vapour pressure may be directly converted into gas (sublimation). By inert gas support and suitable evaporating temperature, a large number of alloys and compounds can be deposited with slight deviation from standard procedure.

Electron beam heating is useful in materials having high melting point and materials which react to support materials. The more energetic the electron beam, greater will be the deposition rate. Controlled kinetic energy from the source is possible and the electron beam can enter into the layers of the material up to any depth.

13.7.3 Top down approach

Two fundamentally different approaches to nanotechnology are graphically termed "top down" and "bottom up" approach. The top down approach is physics friendly and deals with taking a bulk of materials and using molecular beam epitaxy technique to produce the nano structure. The size limit that can be created depends on the tool. Milling, grinding and compounding materials also fall into this class. Ball mills are the common mills in use. These include tumbler mills, attrition mills, vibratory mills and planetary mills. In high energy ball milling, the grain size in a powder sample is reduced to nano size by mechanical deformation. Hard metals with a bcc structure (Cr, Nb, W) or an hcp structure (Zr, Hf, Ru) are subjected to high energy ball milling for nano particle synthesis. The advantages are high production rate. It permits several kg of materials in times up to 100 hours. The disadvantages in this method include (i) imperfection or defects in the crystal, (ii) crystallographic damage, (iii) surface contamination and the like.

13.7.4 Bottom up approach

In the bottom up approach, individual atom or molecules are assembled or self assembled to the desired size. However, to obtain an atomically perfect nano structure is a difficult task. By sophisticated techniques, it has been possible to arrange atomic or molecular building blocks in electronic devices like molecular switches, molecular wires. Similarly, biologists develop bio sensors. All living beings in nature grow to bigger size by this method. When the precursors are at the atomic or molecular stage, this approach is favourable to obtain nano structures. This is analogous to the formation of ice crystals on a cold window pane or a snow flake.

The advantage of the bottom up approach is the structural repeatability of a complex component that forms predictable structure motifs organized over a large area or volume.

Molecular beam epitaxy (MBE) is a method of laying down layers of materials with atomic thicknesses on to substrates. This is done by creating a "molecular beam" of a material which impinges on to the substrate. The resulting "superlattices" have a number of technologically important uses including quantum well lasers for semiconducting systems and giant magneto-resistance for metallic systems.

Ionized cluster beam (ICB) deposition is an ion-assisted film deposition technique by which high quality films of metals, dielectrics and semiconductors can be formed at a low substrate temperature. In the ICB process, film materials are vaporized from a confinement crucible under conditions which result in the formation of aggregates of atoms (clusters). Clusters are ionized by electron impact and subsequently accelerated by high potentials. Through selection of available parameters, it is possible to control the average energy of depositing species over the range from thermal to above 200 eV per atom, which make possible well controlled crystalline film deposition and epitaxy.

The pulsed laser deposition (PLD) method of thin film growth involves evaporation of a solid target in an ultra high vacuum chamber by means of short and high energy laser pulses. In a typical PLD process, a researcher places a ceramic target in a vacuum chamber.

A pulsed laser beam vaporizes the surface of the target, and the vapour condenses on to a substrate.

Sol–gel process which is a standard technique in its own right falls into this category. This is a process of evolution of inorganic networks through the formation of colloidal suspension (sol) and gelation of sol to form the network of the liquid phase (gel). A sol is a dispersion of the solid particles (~ 0.1–1 mm) in a liquid where only Brownian motions suspend the particles.

A gel is a state where both liquid and solid are dispersed in each other. It presents a solid network containing liquid components. The desired colloidal particles are dispersed in a liquid to form a sol. The deposition of sol solution produces coatings on the substrates by spraying, dipping or spinning. The particles in the sol are polymerized through the removal of the stabilizing components and produce a gel in a state of continuous network. The final heat treatments pyrolyze the remaining organic or inorganic components and form an amorphous or crystalline coating

13.8 Preparation of Solid Thin Films

Solid films of less than one micron thickness are commonly called thin films. Thin films are two-dimensional solids. The third dimension is negligibly smaller than the other two. If the thickness of the film is more than one micron, then it is called a thick film. The history of thin film dates back to 1838, when a thin film was formed by electrolysis. Bunsen and Grover prepared a solid metal film in 1852 by chemical reaction. However, it was Faraday who obtained a metal film by thermal evaporation. Thin films in large quantities are in demand in microelectronic, optoelectronic and photovoltaic industries. The applications include sensor, storage memory, energy harvesting, microelectronic, drug delivery, MEMS (micro-electric mechanical systems), semiconducting devices, photovoltaic and optoelectronic devices, radiation detectors, solar energy convertors and so on.

Thin films have merits over their bulk counterpart due to more surface atoms, high purity, high current density, surface dependent properties like wear friction, corrosion and oxidation. Other exceptional properties include: smaller size, material savings, ability to allow emission of visible light, increased band width, low cost and light weight. Films are normally grown on a substrate. Hence, film + substrate = thin film. Subtracts may be orientation stable, conductive, transparent and chemically stable. Examples: Glass, Si, GaAs, GaN, sapphire, quartz, lithium niobate and steel.

Thin film deposition techniques are classified as (i) physical vapour deposition (PVD) techniques or (ii) chemical vapour deposition (CVD) techniques.

In physical vapour deposition, small clusters of atoms are removed from the source. These atoms travel in vacuum reaching the substrate to form film. Steps involved: emission of particle, transport of particle, deposition or condensation and annealing.

In all physical deposition methods, vacuum is essentially needed. The mean free path of the molecule should be more than the distance between the substrate and the target. Without vacuum, the molecules get scattered and contaminated. Atomic collisions caused by finite mean free path may stop the evaporating atom at the surface. Vacuum (i) lowers the melting point, (ii) increases the mean free path, (iii) reduces the vapour pressure and (iv) increases the purity of the film

On the other hand, in CVD, there is chemical reaction of the constituting particles as gas or vapour (pyrolysis, hydrogen, reduction, oxidation and hydrolysis).

Another chemical deposition technique is the aqueous chemical growth (ACG) technique which involves spinning coat, dipping coat, spraying pryrolysis, electroplating and electro less plating.

13.8.1 Physical vapour deposition (PVD)

According to the deposition mechanism, physical deposition techniques are classified into

i.   Thermal evaporation

ii.   Electron beam evaporation

iii. Pulsed laser deposition

iv.   Molecular beam epitaxy

i. **Thermal evaporation**   It is a well-known method of preparing solid thin films from solid (metals, semiconductors or dielectrics). Thermal evaporation is a thermally activated process. It involves heating the bulk form of the material with a resistive heater (RF) to convert it into vapour form. The heating is done in a chamber where sufficient vacuum is created. The emission of the particle in vapour form is supported by a large mean free path (greater than the target and substrate distance). An appreciable amount of evaporating atoms are possible at high temperature. For conducting film, evaporation occurs from the surface and for dielectric and insulating film, evaporation may take place below the layer. Target material may be taken in the form of a foil, wire or ingot. Materials having high vapour pressure may directly convert into gas (sublimation).

To achieve uniform thickness, the substrate has to be rotated in a manner such that each point of the substrate will receive the same amount of materials during deposition.

The advantages are that it is very simple and cost effective. Films are obtained in extremely pure form. The disadvantage is that this method is limited to solids of low melting point. High melting point solid like dielectrics cannot be easily evaporated. The films so prepared by this method have poor density and adhesion.

ii. **Electron beam evaporation**   Here heat energy is supplied by the bombardment of accelerated electrons on the target. An electron beam is accelerated through a potential of 5 to 10 kV and focused on the material. The accelerated elections lose their kinetic energy mostly as heat. Temperature at the target may reach 3000°C and thus, it vaporizes materials of even high melting point. The vapour is transported and condensed on the substrate as a solid thin film. The more energetic the electron beam, greater will be the deposition rate. This method is useful for materials having high melting point and materials which react with the support materials. Focusing of electrons can be done in two ways: (i) by magnetic and electrostatic field where the path of the electron beams is straight and (ii) by bending beams.

Advantages: (i) It is very versatile and is highly efficient. (ii) It can be used for materials with high melting points. (iii) Supporting material can be eliminated. (iv) There is better control of structure and morphology of the films. (v) It has a high deposition rate. This method is best suited for wear resistance thermal barrier coating, electronic and optical thin films and semiconducting materials.

Disadvantages: However, this technique is very complicated to install and very expensive. It requires more space. This process is not suited for coating inner surface of complex geometrical surfaces. There is a potential risk of filament degradation which causes non-uniform evaporation rate

iii.   **Pulsed laser deposition (PLD)**   This is a laser induced vaporization. A powerful laser beam strikes the target producing considerable vapour in the PLD process. Due to the high power laser in a small area, the evaporating materials attain plasma state. Plasma consists of energetic atoms, ions and molecules. The material in plasma state reaches the target in a broad energy distribution of the order of 0.1 to 15 eV. The major physical parameters that control the PLD is target–substrate distance, growth rate, background gas, substrate temperature and so on. The films prepared are of high conductivity, with high infrared transmittance coefficient, excellent substrate adhesion, better hardness and chemical inertness. Normally, Eximer laser beam is used. The characteristics of the film are controlled by laser parameters as well as the substrate medium. The high vapour pressure and low mean free path ensure good quality films. The background gas is responsible for the optimal thickness in the films. This technique is used to prepare high T_C superconducting, ferroelectric and giant magneto resistance (GMR) films.

Advantages: This technique is flexible and easy to implement in any environment. It is possible to have epitaxial growth at low temperatures. Atoms arrive in bunches, allowing for much more controlled deposition. This method lowers the substrate temperature requirement. It is a clean, low cost method.

Disadvantages: The laser ablation cross-section is small. Thickness monitoring is difficult.

iv.   **Molecular beam epitaxy (MBE)**   This is a well-defined crystallization technique where a beam of atoms or molecules travels in a collision free manner. There is a layer by layer atomic growth of special material, characterized by continuation of crystal structure from the substrate. The technique needs a clean room and an ultra high vacuum (UHV) chamber. The UHV environment minimizes contamination. Materials should have low vapour pressure and low sticking coefficient.

13.8.2 Chemical vapour deposition (CVD)

In this process, the substrate is placed inside a reactor to which a number of gases are supplied. The fundamental principle of the process is that a chemical reaction takes place between the source gases. Preparation of materials in CVD involves the dissociation and/or chemical reactions of gaseous reactants in an activated (heat, light, plasma) environment. CVD involves the following steps as shown in Fig. 13.1.

• Mass transport of reactant gaseous species to the vicinity of the substrate

• Diffusion of reactant species through the boundary layer to the substrate surface or homogeneous chemical reactions to form intermediates

• Adsorption of reactant species or intermediates on the substrate surface

• Surface migration, heterogeneous reaction, inclusion of coating atoms into the growing surface, and formation of by-product

• Desorption of by-product species on the surface

• Transport of by-product gaseous species away from the substrate (exhaust)

Due to its process parameter, CVD has got a number of advantages in terms of preparation of nano materials. The advantages are that it is a simple method, and that it does not require high temperature and pressure for the synthesis of nano materials. Furthermore, CVD can be used in a wide variety of precursors such liquid precursors, gaseous precursors, solid precursors, which include halides, hydrides, metal-organic compounds, organic compounds, and so on. The precursor materials fall into a number of categories such as, halides, hydrides, metal organic compounds, metal alkyles and the like.

CVD enables the usage of a variety of substrates and allows material growth in a variety of forms, such as powder, thin or thick films, aligned or entangled, straight or coiled nanotubes. It also offers better control on the growth process of materials. CVD allows proper control of the deposition rate and pressure temperature so as to prepare and maintain the desired structure, composition and size of materials.

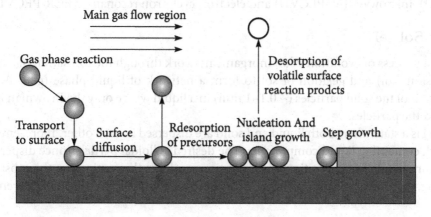

Figure 13.1 | Illustration of chemical vapour deposition steps

Due to its versatile nature and ability to use a wide variety of precursors (source materials), CVD is a potential technique for industrial application and laboratory research. The technique has been used in the coating industry for making wear resistance, corrosion resistance layers, high temperature protection layers, erosion protection layers and many more. The CVD technique has also been used in the semiconductor industry for the making of integrated circuits, sensors, optoelectronic devices, solar cells and so on. It is very difficult to fabricate dense structural part according to requirements in many other preparation methods. In this regard, CVD can be used to produce components that are difficult or uneconomical to produce by using conventional fabrication techniques. Furthermore, CVD has been a technique for production of composites. Using CVD techniques, ceramic matrix composites such as carbon–carbon, carbon–silicon carbide and silicon carbide– silicon carbide composites can be synthesized.

CVD can be modified according to how much energy is needed to ignite or to activate the reaction inside the reaction chamber. High temperature CVD is predominantly used for structural materials. Low temperature CVD is used where the substrate cannot sustain high temperatures. In terms of process control, CVD can be modified to continuous, discontinuous and pulsed CVD (P-CVD). In terms of source of activating the chemical reaction, CVD can be modified to plasma enhanced CVD (PECVD), laser induced CVD (LCVD), photo CVD (PCVD).

Plasma-enhanced chemical vapour deposition (PECVD) is similar to chemical vapour deposition (CVD). The important difference is that in CVD, thermal energy is used to activate the gas and in PECVD, the molecules are activated by electron impact. The main purpose of using plasma enhancement is to reduce the activation energy for a deposition process. It has been recognized that one of the most important and unexpected benefits of PECVD growth is the alignment growth of nano materials due to interaction with the electric field.

A variety of plasma sources have been successfully used for the deposition of nano materials. These sources include direct-current (dc PECVD), hot-filament dc (HF-dc PECVD), magnetron type radio frequency (rf PECVD), inductively coupled plasma (ICP-PECVD), microwave (M-PECVD) and electron cyclotron resonance (ECR-PECVD).

13.8.3 Sol gel

This is a process of evolution of an inorganic network through the formation of a colloidal suspension (sol) and gelation of sol to form a network of liquid phase (gel). A sol is a dispersion of the solid particles (~ 0.1–1 mm) in a liquid where only the Brownian motions suspend the particles.

A gel is a state where both liquid and solid are dispersed in each other. It presents a solid network containing liquid components. The desired colloidal particles once dispersed in a liquid form a sol. The deposition of sol solution produces the coatings on the substrates by spraying, dipping or spinning. The particles in sol are polymerized through the removal of the stabilizing components and produce a gel in a state of continuous network. The final heat treatments pyrolyze the remaining organic or inorganic components and form an amorphous or crystalline coating.

13.8.4 Ball milling

This is a top down approach. Milling, grinding and compounding materials also fall into this class.

It is well known that finer particles are more reactive than larger particles. High reactivity results from availability of greater surface area by reducing the size either by grinding, pounding and milling. The readers may refer Section 13.4.

13.9 Few Wonder Nano Materials

Carbon is found abundantly in nature and is the backbone of life, both animals and plants. Tiny algae to big trees, small cats to big elephants, DNA, blood cells, sugar, protein and cell membranes are all due to the structural diversity of carbon.

The commonly used pencil lead material is going to have its place in ultra modern gadgets. Carbon is present in organic and inorganic compounds, the atmosphere, the earth's crust and even the stars.

Carbon atom with an electronic configuration of 1s 2s2p, has four valence electrons, viz., two each from the 2s and the 2p sub-shells. Its covalent bonding capacity and reactivity with itself and hydrogen, oxygen, nitrogen and phosphorus helps to produce millions of known compounds. Carbon has the distinction of having more compounds alone than compounds of all the elements taken together.

The three well-known allotropes are amorphous carbon (charcoal and lampblack), graphite and diamond. Graphite is a planar hexagonal structure.

Some exotic allotropes in the form of cluster and nano materials recently discovered are fullerene, carbon nano tube and grapheme.

13.9.1 Fullerenes

Fullerene is a carbon allotrope composed entirely of carbon atoms ranging from 16 to 100. The first fullerene discovered was the Buckminister fullerene containing 60 carbon atoms C_{60}. It is also called the Bucky ball. The discovery was accidentally discovered in 1985 by Harold Kroto of UK and Richard Smalley of USA. Fullerenes are a class of big molecules or clusters that evolve out of the three-dimensional bonding of carbon atoms in a hollow closed structure, like a cage as shown in figure 13.2. The major isomers of fullerenes are C_{60}, C_{70}, C_{76}, C_{78}, C_{80}, C_{82}, C_{84} and so on.

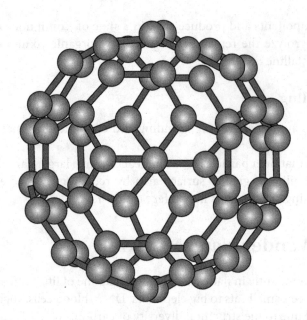

Figure 13.2 | Buckminister fullerene, also called the Bucky ball, containing 60 carbon atoms

The building block of C_{60} molecule is a truncated icosahedron with 60 vortices and 32 faces. Of the 32 faces, 12 are pentagons and 20 are hexagons. All the rings are fused and the double bonds are conjugated. The combination of pentagon and hexagon is very necessary as only hexagons like graphite cannot be folded up into a sphere. The basic requirement here is that the structure must be closed perfectly without gaps and overlaps.

The Buckminister fullerence has high symmetry. There are 120 point group symmetry operations such as rotation; reflection can map the molecule onto itself, and thus, it has icosohedral point groups making it one of the most symmetric molecule synthesized. The highest symmetry axis is a 5-fold axis and there are six 5-fold axes.

It is very easy to produce and most popularly applied in a variety of fields ranging from physics, chemistry, biology and mathematics.

The new field of carbon chemistry emerged from the discovery of C60. The other two cousins of zero-dimensional (0D) fullerene are one-dimensional (1D) carbon nano tube (CNT) and two-dimensional (2D) graphene.

13.9.2 Carbon nanotube (CNT)

CNT are fullerenes that are extended in one dimension with a high aspect ratio, acquiring a cylindrical structure as shown in Fig. 13.3. The properties of CNT vary due to their unique atomic arrangement. It was discovered by S.Iijima in 1991. CNT is a grapheme sheet wrapped seamlessly to form a hollow tube with diameter of the order of 10^{-9} m; it can be as long as a few millimeters. CNTs are composed of four carbon atoms arranged in a three-dimensional cage like structure. The tube can sustain a stress of 150 GPa. As an example of getting some sense on the scale of the world we live in, this means that

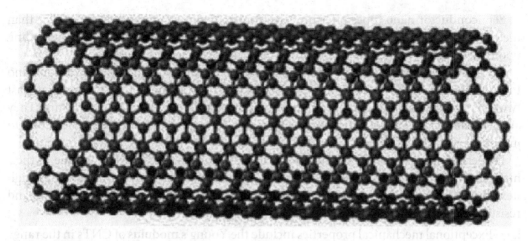

Figure 13.3(a) | Single walled carbon nano tube (SWCNT)

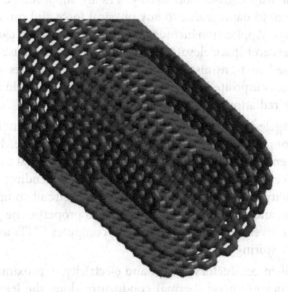

Figure 13.3(b) | Multi walled carbon nano tube (MWCNT)

a one millimeter string can hold a weight of 20 tons. In CNT, the carbon allotropes are bonded by sp2 carbon atoms. CNTs are ultrathin carbon fibers with nanometer size diameter and micrometer size length. Apart from the mechanical property, the CNT has peculiar electronic properties too. CNTs are physical realizations of a one-dimensional electronic system. Depending on the way the sheets are folded, CNTs are either metallic or semiconducting. Carbon atoms are arranged in hexagons. If the rows of hexagons are straight, they act as an electrical conductor; if they are wound around in a helix, they act as a semiconductor.

Semiconductor nano tubes offer tiny electronic circuits a thousand times smaller than the present technology. Generally, graphene sheets are curled to form CNT. Depending upon the direction of the curl, CNTs are either metallic or semiconducting.

CNTs may be a single walled carbon nano tube (SWCNT) or a multi walled carbon nano tube (MWCNT) based on the number of layers constituting them. CNTs contain at least two layers, often many more ranged in the outer diameter from about 3 nm to 30 nm. They are invariably closed at both ends. A new form of CNTs are also discovered which contain just one layer.

Properties and uses of CNTs: The properties and application of CNTs are due to their high tensile strength and Young's modulus, high electrical and thermal conductivities, high flexible and aspect ratio (length = 1000 × diameter). Low thermal expansion coefficient and resistance to chemical reaction are other gifted properties.

i.   Exceptional mechanical properties include the Young's modulus of CNTs in the range of 1–5 TPa for single walled CNTs and a tensile strength of around 150 GPa for multi walled CNTs. These values are much larger than the best known mechanical materials and comparable with high-carbon steel. CNTs are supposed be high strength fiber. They are resistant to damage due to any physical force and can recover from severe structural damage. Application includes everyday items such as clothes and sports gear to combat jackets and space elevators. CNT composites, because of their exceptional strength are used in transmission line cables, woven fabrics and strain resistant textiles. A plastic composite of CNT is used as lightweight shielding materials against electromagnetic radiation.

ii.   Many fascinating electrical and electronics properties are due to the intimate connection between electronic band gaps and the geometrical structure. It is found that the energy gap of semiconducting CNTs is inversely proportional to the diameter of the tube. CNTs can either be metallic or semiconducting depending on the diameter and helicity or chirality of the tube. Applications include circuit components, transistors, digital switching and sensors. Change in electrical properties helps to design sensors for sensing ppm level of O_2, NO_2, NH_3 etc. In computer CNTs are used as switching devices and data storing devices

iii.   CNTs are excellent conductors of heat and electricity, approximately 20 times more than copper. They are good thermal conductors along the length; they also show ballistic conduction along the length. Along the lateral direction, it behaves like an insulator. CNTs are very temperature stable in air; the temperature stability increases in vacuum.

iv.   CNTs are resistant to chemical reactions and can be used as catalysts in chemical reactions.

13.9.3 Graphene

Graphene is nothing but a sheet of crystalline carbon that is one atom thick as shown in Fig. 13.4. It is a single layer of graphite, which is prepared by peeling the top layer of graphite. Many initial efforts to prepare stable grapheme ended in failure. The principles of preparation of graphene before the present sophisticated technique of epitaxial growth were purely mechanical. These include

i.   Splitting graphite crystal into progressively thinner wafers by scrapping or rubbing them against another surface. This is known as micromechanical cleavage.

ii. Using an adhesive tape to rip off sheets of carbon from graphite. This is known as Scotch tape method.

It is a 2D nanomaterial possessing several distinct properties. Graphite can be viewed where a number of 2D graphene crystals are weakly coupled together. Here carbon atoms are arranged in planner and hexagonal form. It has two atoms per unit cell. Carbon carbon bond length in graphene is 1.43A0. It has a honeycomb structure and it acts as a precursor for other members of carbon family like graphite, CNT and fullerne.

Graphene sheets are held together by van der waals forces. Graphene has high spring constant of the order 1-5 N/m and its young modulus is of 0.5 Tpa. These high values make graphene strong and rigid and helps in using grapheme as pressure sensor and resonators. The thermal conductivity of grapheme is 100 times larger than that of graphite. What more important and fascinating is the relativistic effect.

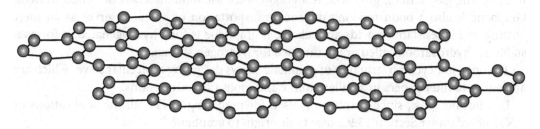

Figure 13.4 │ A typical graphene sheet

In graphene, electrons obey linear dispersion relation in energy i.e. E is proportional to wave vector ($E = h/2\pi \, kv_F$). This is the characteristics massless Dirac fermions. It is a gapless semiconductor with a linear energy spectrum. In grapheme, electrons move with a velocity of the order of 10^6 ms^{-1}. This is of course 300 times slower than the velocity of light. The electrons in graphene strongly interact with the lattice and can also be manipulated through the external applied electromagnetic field.

Graphene has high carrier mobility at room temperature and its mobility is not influenced by dopants even at a higher order of concentration. Contrary to ordinary electrons that suffer scattering due to impurities in the crystal, electrons in graphene are not affected by the impurities there. These electrons can travel large distances without suffering collision with impurities. This property helps in manufacturing high speed electronic switches. The resistance offered by graphene is even less than the least resistive metallic silver.

The electrons of graphene are in the conduction band. It has no band gap in the absence of the applied field. However, bi-layer graphene has a tunable band gap which is useful for the band gap engineers.

Graphene shows anomalous quantum Hall effect (QHE) even at room temperatures; other materials show this effect at absolute zero. The Hall affect is the phenomenon in which a material with current flowing in a transverse magnetic field shows a potential drop in a direction at right angle to the current and the field. The ratio of the potential drop to the current is called Hall resistivity. It is found that at absolute zero temperature, the Hall resistivity is quantized.

Applications: Graphene layer possesses high carrier mobility and low noise, high area to mass ratio, high electrical conductivity and optical transparency. It is resistant to acids and alkalis. These properties help to make many noble applications.

Graphene can be used to make excellent transistors. Electrons travel faster over sub-micron distances. Graphene based transistors can run at higher frequency than silicon based transistors. One can dream of inventing the world's smallest transistor using graphene and integrated circuits.

High electrical conductivity and optical transparency can be properly exploited for designing transparent electrodes. The benefitted device includes: touch screen phones, handheld computers, organic photovoltaic and organic light-emitting devices and nano electronics.

Gas molecules that land on graphene largely affect its electronic property. This helps in designing gas sensors; graphene is a good choice for solid state gas detection devices. Graphene is also a boon for long distance transportation of gases. It serves as an inert coating and is resistant to acids and alkalis. If graphene is fully hydrogenated, it forms a stable 2D hydrocarbon called graphane used for hydrogen storage

The open honeycomb structure of graphene allows designing molecular sieves which are analogous to filter papers in smaller atomic and molecular dimensions.

Last but not least, since graphene is used for preparing CNTs, all the applications of CNTs noted under Section 13.9.2 owe their origin to graphene.

Questions

13.1 Why is nano so important?

13.2 What is the meaning of phrase "There is plenty of room at the bottom".

13.3 Discuss the role of nanotechnology in all possible areas.

13.4 What are the properties that change at the nano scale? Give examples from each.

13.5 Give some examples to visualise nano.

13.6 What is nano science and nano technology?

13.7 At the nano scale, the properties of materials changes. Do you agree? Justify with some examples.

13.8 Nature also mimics some nano products. Do you agree? Give examples.

13.9 From the quantum confinement concept, how do properties change at the nano scale?

13.10 Discuss how surface to volume ratio play a role in changing the properties at nano scale.

13.11 CdSe changes colour from orange in bulk to yellow and white in different nano states. Justify the reason.

13.12 Do you mean gold is always yellow? Justify your answer.

13.13 What is nano cluster? Do you mean to say that clusters with any number of atoms are stable?

13.14 What is the magic number in cluster formation? Give examples.

13.15 What are nano clusters? Discuss magic numbers with reference to nano clusters.

13.16 What is the different classification of nano fabrication? Are the classifications sharp?

13.17 Distinguish between physical vapour deposition (PVD) and chemical vapour deposition (CVD)

13.18 Distinguish between gas phase and condensed phase classification of nanos.

13.19 Discuss size effect in CdSe and gold (Au).

13.20 What is a sol and gel? How can you prepare a thin film by the sol–gel technique?

13.21 Distinguish between top down and bottom up approach of nano fabrication.

13.22 What are the different physical vapour deposition (PVD) techniques for thin film preparation with their relative merits and demerits?

13.23 Mention the different cluster forms of allotropes of amorphous carbon.

13.24 What is a Buckminister fullerene? Mention the major isomers of fullerenes.

13.25 How many point group symmetry and highest symmetry axis does the fullerene have?

13.26 Why is the carbon nano tube (CNT) so named?

13.27 Under what condition will a carbon nanotube (CNT) will behave like (a) a semiconductor; (b) a conductor.

13.28 Distinguish between a single walled carbon nano tube (SWCNT) and multi walled carbon nanotube (MWCNT). What is the ball milling process? How does it work for the preparation of a nano particle? Mention its merits and demerits.

13.29 With a neat sketch, explain crystallographic structure, symmetry, uses and applications of Fullerene.

13.30 With a neat sketch, show the carbon nano tube (CNT) and write its exceptional mechanical properties.

13.31 Narrate the exceptional electrical, electronics and chemical properties of a carbon nano tube (CNT).

13.32 With a neat sketch, show graphene and how it is prepared.

13.33 Narrate the different applications of graphene.

13.34 Mention one property of graphene that helps to design high speed electronic switches.

13.35 Mention a few gifted properties of graphene.

Multiple Choice Questions

1. "There is plenty of room at the bottom" was the vision of
   (i) Feynman
   (ii) Einstein
   (iii) C. V. Raman
   (iv) None of the above

2. One nanometer is
   (i) 0.1 Å
   (ii) 100 Å
   (iii) 10 Å
   (iv) None of the above

3. Gold nano particle of size 3 nm is
   (i) Yellow
   (ii) Red
   (iii) Orange
   (iv) Blue

4. Widespread use of nano technology is due to
   (i) Small scale miniaturization
   (ii) The fact that it is faster and cheaper
   (iii) Its lower cost
   (iv) All the above

5. The allotrope of carbon having 2D structure and 1D confinement
   (i) Fullerene
   (ii) Carbon nano tube (CNT)
   (iii) Graphene
   (iv) Graphite

6. The allotrope of carbon having 1D structure and 2D confinement
   (i) Carbon nanotube (CNT)
   (ii) Graphite
   (iii) Graphene
   (iv) Fullerene

7. The allotrope of carbon having 0D structure and 3D confinement
   (i) Fullerene
   (ii) Carbon nanotube (CNT)
   (iii) Graphene
   (iv) Graphite

8. For a smaller cluster structure, the magic number is
   (i) 3,5,7
   (ii) 1,3,5
   (iii) 3,5,1
   (iv) 1,5,7

9. Ball milling is
   (i) Gas phase
   (ii) Top down approach
   (iii) Bottom up approach
   (iv) Condensed phase

10. The building block of C_{60} molecule has
    (i) 90 vortices and 32 faces
    (ii) 32 vortices and 60 faces
    (iii) 60 vortices and 32 faces
    (iv) None of the above

11.   The number of point group operation in fullerene is

   (i)   120   (ii) 32

   (iii) 240   (iv) None of the above

12.   The carbon–carbon bond length of graphene is

   (i)   1.34 Å   (ii) 1.43 Å

   (iii) 2.43 Å   (iv) None of the above

13.   The band gap of graphene is

   (i)   0 eV   (ii) 1.1 eV

   (iii) 0.6   (iv) None of the above

14.   Grapheme shows

   (i)   Anomalous quantum Hall effect   (ii) Hall effect

   (iii) Quantum Hall effect   (iv) None of the above

15.   The largest cluster of carbon atoms in Bucky balls known till today consists of

   (i)   60 carbon atoms   (ii) 75 carbon atoms

   (iii) 180 carbon atoms   (iv) 540 carbon atoms

Answers

1 (i)	2 (iii)	3 (iii)	4 (iv)	5 (iii)	6 (i)	7 (iv)	8 (i)
9 (iii)	10 (iii)	11 (i)	12 (ii)	13 (i)	14 (i)	15 (iv)	

Bibliography

Beiser, A. 1995. *Concepts of Modern Physics*. 5th Edition. New Delhi: Tata McGraw-Hill Publishing Company Limited.

Callister, Jr., William D. and David G. Rethwisch. 2012. *Fundamentals of Materials Science and Engineering: An Integrated Approach*. 4th Edition. New York: John Wiley & Sons. Inc.

Cullity, B. D. 1956. *Elements of Ray X-Ray Diffraction*. 3rd Edition. Massachusetts: Addison-Wesley Publishing Company, Inc., Reading

Kittel, C. 1995. *Introduction to Solid State Physics*. 7th Edition. New York: John Wiley & Sons. Inc.

Kwan, Chi Kao. 2004. *Dielectric Phenomena in Solids*. Amsterdam: Elsevier Academic Press.

Pillai, S. O. 2008. *Solid State Physics*. 4th Edition. New Delhi: New Age International (P) Limited Publisher.

Raghavan, V. 1998. *Materials Science and Engineering*. 4th Edition. New Delhi: PHI Private Limited.

Vijaya, M. S. and Rangarajan, G. 2014. *Materials Science*. New Delhi: TMH Publishing Company Limited.

Index